Calculus

A New Approach for Schools that Starts with Simple Algebra

R. Michael Range
State University of New York at Albany, USA

World Scientific

NEW JERSEY · LONDON · SINGAPORE · BEIJING · SHANGHAI · TAIPEI · CHENNAI

Published by

World Scientific Publishing Co. Pte. Ltd.

5 Toh Tuck Link, Singapore 596224

USA office: 27 Warren Street, Suite 401-402, Hackensack, NJ 07601

UK office: 57 Shelton Street, Covent Garden, London WC2H 9HE

Library of Congress Control Number: 2025019200

British Library Cataloguing-in-Publication Data
A catalogue record for this book is available from the British Library.

CALCULUS
A New Approach for Schools that Starts with Simple Algebra

ISBN 978-981-98-0197-8 (hardcover)
ISBN 978-981-98-0544-0 (paperback)
ISBN 978-981-98-0198-5 (ebook for institutions)
ISBN 978-981-98-0199-2 (ebook for individuals)

For any available supplementary material, please visit
https://www.worldscientific.com/worldscibooks/10.1142/14075#t=suppl

Preface

A) Why do we need a new approach?

Calculus is one of the great creations of the human mind. Its main geometric problems, dealing with areas and volumes, as well as with tangent lines to familiar curves, have a long history, going back to Greek philosophers and geometers well over two thousand years ago. It then vanished into the background for many centuries, until the Renaissance brought about renewed interest and fresh minds. Many philosophers and scientists in the 16th and 17th centuries contributed major ideas, though it is generally recognized that Isaac Newton (1642–1727) and Gottfried Wilhelm Leibniz (1646–1716) made the fundamental progress that allowed Calculus to grow into a phenomenal mathematical tool that has become indispensable for understanding the physical world around us. In particular, these tools continue to be central in modern times, for example, they provide the theoretical foundation for traveling in space, which led man to first step onto the Moon in 1969, and most likely onto Mars in the not-too-distant future. No wonder that calculus continues to be the core mathematical topic at the beginning of college, and that for decades it has reached down deeply into the high school curriculum.

Mathematicians in the 18th and 19th centuries, prompted by numerous subtle phenomena, questions, and inconsistencies, vastly extended and strengthened the theoretical foundations of calculus, thereby creating what has become known as *Analysis*, one of the traditional three areas of mathematics, next to *Geometry* and *Algebra*. In particular, it was recognized that it was necessary to develop a precise formulation of the concept of *limit*, whose intuitive roots go back to the 17th century. In turn, this required a deeper understanding of the *continuum* of points on the number line. More precisely, this led to a vast extension of the number system from the familiar rational numbers to the much more intricate and elusive *real numbers*, whose most familiar representation involves *infinite* decimal expansions. Adherence to logical order, so central to the thinking of mathematicians, then required to begin the teaching of calculus with an extensive investigation of real numbers and of limits. And that's the way it has been for nearly two centuries. While logical order is desirable—and we certainly shall rely on it in this book—there are different ways to implement it. Unfortunately, the efforts to build a solid foundation for the historical developments and ideas of the late 17th century, resulted in placing the new and most difficult sophisticated concepts right at the beginning, and this presents a major hurdle for the student. Also, it makes the whole subject appear somewhat disjointed from the elementary geometric and algebraic concepts students are learning about in middle school and early high school, and as we shall see, it makes matters look much more complicated and mysterious than necessary.

These difficulties have of course been noticed for a long time, starting at least as early as in the 1960s, when an increasing number of students—far beyond the traditional clientele of mathematics and physics majors—were required to learn the basics of calculus. Most medical programs in the United States even made a good grade in calculus a necessary prerequisite for admission. A climax was perhaps reached in the late 1980s, when "Calculus Reform" became a hot topic in the United States, leading to new revised editions of traditional calculus textbooks. In particular, the introduction of computers and graphing calculators facilitated using numerical and graphical methods to make key concepts more visible, and hopefully more understandable, to the student.

Unfortunately, nothing was really done about changing the order in which calculus is taught. The basic logical order which places deep theoretical concepts at the beginning, continues to dominate. Worse, the mathematician's love for logical order has inflated the "prerequisites" for learning calculus, leading to a calculus curriculum that now includes a year-long "Pre-Calculus" course in high schools and colleges that makes the whole subject look even more complicated, and that moves the fundamental problems and concepts even further into the distance for the student who wants to understand and learn the basics. Why, for example, must the student learn a substantial amount of trigonometry before being exposed to the basic ideas involving tangents, differentiation, and integration? The state of affairs is clearly visible in the recently published three volume sequence by the noted complex differential geometer Hung-Hsi Wu that covers the secondary school curriculum in mathematics,[1] especially in Volume 3 devoted to Pre-Calculus and Calculus. Wu has been motivated by the laudable goal to be "respectful of *mathematical integrity* as well as the standard school curriculum", in other words, to place Textbook School Mathematics on a solid mathematical foundation. Still, nothing has changed with the order to help the student.

A recent book by the well known mathematician David Bressoud[2] makes explicit these concerns about the standard calculus sequence that has been in use for at least 150 years. He points out that this sequence is really the reverse of the historical order, which long ago first focused on key ideas of areas and volumes, later on (instantaneous) rates of change, such as velocity, and eventually on infinite series. The fourth topic, *limits*, only appeared much later in the 19th century. Bressoud's intent is "to use the historical development of these four big ideas to suggest more natural and intuitive routes into calculus." (op. cit., Preface) Following the historical order would provide a context for students to understand how these ideas developed. Furthermore, and most relevant for all who either teach or want to learn calculus, Bressoud states that the current syllabus is not pedagogically sound, and that very few of the large number of students that have to learn calculus today profit from the standard logically correct sequence.

*The purpose of this book is to propose a substantial reordering of the calculus curriculum, one that differs in key points from the standard curriculum, as well as from the historical sequence, as outlined, for example, by David Bressoud. The **guiding principle** is to develop the subject in such a way that it naturally ties in with elementary school topics, and to introduce new and more advanced topics gradually as the need for more sophisticated methods becomes clearly visible to the student.*

This new approach has been inspired and made possible by the discovery that René Descartes' (1596–1650) long forgotten algebraic *double* root method for finding tangents has a most elementary implementation that can be explained to the student as soon as she/he has learned the basics of

---

[1] Hung-Hsi Wu, Vol. 1: *Rational Numbers to Linear Equations*, Vol. 2: *Algebra and Geometry*, Vol. 3: *Pre-Calculus, Calculus, and Beyond*. American Mathematical Society, Providence, Rhode Island, **2020**.

[2] David M. Bressoud, *Calculus Reordered*, Princeton University Press, 2018. The interested reader might want to first read the review of this book by Fernando Gouvea, The American Mathematical Monthly **127** (2020), 470–477.

linear functions and of the quadratic function and equation. This allows a natural transition from simple algebra to key calculus concepts and techniques without the need to first introduce *real* numbers, limits and related matters. Even more remarkable is the realization that the main algebraic tool—a simple, yet central factorization result—when combined with a most basic estimate, naturally leads to the idea of *continuity*, which is much more elementary than the general concept of limit. Most importantly, it suggests how algebraic derivatives can also be captured by a *non-algebraic* approximation process, and thereby it ties the elementary algebraic approach to the amazing new ideas developed in the late 17th century. This discovery initiates the transition to the fundamental new concepts that are the heart of analysis, and that are indispensable for treating non-algebraic functions, such as exponential functions. To summarize, students get naturally guided from simple familiar algebra to the idea of tangents and derivatives, and then discover a new, non-algebraic, process that will open the door to the heart of analysis.

As for the historical context, we should keep in mind that Greek philosophers over 2000 years ago did not just investigate how to approximate areas and volumes, but they also introduced tangents to simple curves, such as the conic sections, and developed techniques to describe them in special cases. And of course Descartes' fundamental idea to identify tangents algebraically, which is central to our new introduction to calculus, predates by nearly half a century the work of Newton and Leibniz. It is an unfortunate twist in history that attempts to implement Descartes' double point method in the 17th century led to formidable algebraic complications, so that this approach was eventually abandoned and ultimately forgotten, until it was rediscovered in a new framework a few years ago. Therefore, to focus on the tangent problem in order to introduce students to the central ideas of analysis, does indeed respect, at least partially, the historical roots. Furthermore, I believe that the more complicated technical details involved in studying areas, volumes, and so on, something that Bressoud states at the beginning, would be a major obstacle for the student, and therefore it seems reasonable to postpone their discussion to a later point. Again, our main purpose is to help the students by presenting topics in a sequence that moves from simple to deeper and more complicated concepts, and by explaining why new tools and deep ideas are eventually needed in order to understand the physical world around us.

A fairly complete modern implementation of Descartes' method, extended to handle all algebraic functions and formal power series (the most general type of function considered in the 17th and 18th century) had been published already in 2011,[3] and subsequently this approach was expanded into a book in 2015.[4] A few years later this author published a brief outline of this "reordering", emphasizing the quadratic equation as the starting point.[5] However, it was then realized that in order for these ideas to not just circulate among mathematicians and scientifically minded readers, it would be necessary to write a detailed textbook that shows how the new ordering could be implemented in a revised high school curriculum. Wu's books mentioned above, especially the third volume, have been our guide to the current standard curriculum. However, aside from starting off with Descartes' algebraic definition of tangents, it has been my choice to not just change the order and introduce a number of simplifications, but also to omit various topics and details that I do not view as relevant for guiding the student into the heart of calculus. Of course, some of these topics will eventually be discussed in college level courses, at a time when it will be clear to the student why they are needed.

The hope is that such a detailed outline of a possible new high school curriculum would initiate and stimulate extensive discussions among the many stakeholders who are aware of the pedagogical hurdles in the standard curriculum. The goal, of course, is to agree on changes, to test them in the classroom, and to get feedback from many teachers and students, ultimately resulting in what could become a "new standard calculus curriculum" for high schools. Again, the guiding principle should be to help our students to understand and learn this most important subject.[6]

B) The new approach

Let me outline the proposed new ordering in more detail. Along the way I will highlight a few of the other more notable deviations from the standard curriculum. We begin with a fairly detailed discussion of the rational numbers, the most widely known and used numbers. These will underly

---

[3] R. Michael Range, *Where are Limits Needed in Calculus?*, Amer. Math. Monthly **118** (2011), 404–417.

[4] R. Michael Range, *What is Calculus? From Simple Algebra to Deep Analysis*, World Scientific Publ., Singapore, London, and New York 2015.

[5] R. Michael Range, *Using high school algebra for a natural approach to derivatives and continuity*, The Math. Gazette **102** (2018), 435–446.

[6] A brief outline of the new introduction to calculus presented in this book has been published in: *Calculus: A new approach for schools that starts with simple algebra*, European Math. Soc. Magazine **124** (2022), 42–48.

our discussion as long as we deal with algebraic functions, including differentiation and continuity. In particular, in order to prepare the student for the more abstract *real* numbers that will ultimately be needed, we emphasize how the basic rules (later called *axioms*) are the key tools that guide us to extend familiar properties from the counting numbers to more general numbers, such as *integers*, *rationals*, and—much later—the intriguing *real* numbers. We assume basic number facts, such as techniques to add and multiply positive integers, factorization into prime numbers, the division process, and so on, as well as basic facts from elementary geometry, such as similarity of triangles, parallel lines, area formulas for triangles and quadrilaterals, and so on.

Next, we introduce the general concept of a function and of its related graph. Functions are the basic tool used in applications to express the relationship between the different quantities under investigation. And calculus investigates special properties of functions that are relevant for applications, such as the fundamental notion of *rate of change*. This early emphasis on functions differs from the standard school curriculum that focuses more on equations and (algebraic) formulas. The most elementary (mathematical) examples are linear and quadratic functions, which are discussed in detail, including their relevant geometric properties. The notion of a tangent is introduced in the context of the quadratic function, where it is most easily connected directly to the concept of double root of a quadratic equation, building upon the 400 year old deep insight of René Descartes. We also review the ancient formulation of tangents introduced by Greek philosophers and explain how it is made precise by Descartes' idea. Important applications of the formula for tangents are the (mathematical) verification of the physical reflection properties of a parabola, as well as the idea of velocity at a single point in time that first appeared in Galileo's investigations in the early 17th century. Furthermore, since the vertex of a parabola—described in standard form as the graph of a quadratic function—is that point on the graph where the tangent is horizontal, the formula for tangents allows to find the vertex, and consequently graph the parabola described by an arbitrary quadratic function, very easily. It also provides a first example of the relationship between tangents and local maxima and minima of a function.

These elementary ideas are then applied to the class of polynomial functions. An important new result is the "Chain Rule" for derivatives of a *composition* of (polynomial) functions. In contrast to traditional approaches, where this rule usually comes after product and quotient rules, we highlight the simplicity of this most natural operation on functions, and of the completely straightforward proof of the relevant rule that is much simpler than any of the standard proofs. Our goal is to familiarize the student with the most basic ideas and rules in the simplest context of polynomials over the rational numbers, before discussing the more complicated product and quotient rules for differentiation. Furthermore, the simple proof of the Chain Rule will carry over to the general class of differentiable functions (see further down), once one has formalized the idea of continuity and its basic natural properties.

Next we introduce *rational functions*. The collection $\mathbb{Q}(x)$ of all such functions provides another important example to highlight the familiar rules that govern the *field* of rational numbers $\mathbb{Q}$. Most importantly, we show how tangents and the rules of differentiation, augmented by product and quotient rules, carry over most naturally to this more general setting. The key lesson is that no really new ideas are required in order to extend the central concepts from the simple quadratic function to more general functions that are constructed by algebraic techniques.

Things are very different, however, once one goes beyond algebraic functions. The study of these functions requires a new approach to tangents and derivatives. What is most surprising is the realization that the all-important algebraic factorization is not only the key to tangents and derivatives for algebraic functions, but that it most naturally guides us to a new and more powerful way to capture derivatives. More precisely, we combine the (algebraic) factorization with a standard estimate that reveals the concept of *continuity*, a most fundamental property that makes precise that most natural processes do not make abrupt changes. This estimate, in fact, provides a simple direct proof of the continuity of all polynomials, and, more generally, of rational functions, that is much simpler than any of the standard proofs in traditional textbooks. In particular, it shows that the simple algebraic derivative can also be captured by an approximation process. We discuss this process in some detail, including some properties of average rates of change that will be needed later, and leading us to the idea of *instantaneous rate of change* that is fundamental in calculus.

We are now ready to study the tangent problem for exponential functions. We begin by introducing these functions in a preliminary elementary form. These are the most important and most widely known functions in numerous applications, such as compound interest in finance, radioactive decay, and exponential growth, the latter one having appeared extensively in recent discussions of the spread of Covid-19. However, attempts to apply the elementary algebraic techniques to study the tangent problem, quickly reveal the limits of the simple techniques used up this point. Fortunately, the idea of continuity, discovered earlier in the context of rational functions, suggests a new different method, namely, to capture the algebraic derivative by approximation. However, attempts to apply this non-algebraic process to the exponential function reveal surprising new phenomena and fundamental difficulties. In particular, one recognizes that a successful implementation of these ideas forces us to go beyond the rational numbers and to introduce the deep and crucial property of *completeness* of the number line that is at the heart of the *real* numbers $\mathbb{R}$. Again, this approach should help the student to understand why these more sophisticated ideas that transcend the familiar algebra need to be introduced.

We have thus reached the point where it is necessary to introduce the *real* numbers and to study them carefully. As we just mentioned, the critical new property (or axiom) that distinguishes the real numbers from the familiar rational numbers is known as *completeness*. There are different equivalent formulations of this property, and we choose the so-called *Least Upper Bound* property. Not only does this version make precise something that is intuitively quite obvious, but it also provides perhaps the easiest way to prove the existence of all the limits that are central to our approach to calculus. Furthermore, beyond providing the new sophisticated tool that is essential for solving the tangent problem for non-algebraic functions, the real numbers also open the door to a completely new and unexpected phenomenon. More precisely, it turns out that *completeness* implies that the set of real numbers $\mathbb{R}$ involves a much higher order of "infinity" than the one we are familiar with from the natural counting numbers. This amazing result, due to Georg Cantor (1845–1918), the German mathematician who created *set theory*, should be compared to the fundamental discovery by Greek philosophers over 2000 years ago that simple geometric quantities, like the diagonal of a square with sides of length 1, can not be measured exactly by the familiar rational numbers. While this surprising property of the real numbers is not essential for the understanding of calculus, we do mention it here and include a fairly simple proof of it, because of the deep impact it has had on abstract mathematics and on our understanding of *infinity*.

Once we have a good understanding of the *real* numbers, it is then possible to carefully extend the domain of exponential functions to all such numbers, and to establish their basic properties in this more general setting.

In contrast to common perceptions,[7] we do not view this as a tedious task, but rather as the natural path to show the student how to apply the central new concepts that are the foundation of analysis. This is the place where it is most useful to introduce precise definitions and properties of limits, both in the context of sequences and series, as well as of functions. As a by-product, we discuss the representation of real numbers by (usually infinite) decimal expansions. The culmination of all this is the thorough discussion and solution of the tangent problem for exponential functions, perhaps the most significant application of limits and their main properties in our introduction to calculus. Here, too, our goal is to guide the student to recognize the importance of limits, to understand the subtlety of their formal definitions, and to apply them to solve a central problem.

This emphasis on exponential functions differs from the historical order, and also from the standard curriculum, as visible, for example, in Wu's books. In the 17th and 18th century the exponential function was not investigated directly, but rather through the back door, either as the inverse of logarithms, that were widely used already in the 17th century as a computational tool, or by searching for a solution of the differential equation $y' = y$. Since in those times formal power series were accepted and handled just like polynomials, without any concerns about convergence, Leonhard Euler (1707–1783), widely recognized as the greatest mathematician of the 18th century, readily was able to find such a series representation for the solution and establish its basic properties. More recently, many books written for non-math majors, recognizing the importance of exponential functions for applications, introduce their derivatives early on, although in a rather intuitive way suitable for the intended audience. In contrast, the standard curriculum, after introducing limits and derivatives, typically moves on to areas and integrals, defines the *natural logarithm* by integration, and at last introduces the *natural* exponential function as the inverse of that logarithm. Again, this logical sequence pleases experienced mathematicians, but surely it involves a very long detour that doesn't help the student who needs to use exponential functions in applications. It also does no justice to the fundamental importance of exponential functions, not just for applications, but even more for *pure* mathematics. After all, these functions are the "eigenfunctions" of the differentiation operator,[8] and therefore Euler's

---

[7] "...the laws of exponents for arbitrary *real* exponents are extremely tedious to prove if we adopt the approach of...extending them step by step..." (H. H. Wu, op. cit, vol. 3, p. 363.)

[8] An *eigenfunction* for a linear operator L acting on functions is a function f with the property that $L(f) = \lambda \cdot f$, where λ is a constant (an *eigenvalue* of L).

approach to exponential functions mentioned above, supported by relevant convergence theorems, would be much more appropriate than introducing them as the inverses of logarithms. Furthermore, let us remind the experts of their fundamental role in analysis, as visible, for example, in the *Fourier transform*.

Once exponential functions are well understood, it is then an easy step to introduce the general concept of "differentiable function". Here, too, we deviate from the standard approach by using a formulation that was introduced by Constantin Carathéodory (1873–1950) at least as early as 1950 in his classic text *Funtionentheorie* (Birkhäuser, Boston 1950), as follows.

A function f of a real variable is differentiable at the point a, if there is a factorization

$$f(x) - f(a) = q(x)(x - a),$$

*where the function q is continuous at a. The value q(a) is called the **derivative** of f at a and is denoted by D[f](a), or more briefly by f'(a).*

This definition was already used in courses and in a number of textbooks in Germany beginning in the 1960s, and we also used it systematically in our 2015 book *"What is Calculus?"*, which we shall refer to as [WiC?][9] from now on.

It is obvious (to a person familiar with calculus) that this definition is equivalent to the standard one in terms of limits of difference quotients. More significant for our purposes is the fact that this formulation is just the natural generalization, enhanced by continuity, of the central factorization in the algebraic setting. Furthermore, it can be viewed as a precise version of the analogous formula $df = f'(a)dx$ involving differentials that goes back to Leibniz, and that is at the core of applications related to rates of change. Moreover, this definition is just a simple reformulation of the fundamental equivalence of standard differentiability with the more modern view point of *good linear approximation*, as follows. Just rearrange the above equation into

$$f(x) - [f(a) + q(a)(x - a)] = [q(x) - q(a)](x - a),$$

a form that identifies the difference between the function and its tangent line. Clearly the continuity of q at a is equivalent to $\lim_{x \to a}[q(x) - q(a)] = 0$, so that the error term $[q(x) - q(a)](x - a)$ is what is commonly known as

---

[9]World Scientific Publishing, Singapore, London, and New York, 2015.

$o(x - a)$,[10] which is the critical property for the error between the graph of a differentiable function and its tangent line.

And finally, on the practical side, this definition of differentiability allows the most elementary and natural proof of the chain rule for differentiation, in complete analogy to the proof for polynomials discussed earlier, as well as of all other rules we had established earlier for rational functions, by reducing technical details to natural properties of continuity. Moreover, it readily generalizes to functions and maps of several variables, once the appropriate linear algebra has been introduced.

Next we discuss trigonometric functions such as *sine* and *cosine*, another important class of functions with a wide range of applications. In the context of triangles and measurements in astronomy and navigation, these functions have a long history, going back thousands of years. In the 17th and 18th century they played a central role in order to describe periodic phenomena, such as the motion of planets around the sun, or waves in many different forms. For the purposes of calculus, we introduce *sine* and *cosine* functions on the unit circle, which is the natural setting that directly exhibits their periodicity. We easily establish their connection to triangles and geometry, but we do not dwell on these aspects. Instead, we focus on proving their differentiability via a geometric argument that directly uses the definition of these functions on the circle. The key analytic ingredient is the verification that $\lim_{t \to 0} \sin t/t = 1$, the familiar limit that is central in all investigations of trigonometric functions. In particular, we avoid the addition formulas for the *sine* and *cosine* functions that are typically required in standard proofs, and whose proofs via basic trigonometry are quite non-trivial. Incidentally, we obtain these formulas later on as an easy application of the differential equation that characterizes the trigonometric functions.

As for additional theoretical results, one of the important facts about differentiable functions in the traditional curriculum is the *Mean Value Theorem*. The standard proof of this result involves a significant detour, something that may be quite puzzling to many students. Again, to help the student to understand better what is going on, we follow a different route that involves a more natural approach that focuses on the key fact that a differentiable function whose derivative is zero at all points of an interval I is necessarily *constant* on I.[11] In particular, it becomes evident that

---

[10]A function $k(x)$ is said to be $o(x - a)$ (read *little oh...*) if $k(x)/(x - a) \to 0$ as $x \to a$.
[11]This was published in a brief note some years ago: R. Michael Range, *On Antiderivatives of the Zero Function*, Math. Magazine **80** (2007), 387–390.

the proof of this intuitively obvious statement requires the completeness of $\mathbb{R}$. The resulting *Mean Value Inequality* is then our main tool to relate geometric properties of the graph of a differentiable function, like increasing or decreasing, and local extrema, to its derivative. Incidentally, this result easily implies the standard Mean Value Theorem for differentiable functions with *continuous* derivative, a mild restriction of concern only to theoretical mathematicians.

The last chapter presents some simple, yet fundamental applications of derivatives, centering on exponential and trigonometric functions. However, this is far from the end of the story. As the expert knows, the next major step involves reversing the process of differentiation, and the closely related idea of definite integrals, with their many significant applications. Beyond that, there are the higher order generalization of the approximation by the tangent line, Taylor series representations of the key transcendental functions, techniques for finding antiderivatives, and so much more. Given that our 2015 book [WiC?] thoroughly discusses these topics, building upon the insights gained while learning about tangents, real numbers, limits, and differentiation, there is no need to include these topics here, as the reader can simply continue the story by referring to that book. In fact, the present book can be viewed as a much more detailed version of the first half of our earlier book, so it naturally prepares the student for the second half of that book. Besides expanding the discussion of the foundations, the main new contribution of the book before you makes explicit and carefully builds upon the connection to the quadratic equation that is a central topic in high school algebra.

Our hope is that this novel introduction to calculus will make it easier for students to learn this important topic, and that it will help them to better understand and appreciate some of the great ideas that have become a central tool in our understanding of the world around us.

Regarding references in text: A reference to, for example Section 4, refers to Subsection 4 in the same section, while Section 2.4 refers to Subsection 4 in Section 2 of the same chapter. If the reference is to an item in a different chapter, the roman numeral of that chapter is included as well, e.g., Section III.1.4.

About the Author

R. Michael Range was born in Germany and grew up in Milano, Italy. He earned his *Diplom in Mathematik* at the University of Göttingen in 1968, where lectures of Hans Grauert, one of the leading experts in *multidimensional complex analysis* of that time, got him hooked to this research area. A Fulbright Fellowship brought him to the United States, where he earned a Ph.D. at UCLA in 1971. He has held academic positions at Yale University, the University of Washington, and the State University of New York in Albany, as well as research positions at institutes in Bonn, Stockholm, Barcelona, and Berkeley. Range has published numerous research articles, and is the author of the book *Holomorphic Functions and Integral Representations in Several Complex Variables* (Springer 1986 and 1998), a widely used introduction to the field. He has lectured in many countries, including mini-courses at the Chinese Academy of Sciences in Beijing, POSTECH in South Korea, and the University of Trondheim, Norway.

More recently he has written historical and expository articles in multidimensional complex analysis, one of which (*Complex Analysis: A Brief Tour into Higher Dimensions*, Amer. Math. Monthly 2003) was recognized with the Lester R. Ford award of the Mathematical Association of America in 2004. He has also spent much time thinking about the calculus curriculum, with the goal of making things easier for the ever increasing number of students with very broad interests. This resulted in *What is Calculus? From Simple Algebra to Deep Analysis*, a book published in 2015 by World Scientific. Range retired from SUNY at Albany in 2015, though he has remained active mathematically. In particular, he gave a lecture in Mexico City in 2021 (via Zoom, because of Covid), *From Cauchy, via Martinelli-Bochner and Leray, to the Henkin-Ramirez Kernel* (Bol. Soc. Mat. Mex. 2023, 29:102). The present book represents his latest efforts towards making Calculus more accessible.

Contents

Chapter I

The Main Characters

I.1 The Rational Numbers

I.1.1 *From Counting Numbers to Integers*

Numbers are the most commonly used tool to count and measure quantities that occur in everyday life as well as in numerous fields of study, such as time, distance, velocity, population size, blood pressure, account balances, profits, rate of inflation, inventories, etc. Relationships between different quantities are then expressed by *functions* of one or several variables, where each of the input variables, as well as the output of the function takes on numerical values.

Thus *numbers* are a fundamental concept, and we need to have a solid understanding of their basic properties. Most people typically first encounter numbers as tools for counting, say counting the fingers on one hand: $1, 2, 3, 4, 5$. Of course, the counting numbers don't stop at 5, or at any other place, as you can always add one more item. Thus the collection of *counting* numbers $1, 2, 3, ...$, also known as the set of *natural* numbers $\mathbb{N}$, is an *infinite* (that is, non-ending) collection or set.

We have learned in elementary school about addition and multiplication of natural numbers, which result again in a natural number, and which enjoy certain very basic properties. Typically, students accept these rules as something obvious or natural, and—especially in the early stages—not much emphasis is given to them. On the other hand, these rules become the guiding principle when we introduce more general numbers, and they determine how the basic operations need to be generalized. It therefore is important that we spell out these rules, so that we can refer to them as needed.

Basic Rules for Addition and Multiplication of Natural Numbers:

i) *commutative*: If $m, n \in \mathbb{N}$, then $m + n = n + m$, and $mn = nm$,

ii) *associative*: if $l, m, n \in \mathbb{N}$, then $(l + m) + n = l + (m + n)$, and $(lm)n = l(mn)$,

iii) *multiplicative identity*: $1n = n1 = n$ for each $n \in \mathbb{N}$,

iv) *distributive*: if $l, m, n \in \mathbb{N}$, then $l(m + n) = lm + ln$.

The associative property ii) shows that brackets, used to signal that the operation inside needs to be carried out first, do not really matter for addition or multiplication separately. In contrast, when the two operations are combined, as in iv), brackets do matter: the expression $l(m + n)$ is not equal to $lm + n$, except in case $l = 1$. This rule is somewhat less natural, and it is important to understand it well and follow it diligently. Unless brackets tell us to carry out addition first, as stated in rule iv), the convention is that multiplication has priority over addition. So, without brackets, $lm + n$ always means $(lm) + n$.

Note that we have not used any particular symbol to denote multiplication: if m, n are two numbers, then mn denotes their product, that is, the number that results by multiplying m and n. This is just a matter of convenience when numbers are identified with specific letters or symbols. However, in order to avoid any possible confusion, it is best to insert a symbol for multiplication, either $\cdot$, or $\times$. For example, if we want to write the product of 3 and 4, we must write $3 \cdot 4$ (or 3×4), since, again by convention, 34 is the shorthand notation for the completely different number $3 \cdot 10^1 + 4 \cdot 10^0$.

Again, these very basic rules i)–iv) carry over to more general numbers that we shall consider later, such as the *integers, rational, real,* or *complex* numbers, as well as to more complicated structures, such as the set $\mathbb{Q}[x]$ of polynomials with rational coefficients (introduced in Section II.4.1). So it is important that we become aware of them early on.

The origins of the natural counting numbers are lost in antiquity. They occur, in different versions and notations, in all ancient societies. First extensions of the natural numbers are much more recent. For example, negative numbers were used in China over 2000 years ago, and in India somewhat later. In Europe, one had to wait until the 16th century, when negative numbers appeared in financial transactions. For example, while keeping track of a person's account balance by counting the units of currency, one could get to a situation that suddenly no money is left (a *zero* balance), or worse, that amounts are being owed, as the individual borrowed currency to make a purchase. How to record this conveniently? That is how the number zero was created, as well as negative numbers, to record when a person had no money or was owing money. Another natural instance came up when people started to record temperatures. For example, in the Celsius scale, the freezing temperature of water was defined to be 0^0 (the superscript 0 identifies degrees), referred to as 0 degrees Celsius (0^0 C), and thermometers showed negative numbers to record temperatures colder than 0^0.

Eventually, zero and the negative numbers were formalized, making sure that the rules i)–iv) known for natural numbers continue to hold, leading to what has become known as the set of *integer numbers*, which is denoted by $\mathbb{Z}$. We shall go through this process in detail in order to show the student in the simplest case how "following the natural rules" determines exactly how to proceed. In other words, the resulting properties, such as "negative $\times$ positive = negative", and so on, are not at all arbitrary conventions, but are the consequence of applying the most natural rules known from the counting numbers $\mathbb{N}$ consistently. This process will be repeated later on as we advance to more complicated number systems and more general structures. While the details may seem tedious and boring, we ask the student to be patient and try to understand the process.

We begin by introducing a new number 0 (zero) and we define its addition with a natural number $n \in \mathbb{N}$ by $0 + n = n$, and of course we also set $n + 0 = n$, so that i) continues to hold. In analogy to iii), we call 0 the *additive identity*. Next, for each $n \in \mathbb{N}$ one introduces a new number $-n$, called the *additive inverse of* n, whose critical defining property is that the addition of $-n$ and n satisfies

$$n + (-n) = (-n) + n = 0.$$

The additive inverse $-n$ of a natural number n is called a *negative* number, while the natural numbers themselves are called *positive*. The number 0 is neither positive nor negative. Note that the defining equation can be used to also introduce the additive inverse of a negative number $(-n)$. In fact, that equation tells us that $n \in \mathbb{N}$ satisfies the critical property that makes n the additive inverse of $-n$, so we have $-(-n) = n$. By adding to $\mathbb{N}$ the new numbers 0 and all the additive inverses of the natural numbers, one obtains the set of *integer* numbers $\mathbb{Z}$. Each $m \in \mathbb{Z}$ has an additive inverse $-m$, which satisfies $(-m) + m = m + (-m) = 0$. Clearly $-0 = 0$.

One then extends addition in a natural way from $\mathbb{N}$ to $\mathbb{Z}$, so that the properties i) and ii) are preserved, and so that 0 is the additive identity in $\mathbb{Z}$, that is, the property

v) *additive identity*: $0 + m = m + 0 = m$ for each $m \in \mathbb{Z}$
holds.

This procedure is best visualized by writing $\mathbb{Z}$ as a set that extends indefinitely on both sides, i.e.

$$\mathbb{Z} = \{..., -4, -3, -2, -1, 0, 1, 2, 3, 4, ...\}.$$

We know that if m and n are natural numbers, then $m + n$ is the natural number that is obtained by starting at m and moving n places to the right. We do the same thing if m is an arbitrary integer, so, for example, $-4 + 6 = 2$, since 2 is 6 places to the right of -4. We add the *negative* number $(-n)$ by moving n places to the *left*, which is of course consistent with $n + (-n) = 0$. One then must check that properties i) and ii) hold for addition of numbers in $\mathbb{Z}$, but we shall skip these uninspiring details. We note that in every day life, the expression $m + (-n)$ is usually denoted by $m - n$; this "subtraction" is not really a new operation, it is just a convenient notation for a special type of addition.

The rules imply that the additive inverse of every $m \in \mathbb{Z}$ is *unique*. In fact, if two integers p and q are additive inverses of m, that is, if they satisfy $p + m = 0 = q + m$, then adding $(-m)$ to both sides gives

$$(p + m) + (-m) = (q + m) + (-m),$$

and hence, by ii),

$$p + (m + (-m)) = q + (m + (-m)).$$

This last equation then implies

$$p + 0 = q + 0, \text{ that is, } p = q \text{ by rule v).}$$

We can thus add the following property or rule to our list:

vi) *additive inverse*: for each $m \in \mathbb{Z}$ there is a unique number $-m$ in $\mathbb{Z}$ which satisfies $m + (-m) = 0$.

Note that if $m \neq 0$, then $-m \neq 0$ as well, because otherwise we would have $m = m + 0 = m + (-m) = 0$, which contradicts the assumption $m \neq 0$.

To see how multiplication must be defined in $\mathbb{Z}$, subject to the rules i)–vi), we first show that iii) extends to all numbers in $\mathbb{Z}$.

Lemma 1 *The number 1 is the multiplicative identity for all integers in $\mathbb{Z}$.*

Proof. We begin by analysing what the rules imply for $1 \cdot 0$. Note that

$$1 \cdot 0 + 1 = 1 \cdot 0 + 1 \cdot 1 = 1(0 + 1) = 1 \cdot 1 = 1,$$

where we have used iii) and iv). Now add (-1) to both sides of $1 \cdot 0 + 1 = 1$ to obtain $1 \cdot 0 + 0 = 0$, so that $1 \cdot 0 = 0$. Next, if $n \in \mathbb{N}$, we already know that $1n = n$ by iii). So we are left with examining $1(-n)$. Note that

$$1(-n) + n = 1(-n) + 1n = 1[(-n) + n] = 1 \cdot 0 = 0,$$

where we again have used the distributive rule iv). This last equation shows that $1(-n)$ is the additive inverse of n, that is, we have verified that

$$1(-n) = -n. \blacksquare$$

Next we show that the rules imply that $0 \cdot n = 0$ for every $n \in \mathbb{Z}$. To see this, consider

$$0 \cdot n + n = 0 \cdot n + 1 \cdot n = (0 + 1) \cdot n,$$

where we have used iii) extended to all integers, and iv). Since by v) and iii) one has $(0 + 1)n = 1n = n$, we see that $0n + n = n$. Adding $(-n)$ on each side, and using ii) gives $0n + 0 = 0$; since the left side equals $0n$ by v), we obtain the desired result.

Lemma 2 *For any integer n we must have*

$(-1)n = -n$, *that is, $(-1)n$ is precisely the additive inverse of n.*

Proof. Note that $(-1)n + n = (-1)n + 1n = [(-1) + 1]n$ by iv) and iii), and the latter expression equals $0n = 0$. So $(-1)n$ is indeed the additive inverse of n. $\blacksquare$

We can now easily see how multiplication must work in $\mathbb{Z}$ in general, starting with the known multiplication for natural numbers. Suppose n, m are natural numbers. We want to see what $(-n)m$ must be. Just note that by the Lemma we just proved, $(-n) = (-1)n$, so that $(-n)m = [(-1)n]m = (-1)(nm) = -(nm)$, by the associative rule ii) and the above Lemma. By i), this of course also shows that $m(-n) = -(mn) = -(nm) = (-n)m$, so that i) continues to hold. Finally, we must consider $(-n)(-m)$. As before, this expression equals $((-1)n)((-1)m) = (-1)[(-1)(nm)]$, where we have used i) and ii). Since $(-1)(nm) = -(nm)$, we see that we must have

$$(-n)(-m) = -[-(nm)] = nm.$$

So by strictly following the rules i)–vi) we have determined how multiplication *must* be defined for arbitrary integers. In other words, if we want to continue following the rules, there is only one way in which multiplication can be extended from $\mathbb{N}$ to $\mathbb{Z}$.

We now turn matters around and use the results we just obtained to *define* for $m, n \in \mathbb{N}$:

$$(-n)m = -(nm),$$

and

$$m(-n) = -(mn).$$

By rule i) applied to $m, n \in \mathbb{N}$, we see that $(-n)m = m(-n)$. Furthermore, we *define*

$$(-m)(-n) = mn = nm = (-n)(-m).$$

If either m or n equals 0, then the above products are all $= 0$, which is consistent with the rules as seen above. Note that the given definitions imply that multiplication remains commutative in $\mathbb{Z}$, that is, rule i) still holds in $\mathbb{Z}$.

In particular, we have seen that the properties

negative times positive equals negative,

and

negative times negative equals positive,

which typically baffle beginners, are not at all weird arbitrary choices, but are the consequence of following the same simple basic rules that we are familiar with from the natural numbers $\mathbb{N}$ in the more general case when we consider all numbers in $\mathbb{Z}$.

There is one important task left. While we followed the rules as we determined how addition and multiplication should work in $\mathbb{Z}$, this does not guarantee that the final result does indeed follow the rules. In particular, one must check that rules i)–iv), known to hold for *natural* numbers, hold for arbitrary *integers* as well. Note that we already saw that multiplication satisfies rule i), and Lemma 1 shows that rule iii) holds in $\mathbb{Z}$ as well. We shall skip the somewhat uninspiring task to verify the remaining rules ii) and iv), but some of the steps are outlined in Exercises 1–4. We also note that the additional rules v) and vi) have been introduced and hold for all integers $m \in \mathbb{Z}$. We therefore have verified the following theorem.

Theorem 3 *Addition and Multiplication for Integer Numbers $\mathbb{Z}$ satisfy the following basic rules:*

Q1) *commutative*: If $m, n \in \mathbb{Z}$, then $m + n = n + m$, and $mn = nm$,

Q2) *associative*: if $l, m, n \in \mathbb{Z}$, then $(l + m) + n = l + (m + n)$, and $(lm)n = l(mn)$,

Q3) *multiplicative identity*: $1m = m1 = m$ for each $m \in \mathbb{Z}$,

Q4) *distributive*: if $l, m, n \in \mathbb{Z}$, then $l(m + n) = lm + ln$,

Q5) *additive identity*: $0 + m = m + 0 = m$ for each $m \in \mathbb{Z}$,

Q6) *additive inverse*: for each $m \in \mathbb{Z}$ there is a unique number $-m$ in $\mathbb{Z}$ which satisfies $m + (-m) = 0$.

We have labeled the rules Q1–Q6, in anticipation of applying them to the set of *rational* numbers $\mathbb{Q}$ that will be introduced in Section 1.3. To place all this in a larger context, a set R in which two operations *addition* and *multiplication* are defined, subject to satisfying the rules Q1–Q6, is called a *commutative ring with identity*. In more advanced abstract algebra, such rings are among the common structures that are investigated, and beyond the larger number systems that we shall consider, there are many other examples that arise naturally in applications.

I.1.2 *Order Properties*

We have already freely used the terms positive and negative to distinguish one side of the number line (where the natural numbers are), from the other side (where all the additive inverses of the natural numbers are). We shall now formalize these concepts by introducing the precise rules that are satisfied in $\mathbb{Z}$, so that we can rely on them when later on we consider rational or real numbers.

Moving from 0 towards 1 on the number line identifies the preferred direction of the line (usually displayed from left to right, or from bottom to top). This direction reflects the natural order in the numbers. Numbers n to the right of zero, i.e., on the side where 1 is, are said to be *greater than zero* (we write $n > 0$), and are also called *positive*, those on the left are said to be *less than zero* ($n < 0$) and are called *negative*. Remember that for the time being the numbers we consider are integers.

One has the following evident property:

O1) *For any number n exactly one of the following three distinct possibilities must hold: either $n = 0$, or $n > 0$, or $n < 0$.*

Given two numbers n and m, we say n *is less than* m (written $n < m$) if $n - m < 0$, and we say n *is greater than* m (written $n > m$) if $n - m > 0$.

By applying O1) to $n - m$ instead of n, one sees that given any two numbers n, m, one of the following must hold: either $n = m$, or $n < m$, or

$n > m$. It is convenient to introduce the notation $n \leq m$ (n *is less than or equal to* m), which can also be written $m \geq n$ (m *is greater than or equal to* n), to mean that either $n = m$ or $n < m$.

The natural ordering interacts with arithmetic operations according to precise rules as follows.

O2) *Given two positive numbers* $n, m > 0$, *then their sum* $n + m > 0$.

O3) *Given two positive numbers* $n, m > 0$, *then their product* $nm > 0$.

Let us mention that these rules imply that

$$\text{if } m \cdot n = 0, \text{then either } m = 0 \text{ or } n = 0;$$

in fact we saw that if both are different from 0, that is, are either positive or negative, then their product is either positive or negative, so different from 0 as well. A fancy way to describe this property is to say that in the integers $\mathbb{Z}$ the number 0 has NO non-zero divisors.

In these rules, as well as in all other rules discussed below, one may replace $<$ by $\leq$, respectively $>$ by $\geq$.

Clearly these rules just formalize familiar basic properties of natural numbers. Other rules then follow. For example, if $a < b$ and $b < c$, then it follows that $a < c$. This is known as the *transitive* property. In fact, since $b - a > 0$ and $c - b > 0$, O2) implies $(b - a) + (c - b) > 0$. By using the basic rules of $\mathbb{Z}$, it then follows that

$$0 < [b + (-a)] + [c + (-b)] = [c + (-a)] + [b + (-b)] = c - a + 0 = c - a,$$

i.e., $a < c$.

Furthermore, assume that $a < b$ and let c be any other number. Then $a + c < b + c$. The proof is left as an exercise. The analogous multiplicative version $ac < bc$ however holds only if c is positive! In fact, by assumption, $b - a > 0$, so if $c > 0$, by O2), we get $(b - a)c > 0$. By Q4 this implies $bc - ac > 0$, i.e., $ac < bc$. However, replacing c with $-c$, one obtains

$$(b - a)(-c) = (b - a)[(-1)c] = (-1)[(b - a)c] = (-1)(bc - ac).$$

This latter number is the additive inverse of the positive number $bc - ac$, and hence it is negative. Therefore $[b(-c) - a(-c)] < 0$, which means that $a(-c) > b(-c)$, so that multiplication by the negative number $-c$ reverses the inequality! For example, multiplication of $2 < 3$ by (-1) on both sides gives $-2 > -3$; note that on the number line the number -2 is to the right of -3.

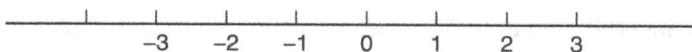

Fig. I.1   Position of −3 and −2 compared to 2 and 3.

We emphasize again that these properties are not at all arbitrary, but they are a consequence of the need to work with numbers in a logically consistent manner, starting with the familiar basic properties of the natural numbers. For example, we had seen that the basic rules Q1–Q6 imply that for an integer n one has $(-1)n = -n$ (= the additive inverse of n, i.e., that unique number that satisfies $n + (-n) = 0$). And we had seen that this property is at the heart of the familiar rules for multiplication

$$negative \times positive = negative, \text{ and } negative \times negative = positive.$$

Finally, we introduce the *absolute value* $|a|$ of a number a; it is defined by

$$|a| = \begin{cases} a & \text{if } a = 0 \text{ or } a > 0 \\ -a & \text{if } a < 0. \end{cases}$$

So $|a| \geq 0$ for any number a. For example, $|-4| = 4$, $|4| = 4$. Geometrically, $|a|$ measures the *distance* between a and 0 on the number line. More generally, if a, b are two numbers, $|a - b| = |b - a|$ measures the distance between the two corresponding points on the number line. One then has the following most useful property, known as the "triangle inequality,[1] which states that

$$|a + b| \leq |a| + |b|. \tag{I.1}$$

This fact is obvious if a and b are both ≥ 0, so that $a + b \geq 0$ as well. Similarly, if both $a, b < 0$, then $a + b < 0$. Hence $|a + b| = -(a + b) = (-1)(a + b) = (-1)a + (-1)b = |a| + |b|$. So the interesting part is when one number, say a, is ≥ 0, and $b < 0$. Suppose first that $a + b \geq 0$; then $|a + b| = a + b = |a| + b < |a|$, since $b < 0$. Finally, if $a + b < 0$, then $|a + b| = (-1)(a + b) = (-a) + (-b) = (-a) + |b| \leq |b|$, since $-a \leq 0$. In either case, $|a + b| \leq |a| + |b|$.

Replacing b with $-b$ results in the equivalent estimate

$$|a - b| \leq |a| + |b| \text{ for all } a \text{ and } b.$$

---

[1]This name originates by considering the distance of points in the plane. In that setting the corresponding inequality states that in a triangle the length of one side is less than or equal to the sum of the lengths of the other two sides.

By applying the estimate (I.1) to $|a| = |(a - b) + b|$, one obtains $|a| \leq |a - b| + |b|$, i.e., $|a| - |b| \leq |a - b|$. This same estimate holds if a and b are interchanged, so that $|b| - |a| \leq |b - a| = |a - b|$. It follows that

$$|a - b| \geq ||a| - |b||.$$

One also has the following property for multiplication:

$$|a \cdot b| = |a| \cdot |b| \text{ for all } a, b.$$

The proof is left as an exercise.

The reader should become familiar with all these rules, as they will be used extensively in all subsequent work.

I.1.3   *Multiplicative Inverses of Integers*

If we carefully look over the rules Q1 to Q6, we notice that, compared with addition, something is missing for multiplication, namely the concept of a "multiplicative inverse", in analogy to the existence of an additive inverse. Given $n \in \mathbb{Z}$, such a multiplicative inverse, that we label n^{-1}, should satisfy $n(n^{-1}) = 1$. Clearly we must have $1^{-1} = 1$ (just like $-0 = 0$) and $(-1)^{-1} = -1$ (Why? Check it out!). But aside from these trivial cases, we can't find any other multiplicative inverses in $\mathbb{Z}$. On the other hand, there clearly cannot exist a multiplicative inverse for 0 if we insist on the rules Q1 to Q6 to continue to hold. In fact, suppose there were a "number" 0^{-1} so that $0(0^{-1}) = 1 = (0^{-1})0$. Let m, n be any two integers. Then $0m = 0 = 0n$. It then would follow that

$$0^{-1}(0m) = 0^{-1}(0n),$$

and hence, by Q2,

$$(0^{-1}0)m = (0^{-1}0)n.$$

The property $(0^{-1})0 = 1$ would then imply $1m = 1n$, that is, $m = n$. Thus any two integers would have to be equal, which is clearly absurd. So we have learned the following:

The basic rules Q1 to Q6 imply that there cannot exist a multiplicative inverse for 0.

Just like in order to "create" additive inverses for natural numbers we had to postulate their existence, and simply add these new "numbers" to $\mathbb{N}$ and $\{0\}$, thereby obtaining the set of integer numbers $\mathbb{Z}$, we now create a new set $\mathbb{Q}$ called the set of *rational* numbers by first adding the set $\{m^{-1} :$

m any nonzero integer}, to $\mathbb{Z}$, and thereafter adding anything else that may be required by the rules. Using the rules Q1 to Q6 we then try to extend the familiar addition and multiplication from $\mathbb{Z}$ to all of $\mathbb{Q}$, making sure that these rules continue to hold. Hopefully, this will lead us to precise answers, just as it happened in the extension from $\mathbb{N}$ to $\mathbb{Z}$.

Before getting into the technical details, let us try to understand why it may be useful to add multiplicative inverses and consider the resulting set of rational numbers. For example, suppose you have 5 dollars, and you want to divide this equally among 5 friends. This is easy if you are holding five \$1 bills: each person gets one bill. But suppose you have one \$5 bill. Then you have to think of what would be one fifth (i.e., 1/5) of your \$5 bill. For example, you could cut up the bill in five equal parts, and hand out the five pieces. (Maybe your friends are lucky, and they find a bank that would trade 1/5 of the \$5 bill for a regular \$1 bill.) In any case 5 times 1/5 of a \$5 bill gives you **one** \$5 bill, so $5 \cdot (1/5) = 1$, so 1/5 qualifies as a multiplicative inverse of 5. Similarly, if you want to equally divide \$1 among 4 people, each one clearly should get 25c, that is one quarter. Note that the corresponding coin is actually called a "quarter", reflecting the fact that it is 1/4 of \$1. Again, 1/4 clearly is a multiplicative inverse of 4. Any measurement with a ruler also involves fractions. Note that 1 inch is 1/12 of a foot. Measurements like $1/4, 5/8$ or 3/16 of an inch are common. If a metric ruler is used, other fractions are more common, say 5 cm, that is, 5/100 of a meter. Another situation arises if you need to divide a whole pizza among four people. It's easy to cut up the pizza into four equal pieces, each piece is then 1/4 of the whole thing, and again the four equal pieces add up to the whole pizza. Note that while in principle the same process applies if you need to divide the pizza among 3 friends, it is quite a bit trickier to divide a pizza into 3 equal pieces. Or if you try to divide \$1 among three people, each one should get a little bit more (to be precise, 1/3 of a cent more) than 33 cents. We see that the multiplicative inverse of 3, while conceptually it is clear what is meant, in practical applications may be somewhat troublesome. Another complication arises when you realize that 2 quarters of a pizza make up one half of the pizza, so $2/4 = 1/2$. We see that while in every day life fractions, i.e., multiplicative inverses, are quite common, even indispensable, their precise formulation is bound to be somewhat complicated. That is probably the reason why most people have trouble working with fractions, or even simply hate them and try to avoid them. But that is not easy: suppose you are a cook and you have a recipe that serves 4 people, but you have to prepare it to serve 6 people.

The preceding discussion suggests that multiplicative inverses are very closely related to fractions, as used in common language, for example we saw that $4^{-1} = 1/4$, or $100^{-1} = 1/100$. In fact, the notations n^{-1} and $1/n$ are equivalent, both refer to the multiplicative inverse of n. So for $n \neq 0$ we write $n^{-1} = 1/n$, although at this point we do not assume any precise knowledge of fractions. The key property of this new object $n^{-1} = 1/n$, regardless of how it may actually be "constructed", is the fact that multiplication of it with n is defined by

$$n \cdot (1/n) = 1 = (1/n) \cdot n.$$

Next, let us observe that the usual rules imply that there can only be one multiplicative inverse. In fact, if $np = 1$, multiplying by n^{-1} gives $n^{-1}(np) = n^{-1} \cdot 1 = n^{-1}$ by Q3, and using Q2 and Q3 then gives

$$n^{-1}(np) = (n^{-1}n)p = 1p = p,$$

so that $p = n^{-1}$. We also note that Q1 and Q2 imply that one must have $m^{-1}n^{-1} = (mn)^{-1}$; in fact, just multiply with mn and use the rules to obtain

$$m^{-1}n^{-1}(mn) = m^{-1}(n^{-1}n)m = m^{-1}1m = m^{-1}m = 1,$$

which proves the claim.

Finally, we need to consider the multiplication of an arbitrary integer m with n^{-1}, where of course $n \neq 0$. This defines a new "number" $m \cdot n^{-1} = mn^{-1}$ (unless m or n equal 1), and in order to satisfy rule Q1, we must require that $mn^{-1} = n^{-1}m$. Using fractional notation, we have

$$m(1/n) = (1/n)m,$$

or

$$m\frac{1}{n} = \frac{1}{n}m,$$

so that we can use $\frac{m}{n}$ unambiguously as a notation for either side, since the two sides are equal. Following customary language, we call the number m on top in the fraction $\frac{m}{n}$ the *numerator*, while the number n in the bottom is called the *denominator*.

As a first attempt, we can thus take as the set of rational numbers $\mathbb{Q}$ the collection of fractions

$$\left\{ \frac{m}{n} : \text{where } m, n \in \mathbb{Z} \text{ and } n \neq 0 \right\}.$$

Note that any $m \in \mathbb{Z}$ can be written as $m = m \cdot 1 = m \cdot 1^{-1} = \frac{m}{1}$, and hence is included in the above set.

As suggested by the earlier motivational discussion, certain fractions really represent the same number, for example $\frac{1}{2} = \frac{5}{10}$, and our first task is to understand this precisely.

Lemma 4 *The usual rules require that for integers m, n, p, q, and $n, q \neq 0$, one has $\frac{m}{n} = \frac{p}{q}$ if and only if $mq = np$.*

Proof. Assume $\frac{m}{n} = \frac{p}{q}$, that is $mn^{-1} = pq^{-1}$. Multiply by nq on both sides and use the rules Q1 and Q2 to rearrange the terms to obtain

$$mq(nn^{-1}) = np(qq^{-1}), \text{ which gives } mq1 = np1, \text{i.e, } mq = np.$$

For the converse, start with $mq = np$ and multiply both sides by $n^{-1}q^{-1}$. Again, using Q1 and Q2, it follows that $(mn^{-1})(qq^{-1}) = (pq^{-1})(nn^{-1})$, which implies $mn^{-1} = pq^{-1}$. ∎

The conclusion then is that the set of rational numbers $\mathbb{Q}$ is a collection of elements, each of which can be represented as a fraction in numerous way, with the above Lemma describing the precise condition for two fractions to represent the same rational number.

This property stated in the Lemma in particular implies the familiar rule that common non-zero factors in numerator and denominator can be introduced or deleted without changing the number represented by the fraction. In fact, suppose $l \neq 0$. Then rearrange $mnl = mnl$ as $m(nl) = n(ml)$; the Lemma then implies

$$\frac{m}{n} = \frac{ml}{nl}.$$

This property is used widely in all work involving fractions.

Let us mention that the unique factorization of natural numbers into products of prime numbers implies that this is indeed the only way that two fractions can represent the same number. In fact, suppose $\frac{m}{n} = \frac{p}{q}$. Let us assume first that all integers are positive. By cancelling common prime factors in numerator and denominator on the left side according to the rule we just mentioned, we can assume that m and n have no common prime factors. By the Lemma, $mq = np$. By assumption, each prime factor of m does not occur as a factor in n on the right side, so must be a factor of p, that is, m itself is a factor of p, i.e., $p = lm$ for some $l \in \mathbb{N}$. So $mq = n(lm)$. Multiplying both sides by m^{-1} and using Q1 and Q2, one obtains

$$(mm^{-1})q = nl(mm^{-1}),$$

which implies $q = nl$. It follows that

$$\frac{m}{n} = \frac{p}{q} = \frac{ml}{nl}.$$

The general case is easily reduced to this special case by using $-m = (-1)m$ and so on, as needed, and keeping track of the factors (-1). We leave the details as an exercise.

After these preliminaries, let us now define addition and multiplication of rational numbers, i.e., fractions. As usual, we shall be guided by requiring that the rules Q1 and Q6 hold.

As for addition, let us begin with two fractions that have the same denominator, i.e., we want to define $\frac{a}{n} + \frac{b}{n}$. We rewrite this as $an^{-1} + bn^{-1} = (a+b)n^{-1}$ by rule Q4. The latter can be written as $\frac{a+b}{n}$. The general case now follows easily as follows. Consider $\frac{a}{n}$ and $\frac{b}{m}$. By changing to the equivalent fractions $\frac{am}{nm}$ and $\frac{bn}{mn}$, and since $nm = mn$, we now have the same denominator. Consequently,

$$\frac{a}{n} + \frac{b}{m} = \frac{am}{nm} + \frac{bn}{nm} = \frac{am + bn}{nm}.$$

While this may look complicated, don't try to memorize the final outcome. Instead, just remember the basic principles that we applied. It also follows that $0 = \frac{0}{1}$ is the additive identity for all rational numbers, and that the additive inverse is given by $-\frac{a}{n} = \frac{-a}{n}$. It is now easy to check that addition for rational numbers satisfies rules Q1, Q2, Q5 and Q6.

Next we consider multiplication. Again, assuming Q1 and Q2, it is clear that we must have

$$\frac{a}{n}\frac{b}{m} = \frac{ab}{nm}.$$

Just rewrite the product on the left as $(an^{-1})(bm^{-1}) = (ab)(n^{-1}m^{-1}) = (ab)(nm)^{-1}$, following the rules. Clearly $1\frac{a}{n} = \frac{1}{1}\frac{a}{n} = \frac{1a}{1n} = \frac{a}{n}$, so that 1 is the multiplicative identity for all rational numbers. Finally, every nonzero number $q \in \mathbb{Q}$ has a unique *multiplicative inverse*. In fact, if $q = \frac{a}{n}$ with $a \neq 0$, then one has

$$\frac{n}{a}\frac{a}{n} = \frac{na}{na} = 1,$$

so that $\frac{n}{a}$ is indeed the multiplicative inverse of $q = \frac{a}{n}$. As usual, we denote this multiplicative inverse of q by q^{-1} or $1/q$.

We thus can add one more rule to our collection Q1–Q6, as follows.

Q7 *Multiplicative inverse in* $\mathbb{Q}$: every rational number $q \neq 0$ has a multiplicative inverse q^{-1} in $\mathbb{Q}$, defined by the property that $qq^{-1} = 1$.

It now follows that if $q_1 q_2 = 0$ for two rational numbers, then at least one of them must be zero. In fact, if one of them, say $q_1 \neq 0$, then by Q7 there is a multiplicative inverse q_1^{-1}. Multiplying both sides of $0 = q_1 q_2$ by q_1^{-1} and applying Q2, one obtains $0 = q_1^{-1}(q_1 q_2) = (q_1^{-1} q_1) q_2 = 1 q_2 = q_2$.

The careful reader may be wondering whether the outcome of addition or multiplication depends on the particular form of the fraction chosen. For example, if $\frac{a'}{n'} = \frac{a}{n}$ and $\frac{b'}{m'} = \frac{b}{m}$, does it follow that $\frac{a'}{n'} + \frac{b'}{m'} = \frac{a}{n} + \frac{b}{m}$? The answer is indeed yes, and similarly for multiplication. The details are left as an exercise.

Finally, it is straightforward to check that multiplication in $\mathbb{Q}$ is commutative and associative. It takes a little bit more work to check that the distributive property Q4 also holds in $\mathbb{Q}$, but the verification is really quite routine, and we leave it as an exercise.

Next, we extend the order properties to $\mathbb{Q}$ by defining that $q = \frac{m}{n}$ is positive ($q > 0$) if both m and n are positive. Since $\frac{m}{n} = \frac{m(-1)}{n(-1)} = \frac{-m}{-n}$, we see that q is also positive if both numerator and denominator are negative. The remaining possibility for $q \neq 0$ is that numerator is positive and denominator is negative, or vice versa. In that case we say that q is negative ($q < 0$). It is then clear that rule O1 holds, and it readily follows that O2 and O3 hold as well in $\mathbb{Q}$. Note that $q < 0$ if and only if its additive inverse $-q$ is positive.

More generally, if p and q are rational numbers and $p - q > 0$ (i.e. $p + (-q) > 0$), we write $p > q$ (*p is greater than q*) or, equivalently, $q < p$ (*q is smaller (or less) than p*). Similarly, $p \geq q$ is equivalent to $p - q \geq 0$.

Once we have become familiar with rational numbers and with fractions m/n, where m and n are integers, it is useful to mention that the various rules we have established for working with such fractions carry over to more general fractions p/q, where p and $q \in \mathbb{Q}$, with $q \neq 0$. The meaning is clear, once we recall that $1/q = q^{-1}$ is just the multiplicative inverse of q. Thus, for example,

$$\frac{p}{q} = \frac{r \cdot p}{r \cdot q} \text{ for any nonzero } r \in \mathbb{Q}.$$

In fact, $\frac{r \cdot p}{r \cdot q} = rp(rq)^{-1} = (rp)(r^{-1} q^{-1}) = (rr^{-1})(pq^{-1})$, where we have used rules Q1 and Q2. The latter expression clearly equals $1 \cdot pq^{-1} = \frac{p}{q}$. The verification of the other rules is equally straightforward, and we shall skip the details.

I.1.4 *Rational Numbers on the Number Line*

The rational numbers $\mathbb{Q}$ that we just introduced in detail, including all the basic rules Q1 to Q7 that are satisfied, are the most basic set of numbers that needs to be considered in every day life. Unfortunately, fractions are quite a bit more complicated than just the natural numbers or even the integers, but as we indicated at the beginning of the discussion, they really occur quite naturally in many situations. Our presentation highlighted the basic rules as the guiding principle in their construction. These rules will continue to hold later on, when we will be forced to consider larger number systems, such as the real numbers $\mathbb{R}$, or the complex numbers $\mathbb{C}$. However, until we reach that point we shall mainly use the rational numbers in order to keep matters as simple as possible.

It is most useful to be able to visualize the rational numbers. The basic tool is the "number line". Essentially, this is just a ruler that extends indefinitely in both directions. We think of the line made up of a huge collection of "points", and the goal is to identify numbers with points on the line. This is most easily done for integers, as follows. Consider the following small section of the line.

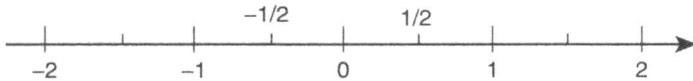

Fig. I.2

The key point is to mark the numbers 0 and 1 on the line (or ruler), usually 1 to the right of 0 (or above it, if the ruler is drawn vertically). Once the length of the segment from 0 to 1 has been fixed, we keep adding that same length on the right side, thus identifying the natural numbers 1, 2 $(= 1 + 1)$, 3 $(= 2 + 1), 4, 5, ...$, and so on, while on the left side of 0 we successively add (-1), that is, we move one segment of the same length as before to the left, to mark the numbers $-1 \, (= 0 + (-1))$, $-2 \, (= (-1) + (-1))$, $-3 \, (= (-2) + (-1))$, -4, -5, ... In essence, we had already displayed the integers $\mathbb{Z}$ in this way earlier, without showing the underlying line (i.e., the ruler).

Once the integers are marked on the line, each rational number (i.e., fraction) is identified with a unique point on the line. Figure I.2 shows the numbers $1/2$ and $-1/2$. The process is clear. For example, to identify the fraction $-3/4$ on the line, we divided the segment from -1 to 0 into 4 equal parts, and starting at 0, we go to the left endpoint of the $3rd$ part: that marks the point $-3/4$ on the line. Here is one more, slightly more complicated example. Consider $11/7$. We note that $11/7 = \frac{7}{7} + \frac{4}{7} = 1 + \frac{4}{7}$. So $11/7$ lies between 1 and 2. We divide the segment from 1 to 2 into 7 equal parts (never mind that this may not be quite simple to carry out exactly); the right endpoint of the 4th segment (beginning at 1) then identifies the point that corresponds to $11/7$.

Let us just remind you of the ordering that we had already introduced earlier. Numbers to the right of 0 (where the natural numbers are) are called *positive*. The statement that q is positive is also denoted by $q > 0$ (q is greater than 0). The numbers on the left side (where the additive inverses of the natural numbers lie) are called *negative*, also denoted by $q < 0$ (q is less than 0).

Rulers that are used in practical work are just finite segments of the number line, typically beginning with 0 on the left side. (But note that thermometers can be viewed as rulers also, and usually show some negative numbers as well.) The actual physical length of the segment from 0 to 1 is determined by the scale used, and by the system of units that are used (typically either metric, or anglo-saxon). Thus that length could be exactly 1 cm, 1 meter, 1 inch, or 1 foot, to give some examples. As for displaying fractions on a ruler, there are key differences: metric rulers highlight fractions with denominators $10, 100, 1000, ...$, while anglo-saxon rulers favor fractions with denominators $2, 4, 8, 16$, and so on. In any case, every rational number corresponds to a specific point on the number line.

We notice that the rational numbers are "dense" on the line, that is, our eyes cannot detect any gaps between them. Between any two *different* rational numbers $q_1 < q_2$, no matter how close they are to each other on the number line, the midpoint $(q_1 + q_2)/2$ always lies in between them, and, in fact, continuing this process, one sees that there actually are *infinitely* many rational numbers between q_1 and q_2. No wonder we cannot see any gaps! This justifies to view the (number) line as just consisting of rational numbers for much of our initial work. The physical visualization does show the number line as being covered by the rational numbers, and for our eyes, at least for the time being, there is no need to wonder if there is anything else on the number line besides the rational points.

On the other hand, the precise reality is much more complicated. It turns out that as one looks more closely at the number line (say with a super magnifying glass), one would see an unimaginable quantity of "tiny holes" scattered among the rational numbers. This "empty" space, i.e., points on the number line that are not rational, is indeed "real", that is, it cannot just be ignored. This fact, first noticed by Greek philosophers in the 4th century B.C., ranks among the great discoveries of the human mind. It shattered the belief that all observed quantities could be measured by integers and ratios between them (i.e., fractions).

Fact: *The diagonal in a square of side one cannot be measured exactly by a ruler that just includes rational numbers.*

We analyse the simple argument. Consider the unit square placed so that one side covers the number line from 0 to 1, as shown in Fig. I.3. By Pythagoras' Theorem, the length d of its diagonal satisfies $d^2 = 1^2 + 1^2 = 2$. The diagonal is rotated onto the number line, thereby identifying a point at distance d from 0.

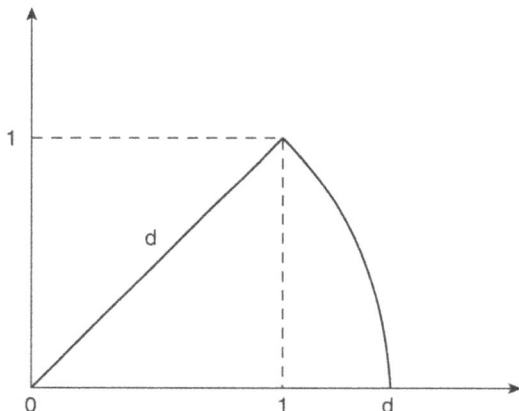

Fig. I.3   Diagonal d in a square of sides 1.

The startling fact, as we shall explain in a moment, is that there is NO fraction (i.e., rational number) $d = m/n$ that satisfies $d^2 = 2$. We are thus forced to conclude that this point d on the number line cannot be represented by a rational number, that is, d is an "irrational" number. *The number line contains more than just rational numbers!*

So how do we see that no rational number m/n satisfies $(m/n)^2 = 2$? Suppose we had integers m, n that satisfy this equation. Clearly we must have $n \neq 0$, and we can also assume that $m, n > 0$. By the properties we established, we can cancel all common factors in numerator and denominator. Suppose that has been done, so that m and n cannot both contain a factor 2, i.e., we can assume that m and n are not *both* even. From $(m/n)^2 = 2$ one obtains $m^2 = 2n^2$, so m^2 is even. Since the square of an odd number is odd (check it!), this implies that m itself must be even, so $m = 2p$ for some integer p. Therefore $2n^2 = (2p)^2 = 4p^2$. After dividing by the common factor 2 on both sides one gets $n^2 = 2p^2$, so n^2 is even. Again, this implies that n itself is even. So the assumption $(m/n)^2 = 2$ leads to the conclusion that both m and n are even, but that had been ruled out at the very beginning by cancelling all common factors! This contradiction shows that it is impossible to find integers m, n with $(m/n)^2 = 2$, that is, the "number" d (i.e., the point d that we identified) that satisfies $d^2 = 2$, which surely exists on the number line (the diagonal of the unit square!), and that is usually denoted by $\sqrt{2}$, is not a *rational* number.

Note that on a *practical* level the matter that $\sqrt{2}$ is not a rational number is not that important. We can always approximate the length of the diagonal by rational numbers to any desired degree of accuracy, say, by 1.414, or 1.41421, and so on. (See Section 5 below for details about these decimal representations.) Any calculator displays $\sqrt{2}$ as a rational number, typically showing 8 or 12 digits, and usually no one worries that this is not "exact". Again, this explains why in everyday life the matter is largely ignored, and why it is alright for us to continue our presentation without carefully analyzing this remarkable phenomenon at this time. After all, mankind, including philosophers, geometers, and mathematicians, basically moved on for about 2000 years without making much of an issue of the existence of irrational numbers. On the other hand, mathematicians in the 19*th* century eventually were confronted by unresolved questions and possible inconsistencies that forced them to come to grips with the reality that the number line contains much more than just the rational numbers. In essence, even though not in detail, we shall follow this historical path and only investigate this phenomenon carefully when our work makes it clear that we can no longer ignore this reality.

Still, let us mention at this time that $\sqrt{2}$ is just one particular example of a number that is not rational. It is a solution of the polynomial equation $x^2 - 2 = 0$. Similar arguments show that there is no rational number that solves the equation $x^2 - 3 = 0$, that is, $\sqrt{3}$ is not a rational number. More generally, an *algebraic number* is a number x that satisfies a polynomial equation (to be discussed more carefully later on)

$$a_n x^n + a_{n-1} x^{n-1} + a_{n-2} x^{n-2} + \ldots + a_1 x + a_0 = 0,$$

where the coefficients $a_0, ..., a_n$ are integers, with $a_n \neq 0$. Every rational number p/q satisfies the equation $q\,x - p = 0$, where p, q are integers, and hence is algebraic. On the other hand, as we just saw, algebraic numbers like $\sqrt{2}$ or $\sqrt{3}$ are *not* rational. From now on, given a number r, we shall use the notation $\sqrt{r}$ (read square root of r) to label a number q, such that $q^2 = r$. Since the square of a nonzero integer is positive, it readily follows that q^2 is positive for any nonzero rational number q. Therefore, in order to consider $\sqrt{r}$ one needs to assume that r is positive (or 0). Furthermore, since $(-q)^2 = q^2$, both q and $-q$ would qualify to be called square root of r. However, by convention, the symbol $\sqrt{r}$ denotes the *positive* solution of $q^2 = r$ for $r > 0$; the other solution is denoted by $-\sqrt{r}$.

Let us describe a simple geometric construction to identify a line segment of length $q > 0$ that satisfies $q^2 = r$ for a given number $r > 0$. In case $r = 0$ we of course know that $(0)^2 = 0$. Let us consider Fig. I.4, which shows a semicircle of diameter $1 + r$.

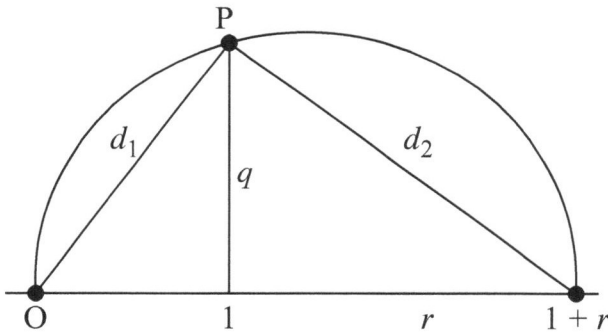

Fig. I.4   Construction of $\sqrt{r}$.

The line through the point 1 and perpendicular to that diameter meets the circle at the point P. We know from elementary geometry that the points $O, P, 1+r$ form a right triangle, and by construction so do the points $O, 1, P$ and $P, 1, (1 + r)$, with hypothenuse d_1 and d_2, respectively. Denote by q the distance of P from the point 1 on the diameter. By Pythagoras Theorem (applied 3 times!) we have

$$1^2 + q^2 = (d_1)^2 \,,\; r^2 + q^2 = (d_2)^2 \text{ and } (d_1)^2 + (d_2)^2 = (1+r)^2.$$

Combining these equations we get

$$(1^2 + q^2) + (r^2 + q^2) = (1 + r)^2 = 1 + 2r + r^2.$$

After cancellations we are left with

$$2q^2 = 2r, \text{ so that } q^2 = r.$$

So this number q, identified geometrically as the length of the line segment from 1 to P, equals $\sqrt{r}$.

At this point we shall not be concerned about proving the "existence" of $\sqrt{r}$, which typically will not be a rational number, aside from the geometric construction we just described, as ultimately this involves enlarging the set of rational numbers. However, we can surely say that if $q \geq 0$ is a rational number, then $q = \sqrt{q^2}$. Going beyond the square root, for a rational number $r = p/q \geq 0$, its *nth* root $\sqrt[n]{r}$, n a positive integer, is a number that satisfies $(\sqrt[n]{r})^n = r$. We note that $\sqrt[n]{r}$, assuming it exists, is a solution of $qx^n - p = 0$, and hence is algebraic, but will not be rational in most cases. Still other algebraic numbers are not even found on the number line: for example, there is no point on the number line that solves the equation $x^2 + 1 = 0$.[2] This latter phenomenon leads us to consider "complex" numbers, but we shall not pursue this at this time.

Unfortunately, algebraic number points on the line still do not capture the vastness of space on the line. In fact, filling in the potholes in the rational line by algebraic numbers hardly makes a dent. This amazing fact was discovered in 1874 by the German mathematician Georg Cantor (1845–1918),[3] and we shall explain it more in detail in Section V.1.6. This discovery is perhaps even more surprising than the discovery of irrational numbers, and yet it is even more removed from everyday life. Furthermore, while it was quite elementary—at least with hindsight—to recognize that $\sqrt{2}$ is not rational, it requires quite a bit more ingenuity to identify specific points on the number line that are not even algebraic. Such points are called *transcendental* numbers. One of the best known examples is the famous number Pi, or π, the ratio between the circumference and the diameter of a circle. Because of its concrete geometric visualization—specifically, π equals the length of the circumference of a circle of diameter 1—this number "exists" just as well as any other number on the line. (Just think of your ruler as a flexible thin wire that you can wrap around a circle with diameter 1.) Approximate values for π, such as $22/7$, were already known in antiquity, and it has been known since the middle of the 18th century that

---

[2]Because of the rule that positive x positive is positive, and negative x negative is also positive, for any nonzero point a on the number line a^2 is positive as well, so $a^2 + 1 > 0 + 1 = 1 > 0$, and hence $a^2 + 1 \neq 0$.

[3]Cantor found a way to distinguish different orders of infinity. He showed that the set of algebraic numbers is "countable" (the simplest type of infinity that is modeled by the set $\mathbb{N}$ of counting numbers), while the set of points on the "complete" number line is "uncountable", that is, it corresponds to a much higher order of infinity.

π is not a rational number. However, it was only verified comparatively recently in 1882 by Ferdinand Lindemann (1852–1939) that π is not even algebraic.

As we shall see, a detailed understanding of the number line, visualized by a geometric line in the plane with the points 0 and 1 identified, is critical for much of more advanced mathematics. In particular, we shall need to learn about the *numbers* that are used to describe it. These numbers are called the *real numbers*, and they are denoted by the symbol $\mathbb{R}$. We shall investigate $\mathbb{R}$ in detail in Chapter V.

For the time being we shall just use "real numbers" as a different name for the "points" on the number line. Similarly, the symbol $\mathbb{R}$ simply denotes the set of all points on the line. As far as precise definitions and applicable rules, at this point we only consider the rational numbers $\mathbb{Q}$, which are a subset of the real numbers $\mathbb{R}$. We remind you that the rational numbers are "dense" in the number line (see above), so that to our eyes the rational numbers are quite sufficient at this time in order to describe the points on the number line.

Finally, we want to mention some convenient notations that rely on the ordering of numbers. Given any 2 numbers a, b, with $a < b$, we define

$$(a, b) = \{x \in \mathbb{R} : a < x < b\}$$

and call it the *open interval* from a to b. Similarly,

$$[a, b] = \{x \in \mathbb{R} : a \leq x \leq b\}$$

is called the *closed interval* from a to b.

Given any number a, and a positive number $\delta > 0$, the set

$$I_\delta(a) = \{x : |x - a| < \delta\} = (a - \delta, a + \delta)$$

is also called the (open) δ - neighborhood of a.

We use the same notations and terminology when we only consider rational numbers, as we mainly do for the time being. It should be clear from the context which numbers are considered.

I.1.5   *First Look at Decimal Expansions*

We are well familiar with the decimal expansion of positive integers. Recall, for example, that the statement $n = 471$ is a shorthand notation for

$$n = 4 \cdot 10^2 + 7 \cdot 10^1 + 1 \cdot 10^0,$$

where $10^0 = 1$. (This is a consequence of properties of powers, see below for a quick review.) Given that every calculator displays numbers, not just positive integers, by using decimal expansions, we will explain the basics of this representation of numbers, leaving a complete discussion that does require the concept of *limits* for later in Section V.2.4.

It is straightforward to extend decimal expansions to negative integers. Just recall that $-n = (-1)n$. By Q4 we thus obtain

$$-n = (-1)(4 \cdot 10^2 + 7 \cdot 10^1 + 1 \cdot 10^0)$$
$$= (-4) \cdot 10^2 + (-7) \cdot 10^1 + (-1) \cdot 10^0,$$

which we could write as $-n = (-4)(-7)(-1)$. But it is much simpler to just note that $(-n) = (-1)471 = -(471) = -471$. And that's the end of the story for integers. What about fractions? The simplest case is the number $\frac{1}{10} = 10^{-1}$. The convention is to continue the standard decimal expansion of positive integers by adding 10^{-1} on the right side, marking the separation between 10^0 and 10^{-1} by the so called decimal point ".". Since $10^{-1} = 0 \cdot 10^0 + 1 \cdot 10^{-1}$, we thus write

$$10^{-1} = 0.1.$$

We could also just write $10^{-1} = .1$, but it is just too easy to overlook the decimal point before 1, so writing 0.1 avoids a possible error. Similarly, $7/10 = 0.7$, and so on. Next we consider fractions $\frac{m}{100}$ with denominator 100. Since $\frac{1}{100} = \frac{1}{10}\frac{1}{10} = 10^{-1}10^{-1} = (10^{-1})^2$, it is tempting to set $\frac{1}{100} = 10^{-2}$. In fact, this is not just a matter of convenience, but, once again, it is the correct definition that needs to be chosen so as to satisfy familiar rules when one extends these to more general numbers. In fact, recall that standard powers of natural numbers are defined by

$$p^n = p \cdot p \cdot \ldots \cdot p, \ n \text{ factors, where } p, n \in \mathbb{N}.$$

It is then readily seen that

$$p^n p^m = p^{n+m} \text{ for } n, m \in \mathbb{N}.$$

It is this rule that turns out to be the crucial property that needs to be preserved when more general numbers are used, both for base and exponent. We will discuss this in detail much later, when we investigate exponential functions, perhaps the most important type of function in analysis and in applications. Here we just want to extend this rule to all integer exponents. First, note that by following this rule, we have

$$p^0 p^n = p^{0+n} = p^n.$$

This shows that p^0 must be equal to the multiplicative identity 1. (Note that if we consider this in $\mathbb{Q}$, then we can just multiply with the multiplicative inverse of p^n on both sides to get this result.) Next, in order to see the correct definition for p^{-n}, where $n \in \mathbb{N}$, note that if we want the rule above to remain correct, then

$$p^{-n}p^n = p^{-n+n} = p^0 = 1,$$

by what we just saw. Thus, p^{-n} must be the multiplicative inverse of p^n, which is consistent with the notation p^{-1} for the multiplicative inverse of p. In particular, this justifies setting $10^{-2} = \frac{1}{100}$. Similarly, $10^{-3} = \frac{1}{1000}$, and, in general, $10^{-n} = \frac{1}{10...0}$ (n zeroes in denominator), i.e., $10^{-n} = 1/10^n$.

We therefore can extend the decimal representation of integers to include negative powers of 10 in a natural way. For example, we write

$$\frac{5831}{10,000} = \frac{5000}{10,000} + \frac{800}{10,000} + \frac{30}{10,000} + \frac{1}{10,000}$$

$$= \frac{5}{10} + \frac{8}{100} + \frac{3}{1000} + \frac{1}{10,000}$$

$$= 5 \cdot 10^{-1} + 8 \cdot 10^{-2} + 3 \cdot 10^{-3} + 1 \cdot 10^{-4} = 0.5831.$$

Conversely, for example, the decimal expansion 75.392 is interpreted as

$$7 \cdot 10^1 + 5 \cdot 10^0 + 3 \cdot 10^{-1} + 9 \cdot 10^{-2} + 2 \cdot 10^{-3} = 75 + \frac{392}{1000}.$$

Unfortunately, when considering general fractions, things get more complicated. The process outlined above works only for fractions whose denominators only contain the prime factors 2 and 5 of the number 10. These are the fractions that can easily be rewritten with the denominator being a power of 10. For example,

$$\frac{1}{4} = \frac{1}{2 \cdot 2} = \frac{5 \cdot 5}{2 \cdot 2 \cdot 5 \cdot 5} = \frac{25}{100} = 0.25.$$

To see the difficulties for more general numbers, let us consider the fraction $\frac{1}{3}$. Note that $\frac{1}{3} = \frac{3}{9} > \frac{3}{10}$, and $\frac{1}{3} - \frac{3}{10} = \frac{10-9}{3 \cdot 10} = \frac{1}{3 \cdot 10}$. So

$$\frac{1}{3} = 0.3 + \frac{1}{3}10^{-1}.$$

Here we can replace $\frac{1}{3}$ on the right side by the formula we just obtained, resulting in

$$\frac{1}{3} = 0.3 + (0.3 + \frac{1}{3}10^{-1})10^{-1}$$

$$= 0.3 + 0.03 + \frac{1}{3}10^{-2} = 0.33 + \frac{1}{3}10^{-2}.$$

After repeating this substitution a total of n times, one obtains

$$\frac{1}{3} = 0.33...3 + \frac{1}{3}10^{-n}, \text{ with } n \text{ digits 3 in the decimal expansion.}$$

(You can prove this precisely by induction.) Obviously there is no end to this process, but we can see that adding more and more 3s in the decimal expansion will get us closer and closer to 1/3, since clearly the difference $\frac{1}{3}10^{-n}$ gets smaller and smaller as n gets larger. All this will be made precise later on, once we have introduced the notion of *limit*. But we really don't need to pursue this right now. Why worry about the problems with the decimal expansion of $\frac{1}{3}$ and most other fractions? The representation of rational numbers by fractions is as simple as it gets, and we have discussed all the relevant operations on such fractions. So there is no need to deal with substantial complications at this time for the sake of using decimal expansions.

At this point it is quite sufficient to understand that fractions with denominators that only contain prime factors 2 and 5 can be written in a decimal form that is a natural generalization of the way we represent natural numbers. This is quite sufficient in many practical applications. Just remember that the numbers represented on metric rulers directly translate into decimal expansions, while the numbers that are highlighted on anglosaxon rulers, such as 3/4, 7/8, 5/16, 9/32, etc. all involve fractions with denominators a power of 2, so can easily be converted into **finite** decimal expansions. For example,

$$\frac{7}{8} = \frac{7 \cdot 5^3}{2^3 5^3} = \frac{7 \cdot 125}{10^3} = \frac{875}{10^3} = 0.875.$$

I.1.6 *Exercises*

1. Verify that addition in $\mathbb{Z}$ is commutative and associative. Hint: Suppose each step taken (forward or backwards) is exactly one foot long. Label your starting point as 0, and note that if $m, n \in \mathbb{N}$, then taking m steps forward followed by n steps backwards takes you to the same place as first taking n steps backwards, followed by m steps forward. Adapt this argument to prove the associative rule for addition.

2. a) Follow the rules to show $n(-1) = -n = (-1)n$ for any $n \in \mathbb{Z}$.

b) Use a) and Exercise 1 to show $(-1)(m + n) = (-1)m + (-1)n$ for $m, n \in \mathbb{Z}$.

3. Suppose $l, m, n \in \mathbb{N}$. Show that $l[(-m) + n] = l(-m) + ln$. Hint: Assume $(-m) + n \in \mathbb{N}$ and use the distributive law valid in $\mathbb{N}$ to show that $l[(-m) + n] + lm = ln$; next add $-(lm) = l(-m)$ (why?) to both sides and simplify. Use Exercise 2a) to handle the remaining case.

4. Use the techniques and results of Exercises 2 and 3 to complete the proof that the distributive rule Q4 holds in $\mathbb{Z}$.

5. In Section 1.1, we proved that if m, n, p, q are positive, then $\frac{m}{n} = \frac{p}{q}$ implies that $\frac{p}{q} = \frac{ml}{nl}$ for some $l \in \mathbb{N}$. Prove this for arbitrary nonzero integers.

6. Prove that addition and multiplication in $\mathbb{Q}$ are commutative and associative, i.e., satisfy rules Q1 and Q2. (You should of course assume these rules for $\mathbb{Z}$.)

7. Prove that the definition of addition and multiplication in $\mathbb{Q}$ given in the text does not depend on the particular form of fractions chosen to represent the rational numbers.

8. Prove that $\mathbb{Q}$ also satisfies the distributive property Q4.

9. Let $p \in \mathbb{N}$ be a prime number. Show that there does not exist a rational number m/n with $(m/n)^2 = p$. More generally, prove that if $p \in \mathbb{N}$ is not a perfect square, that is, $p \neq m^2$ for some $m \in \mathbb{N}$, then p is not the square of any rational number either.

10. Let a, b be two numbers with $a < b$. Let c be any other number. Use the order properties O1–O3 to prove that $a + c < b + c$.

11. Solve the following inequalities. Write the solution set as an interval.

a) $1 - 6x > 2$;

b) $-6 < 5 - 2x < 2$;

c) $|5x - 4| < 4$.

12. Prove that $|a \cdot b| = |a| \cdot |b|$ for all a, b.

13. Simplify the following expressions by eliminating the absolute value sign.

a) $|(-3)(5 - 9)|$;

b) $|(-2)^3 - 2|$;

c) $|(-1)^m + 1|$ for $m \in \mathbb{N}$.

14. Is the statement $\sqrt{p^2} = |p|$ for $p \in \mathbb{Q}$ true or false? If true, give a proof, if false, discuss conditions on p that make it true.

15. Find the decimal expansions of the rational numbers $\frac{3}{625}$ and $\frac{17}{80}$.

16. Find the decimal expansion of $0.327 + 2.773$. Do not just apply a known formula for addition! Use the basic definitions.

I.2   Functions and Their Graphs

I.2.1   *General Concept of Function*

Functions are a very general concept. As used in common language, for example, to say that the speed of a car is a "function" of (i.e., it depends on) the pressure applied to the accelerator pedal, simply suggests that the speed depends on something, namely the input as determined by the pressure on the pedal. This idea is formalized a little bit by saying that a **function** is a rule that assigns a certain output to a given input, where the input is variable (think of the pressure applied by the foot), thereby resulting in possibly different outputs. The "function" buttons on a calculator are a familiar example. For example, enter a number (the input), say 4, and press the "reciprocal" key (the function), typically labelled $\boxed{1/x}$; the calculator than displays the output, which in this case is 0.25. The addition button, usually labeled $\boxed{+}$ is an example of a function whose inputs are a pair of numbers. We enter two numbers, say 7 and 9, separating them by pressing the $\boxed{+}$ key, and then we enter $\boxed{=}$ to tell the function to produce the output, i.e., 16 in this case. A calculator could be programmed differently: for example, one could be asked to enter the function symbol first, and then the two numbers separated by a comma, i.e., $\boxed{+}$ 7, 9, and then press $\boxed{=}$. While the latter format would be consistent with the format typically used in mathematics to label functions of two or more variables, calculators are usually set up so as to follow the order that is most familiar to the user, that is, $7 + 9 =$ in this case.

A function with only finitely many inputs (not too many) is often described by a table that lists the input values and the corresponding output values, rather then trying to explain how the rule that describes the function actually works. Here is a simple example:

Mo, 4/3	Tu, 4/4	We, 4/5	Th, 4/6	Fr, 4/7	Sa, 4/8	Su, 4/9
$52°$	$50°$	$47°$	$49°$	$45°$	$46°$	$47°$

Fig. I.5   Temperature as function of days.

While in applications the inputs and outputs of a function may cover a large number of concepts, the simplest functions considered in mathematics use collections of numbers as inputs, and the output is given by a number as well. In applications, the particular concepts under consideration need to be "translated" into numbers, pairs of numbers, or more, in order to apply the more abstract mathematical functions. For example, temperature is described by a number, where the particular association depends on the temperature scale used (Celsius, Fahrenheit, Kelvin, or something else). Geographic locations on the earth typically are described by a pair of numbers (longitude, latitude), while the position of an airplane requires a triple of numbers (longitude, latitude, altitude).

The most widely used notation for functions uses a letter, or combination of letters, to label the function rule (a common choice is f (*f*unction)), a letter to label the generic input (most commonly x), and another letter, usually y, to label the output, resulting in the symbolisms

$$x \mapsto f(x), \text{ or } f : x \mapsto y = f(x), \text{ or } y = f(x).$$

In any case, the essential information is that the symbol f denotes a specific rule, or "function machine", that assigns to an input x the output $y = f(x)$.

More precise information about a function is usually described by some mathematical formula. The simplest examples are the constant functions. Fix a number c; the constant function f_c assigns to each input $x \in \mathbb{R}$ the output c, in other words, $f_c(x) = c$ for all inputs x.

A less trivial example describes the conversion of degrees Fahrenheit to degrees Celsius that is given by the formula

$$C = C(F) = \frac{5}{9}(F - 32).$$

Here the input F is a temperature measured in degrees Fahrenheit, and the output is the corresponding temperature in degrees Celsius. We note that the notation for input (F) and output (C) is chosen to help to remind the reader of what is involved, while there is no specific symbol for the actual "machine"; instead, the machine is explicitly described by the mathematical formula on the right.

Another familiar example is the area A of a disc, which is a function of its radius r; it is given explicitly by the formula $A(r) = \pi r^2$.

Next, let us mention a simple example from physics that has major historical significance. At the beginning of the 17th century, Galileo Galilei (1564–1642) was trying to understand the motion of a freely falling object. It is said that he dropped stones from the *Leaning Tower of Pisa* and recorded what distance they had fallen in different time intervals. This suggests introducing the function h that to each moment t in time assigns the height $h(t)$ of the stone above ground. More precisely, suppose the tower has height H meters, and we set the clock to measure time (in seconds), so that $t = 0$ corresponds precisely to the moment that the stone is released and begins to fall. Thus $h(0) = H$, and $h(t)$ decreases as t moves forward. At some point t_0 in time the stone hits the ground, i.e., $h(t_0) = 0$. It turns out that this time t_0, which can be measured, combined with sufficient information about the function h, can be used to determine the height of the tower. Based on the evidence collected, and using his creative mind, Galileo was able to describe the function h precisely. We shall discuss Galileo's investigations more in detail in Section II.3.

An example of a function of two variables is the distance $dist(a, b)$ between two points a, b on the number line, defined by

$$dist(a, b) = |a - b|.$$

Sometimes the rule (or machine) is not given by a single formula. For example, the federal income tax $T(x)$ due on a taxable income of \$ x, is described in tax tables; the mathematical formula for $T(x)$ varies according to different income brackets.

We already looked at some very simple "function" keys on a calculator. Scientific calculators handle many functions, usually the ones most widely used in math and science, such as exp, log, the trigonometric functions, and many more. A simple example is the key $\boxed{x^2}$ that identifies the squaring function. After entering a number such as 2.1 (the input) into the calculator, pressing the $\boxed{x^2}$ key results in the calculator displaying the output 4.41 ($= 2.1^2$). Often there is also a key $\boxed{x^3}$ for the cubing function. Note that input and output of calculators are numbers with finite decimal expansion, i.e., rational numbers. So the $\boxed{\sqrt{}}$ key on a calculator (sometimes invoked by $\boxed{inv}$ followed by $\boxed{x^2}$) only provides an *approximation* of the abstract function.

Let us fix some basic terminology and notation. The collection of objects that can be taken as inputs is called the *domain* (of definition) of the function f, and it is denoted by $dom(f)$. The other relevant set is a set $\mathcal{C}(f)$ that contains the outputs of f. We shall call this set the *codomain* of f, but other names, such as *image* or *range*, have also been used, so the reader needs to be careful. The notation $f : \Omega \longrightarrow \mathcal{C}$ is used to refer to a function f with domain Ω and codomain $\mathcal{C}$, implying that to each $x \in \Omega$, the function f assigns a unique output $f(x) \in \mathcal{C}$. The reason that the name for the codomain of a function is not clearly agreed upon is perhaps the fact that the codomain indeed is somewhat vague: note that if $\mathcal{S}$ is any set that contains $\mathcal{C}$, then $f : \Omega \longrightarrow \mathcal{S}$ could be used equally well to refer to the function f, without any need to change anything with the machine f. This of course is not true for the domain. It $\Omega^* \supset \Omega$ is strictly larger, in order to consider Ω^* as domain for f, we must *extend* f from the given domain to the larger one, that is, one has to tell the machine what to do with inputs that are in the larger set, but not in the original domain Ω. For example, consider the square root function $sqr(x) = \sqrt{x}$ with domain $\mathbb{R}^+ = \{x \in \mathbb{R} : x \geq 0\}$. We cannot just simply enlarge the domain without substantial new work. Try to apply the $\boxed{\sqrt{}}$ key in your calculator to the number -1 and see what happens!

A function $f : \Omega \longrightarrow \mathcal{C}$ is called *one−to−one*, or *injective*, if $f(a) = f(b)$ for any two inputs $a, b \in \Omega$ implies that $a = b$. Alternatively, f is injective if $a \neq b$ implies $f(a) \neq f(b)$. f is called *onto*, or *surjective*, if for every $r \in \mathcal{C}$ there is some $a \in \Omega$ with $f(a) = r$.

If the codomain of a function $f : \Omega \longrightarrow \mathcal{C}$ is a subset of the real numbers, the statement that f *is bounded on* Ω means that there is a number $K > 0$, so that

$$|f(x)| \leq K \text{ for all } x \in \Omega.$$

This concept may be applied to any subset S of the domain of f : f *is bounded on* S if the statement above holds just for all $x \in S$.

If $f : \Omega \longrightarrow \mathcal{C}$ is a function, the set of all output values $f(\Omega) = \{f(x) : x \in \Omega\}$ is called the *range* of f, or also the image of f, or, more precisely, the image of Ω under f. Note that $f(\Omega)$ could also be taken as the codomain of f, in fact, it is clearly the smallest set that qualifies as a codomain for f. However, it may be quite difficult to describe this particular set, since in order to verify the statement $y \in f(\Omega)$ one has to show that there exists an input $x \in \Omega$ so that $f(x) = y$. For example, consider the function S with domain $\mathbb{Q}$ defined by $S(x) = x^2$. Then we surely can take $\mathbb{Q}$ as the codomain of S, since $x^2 \in \mathbb{Q}$ for any $x \in \mathbb{Q}$. By the properties of $\mathbb{Q}$, we easily see that the range $S(\mathbb{Q}) \subset \mathbb{Q}^+ = \{x \in \mathbb{Q} : x \geq 0\}$. But is $2 \in S(\mathbb{Q})$? Recall that in Section 1.4 we had seen that the there is NO rational number r that satisfies $r^2 = 2$, so 2 is NOT in the range $S(\mathbb{Q})$. We see that the precise identification of the range may be quite complicated.

It is common practice to denote a function f also by $y = f(x)$, or simply by $f(x)$, thereby blurring the distinction between the *function machine f* and the *output* of f at x. Since x is just a symbol for the *unspecified* variable input, this abuse of notation is not fatal. On the other hand, if a specific value is replaced for x, say $x = 2$, the symbol $f(2)$ definitely should not be used to denote the function f. The symbol $f(2)$ uniquely identifies just a single object, namely the output of f that corresponds to the input 2.

The most important fact to remember about functions is that for every input from an appropriate domain there is **exactly one output**. One is free to choose the notation for the input and output variables. In general discussions mathematicians like to use x and y, although in applications other letters may be chosen to help identify the meaning of the variable. For example, the letter t is often used for a variable that corresponds to time.

I.2.2 *Operations on Functions*

There are several ways in which functions can be manipulated to create new functions. Let us first discuss two most natural operations that can be applied to functions in the most general context, namely, taking the *inverse* of a function, and the *composition* of two functions.

Taking the *inverse* of a function $f : \Omega \to \mathcal{C}$ essentially means putting the function machine in reverse, that is, start with a point $b \in \mathcal{C}$ and consider the machine g, that assigns to the input b the output a, where $f(a) = b$. So if f sends a to b, the *inverse* g sends b back to a. Clearly for this to work out two properties must hold. For once, the codomain $\mathcal{C}$ must be equal to the image $f(\Omega)$ of Ω under f. Thus, if $b \in f(\Omega)$, there definitely exists a point $a \in \Omega$ with $f(a) = b$, that is, a qualifies as the output $a = g(b)$ of the inverse machine g. Furthermore, in order for the inverse machine g that assigns a to the input b to be a *function*, there must be exactly **one output** corresponding to the input b, that is, if a and a^* both satisfy $f(a) = b = f(a^*)$, then we must have $a = a^*$. Recall that earlier we had already identified this property by saying that such a function f is $one - to - one$, or *injective*. To summarize, if $f : \Omega \to f(\Omega)$ is one-to-one, then there exists the inverse function g of f, also denoted by $f^{-1} : f(\Omega) \to \Omega$, which is defined for $b \in f(\Omega)$ by $f^{-1}(b)$ being the **unique** point $a \in \Omega$ (f is one-to-one!) which satisfies $f(a) = b$.

As a concrete example, let us consider the squaring function $S : \mathbb{Q} \to S(\mathbb{Q})$ defined by $S(x) = x^2$. Since $S(x) = S(-x)$ and $x \neq -x$ for all $x \neq 0$, S is clearly not $one - to - one$, so S is not *invertible*, i.e., there is no inverse for S. On the other hand, if we restrict S to the smaller domain $\mathbb{Q}^+ = \{x \in \mathbb{Q} : x \geq 0\}$, that is, if we consider the function $S^+ : \mathbb{Q}^+ \to S^+(\mathbb{Q}^+)$ defined by $S^+(x) = x^2$, then S^+ is one-to-one, and therefore is invertible. Its inverse $(S^+)^{-1} : S^+(\mathbb{Q}^+) \to \mathbb{Q}^+$ is known as the square root function, and is denoted by $\sqrt{\ }$, that is, $\sqrt{y} = (S^+)^{-1}(y)$ for $y \in S^+(\mathbb{Q}^+)$ is the unique number in $\mathbb{Q}^+$ that satisfies $(\sqrt{y})^2 = y$. Similarly, the cubing function $C(x) = x^3$ is one-to-one on $\mathbb{Q}$, and therefore it has an inverse $C^{-1} : C(\mathbb{Q}) \to \mathbb{Q}$ which is known as the cubic root and is denoted by $\sqrt[3]{\ }$, that is, $\sqrt[3]{y} = C^{-1}(y)$ is that unique number that satisfies $(\sqrt[3]{y})^3 = y$.

Next we consider *composition*, a most natural operation that applies to two functions in the most general setting. It simply means to apply one function after the other. There is an obvious requirement for this to be possible, namely the output of the first function must be contained in the domain of the second function. More precisely, if $f : \Omega_f \to \mathcal{C}_f$ and $g : \Omega_g \to \mathcal{C}_g$ are two functions, and if $g(\Omega_g) \subset \Omega_f$, then it makes sense to consider for a given $x \in \Omega_g$ the output $f(g(x))$, that is, we obtain a new function denoted by $f \circ g : \Omega_g \to \mathcal{C}_f$ that is defined by

$$(f \circ g)(x) = f(g(x)).$$

$f \circ g$ is called the composition of f and g. Note that one must be careful about the order: in the composition of f and g the function g is applied first.

As a simple example, let S be the squaring function considered earlier, and let g be defined by $g(x) = 5x + 3$, with domain $\mathbb{Q}$. Then we clearly can consider the composition $S \circ g$ of S with g, which is explicitly defined by the formula

$$(S \circ g)(x) = S(g(x)) = (5x + 3)^2.$$

There is an interesting application of composition in case of an invertible function $f : \Omega \to f(\Omega)$ with inverse $f^{-1} : f(\Omega) \to \Omega$. Clearly we can consider $f^{-1} \circ f$, and by recalling the definition of the inverse, we note that

$$(f^{-1} \circ f)(x) = f^{-1}(f(x)) = x \text{ for all } x \in \Omega.$$

The function we obtained is called the *identity function* on Ω, and it is usually denoted by id_Ω; it is a function with equal domain and codomain Ω, defined by the most simple rule $id_\Omega(x) = x$. Note that if $g : \Omega \to \Omega$ is another function, then one obviously has

$$g \circ id_\Omega = g = id_\Omega \circ g.$$

So if we consider composition as an operation on functions analogous to multiplication on $\mathbb{Q}$, then clearly id_Ω is the analog of the multiplicative identity 1, which explains why we call it the *identity* function.

Similarly, we obtain $f \circ f^{-1} = id_{f(\Omega)}$. The analogy to multiplication on $\mathbb{Q}$ can be carried even further in case of functions with codomain equal to the domain. More precisely, let Ω be a set and consider the collection of functions $F(\Omega) = \{f \mid f : \Omega \to \Omega\}$. Then for any two functions $f, g \in F(\Omega)$, both compositions $f \circ g$ and $g \circ f$ are defined, and we can view $\circ$ as a "multiplication" defined in $F(\Omega)$. Simple examples show that in general

$f \circ g \neq g \circ f$, so this multiplication is NOT commutative. However, note that the associative rule holds:

$$[(f \circ g) \circ k](x) = (f \circ g)(k(x)) = f[g(k(x)] = f[(g \circ k)(x)]$$
$$= [f \circ (g \circ k)](x) \text{ for all } x \in \Omega.$$

Furthermore, id_Ω clearly satisfies

$$f \circ id_\Omega = f = id_\Omega \circ f,$$

and therefore it is the multiplicative identity in $F(\Omega)$. And lastly, if f is both $one - to - one$ and $onto$, then its inverse f^{-1} is again in $F(\Omega)$, so that such functions do have a multiplicative inverse in $F(\Omega)$.

To summarize, composition of functions, under appropriate conditions, is indeed a very general concept that has many formal analogies with the standard multiplication of numbers.

Finally, we want to mention that the familiar operations on numbers extend in a natural way to functions with the same domain, and whose codomain a subset of $\mathbb{Q}$ (or, more generally, of the real numbers $\mathbb{R}$). In detail, suppose $f, g : \Omega \to \mathbb{Q}$. Then one can define new functions $f \pm g$, $f \cdot g$, $\frac{f}{g} = f/g$ in the obvious way, that is,

$$(f \pm g)(x) = f(x) \pm g(x) \text{ and } (f \cdot g)(x) = f(x) \cdot g(x) \text{ for } x \in \Omega,$$

and

$$(\frac{f}{g})(x) = \frac{f(x)}{g(x)} \text{ for all } x \in \Omega \text{ with } g(x) \neq 0.$$

In particular, given such a function f with $f(x) \neq 0$, then $\frac{1}{f}$ is a well defined function with domain Ω, which is the *multiplicative inverse* of f, in the sense that $f \cdot \frac{1}{f} = 1 = \frac{1}{f} \cdot f$, where 1 here denotes the constant function whose output is always the number 1. It is tempting to write f^{-1} for $\frac{1}{f}$, and this is often done. However, one must be very careful and clearly keep in mind the context, so as to distinguish $f^{-1} = \frac{1}{f}$ from the inverse function f^{-1} that we considered earlier in the context of general functions. It is indeed unfortunate that the same notation is used for very different objects, but that's the way it is, and therefore it is very important to keep track of the context.

I.2.3 *Cartesian Coordinate Systems*

Real valued functions of a real variable can be visualized by their graphs. In order to explain this, we first need to introduce the concept of a **rectangular (or Cartesian) coordinate system** in the plane. One fixes a

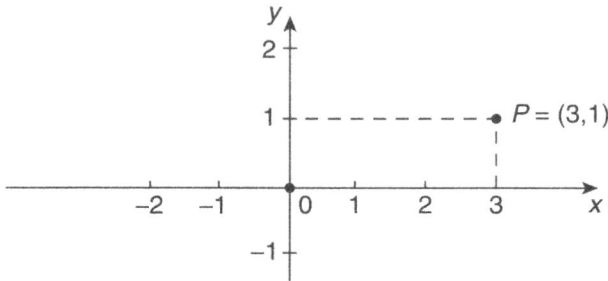

Fig. I.6   A Cartesian coordinate system.

pair of perpendicular number lines in the plane, each one with the number 0 at the point of intersection, as shown in Fig. I.6.

On the horizontal axis (labeled x), numbers are increasing to the right, on the vertical axis (labeled y), numbers are increasing towards the top. According to the figure, each point P in the plane determines an *ordered* pair of numbers (a, b), its *coordinates* with respect to the given coordinate system. Precisely, we draw two lines through P, one parallel to the y-axis, the other one parallel to the x-axis. The first one will intersect the x-axis in precisely one point (why?), at the number 3 in the figure, the other one will intersect the y-axis in one point as well, at the number 1 in the figure. Conversely, by reversing this procedure, every *ordered* pair (a, b) of numbers determines a unique point P in the plane; one writes $P = (a, b)$. By convention, the first number in an ordered pair refers to the horizontal axis. The point of intersection of the two axes has coordinates $(0, 0)$. It is often called the *origin* of the coordinate system and it is denoted by O.

When a coordinate system is introduced in the plane, we shall denote the plane by $\mathbb{R}^2$ to indicate that there are two number lines, that is, two copies of $\mathbb{R}$ that are involved. Also, via the coordinate system, the points in the plane are now identified with pairs of numbers, that is, elements of the set $\mathbb{R}^2 = \mathbb{R} \times \mathbb{R}$.

This looks quite simple. And yet, it was considered a major breakthrough when René Descartes (1596–1650) first introduced these ideas. One has to remember that at that time Euclid's Geometry had dominated the work of philosophers and mathematicians for nearly 2000 years. A central object was the (Euclidean) plane, without any frame of reference to mark specific points, and concepts and results were mainly geometric. The introduction of coordinates changed all that, since it allowed to translate geometric objects and problems into numbers and algebraic problems that can be solved by purely algebraic tools. Conversely, a coordinate system allows us to *visualize* algebraic relations by geometric objects, thereby enhancing our understanding of abstract algebraic relations. We will see this bidirectional relationship at work throughout this book.

Example. The distance between two points P_1 and P_2 in the plane is a geometric notion. To determine it, just connect the two points with a sufficiently long ruler, placing 0 at one of the points. Then the other point marks off on the ruler the number that measures the distance between the two points. By introducing a coordinate system, we can calculate the distance algebraically, provided we know the coordinates of the points. Consider the Fig. I.7.

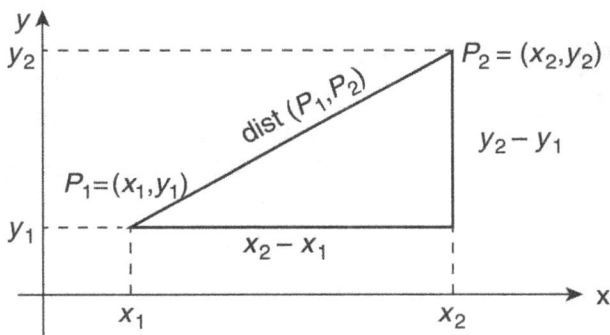

Fig. I.7 Length of diagonal measures the distance $d(P_1, P_2)$.

By applying Pythagoras' Theorem, we see that the distance between the two points $P_1 = (x_1, y_1)$ and $P_2 = (x_2, y_2)$ satisfies

$$[dist(P_1, P_2)]^2 = (x_2 - x_1)^2 + (y_2 - y_1)^2,$$

so that

$$dist(P_1, P_2) = \sqrt{(x_2 - x_1)^2 + (y_2 - y_1)^2}.$$

Example. Suppose we start with the algebraic equation $x^2 + y^2 = r^2$, and we want to describe the set of points $P = (x, y)$ whose coordinates satisfy this equation. The distance formula above, applied to the points (x, y) and $O = (0, 0)$, shows that this set of points is precisely

$$\{P \in \mathbb{R}^2 : [dist(P, O)]^2 = r^2\}.$$

This set is a circle of radius r centered at the origin. (See Fig. I.8.)

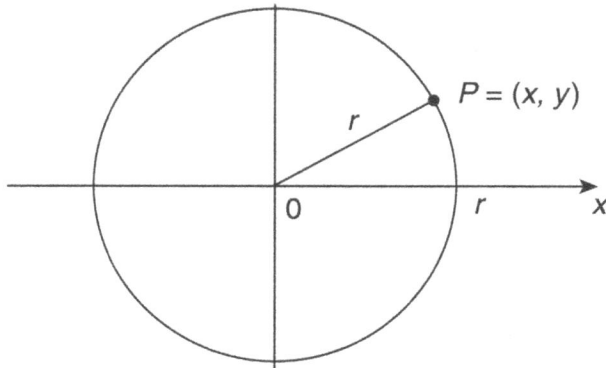

Fig. I.8   Circle of radius r centered at O.

I.2.4   *Graphs of Functions*

The geometric visualization in a coordinate system discussed in the previous section is an important tool in the study of functions. We fix a coordinate system as above. Given a function $f : I \to \mathbb{R}$ defined on a subset I of the number line, its *graph* is the set of points in the plane

$$Graph\,(f) = \{(x, f(x)) \in \mathbb{R}^2 : x \in I\}.$$

Figure I.9 illustrates the concept.

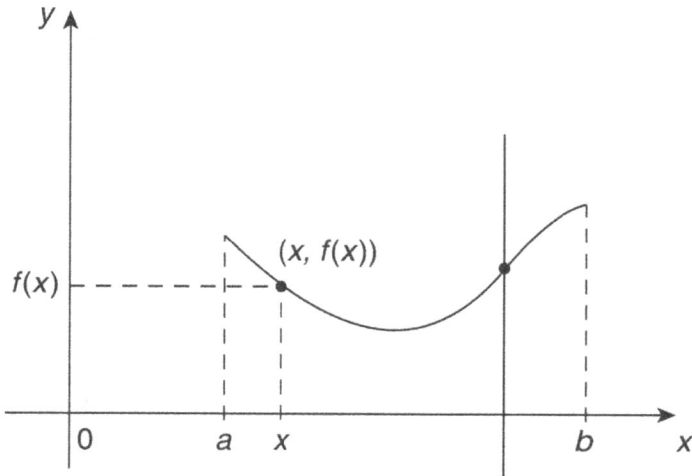

Fig. I.9   Graph of the function $y = f(x)$.

Here the domain of the function f is the interval $I = \{x \in \mathbb{R} : a < x < b\}$. Notice that the graph of f is a "curve" with the distinctive property that every vertical line through a point $x \in I$ meets the graph in exactly one point $(x, f(x))$, corresponding to the fact that a function has exactly one output for each input. This is the so-called *vertical line test* for graphs of functions. Every curve that satisfies the vertical line test can be viewed as the graph of a function. Notice that a (complete) circle fails the vertical line test, and hence is not the graph of a function.

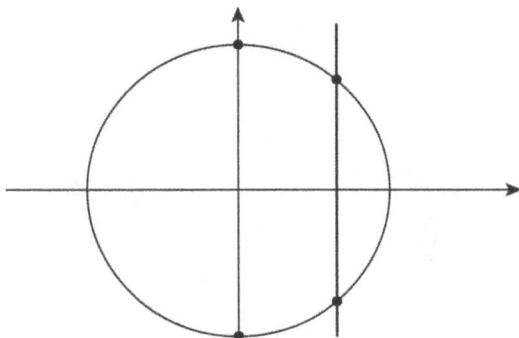

Fig. I.10   A circle fails the *vertical* line test.

I.2.5   *Some Simple Examples*

The temperature conversion function $C = C(F) = \frac{5}{9}(F - 32)$ mentioned earlier is an example of a *linear function*

$$y = f(x) = mx + b,$$

where m and b are constants. By rewriting $\frac{5}{9}(F - 32) = \frac{5}{9}F - \frac{5}{9} \cdot 32$, we see that $C(F)$ fits the above format, with $m = \frac{5}{9}$ and $b = -\frac{5}{9} \cdot 32$. The name reflects the fact that the graph of a linear function is a line. We shall discuss this relationship more in detail in the next section.

Algebraically, the formula for a linear function is given by a *polynomial* of degree 1. We shall discuss polynomials systematically in Section II.4.

Slightly more complicated algebraically are **quadratic functions** (or polynomials of degree 2), that is, functions of the form

$$f(x) = ax^2 + bx + c,$$

where a, b, c are constants with $a \neq 0$. The graphs of such functions are parabolas, as shown, for example, in Fig. I.11.

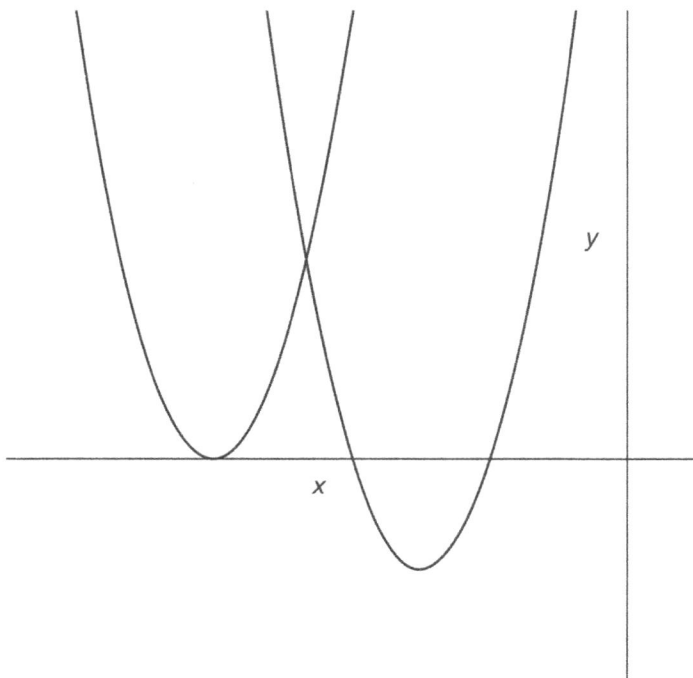

Fig. I.11   Graphs of two quadratic functions.

We note that quadratic functions are NOT bounded on their domain $\mathbb{R}$. However they are bounded over any interval (a, b).

Again, we shall discuss such functions and their graphs in detail in Section II.1.

An important application of quadratic functions arises in the description of a stone dropped from a tower, the phenomenon investigated by Galileo that we mentioned earlier. Galileo determined that if the height of the tower is H meters, the stone falls towards the ground according to the following law: its height $h(t)$ in meters after t seconds is given by the formula, i.e., by the function

$$h(t) = H - 4.9t^2.$$

This relationship can be tested experimentally. See Fig. I.12 for the graph of h corresponding to $H = 50m$.

Often the formula (i.e., function machine) under consideration is meaningful only for inputs that are restricted by certain conditions, so that the domain will be a *proper* subset of the real numbers.

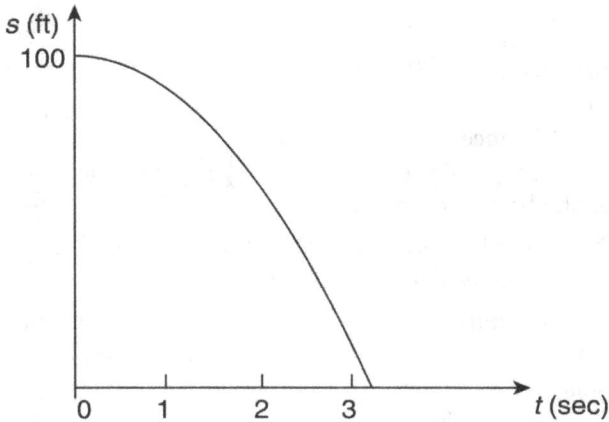

Fig. I.12   Height of falling stone in dependence of time t.

Example. The function $y = \sqrt{x}$ has as its domain the set of nonnegative numbers $\{x \geq 0\}$. Its graph is shown in Fig. I.13.

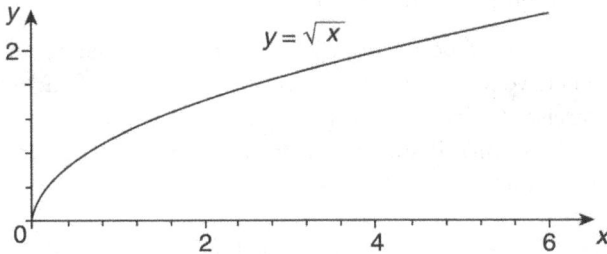

Fig. I.13   Graph of the function $y = \sqrt{x}$ for $x \geq 0$.

Recall that as long as we only consider rational numbers, the square root function is not really defined at every point $x \geq 0$. To be safe, we shall, for the time being, limit the domain of this function to the set $\mathbb{Q}^{sq} = \{x : x = r^2$ for some rational number $r\}$. Since this set is dense in the positive number line (recall that $\mathbb{Q}$ is dense in $\mathbb{R}$), the picture above gives a faithful representation of the graph for our eyes.

I.2.6   *Exercises*

1. a) Identify the points $P = (-3, 4), Q = (0, 3)$ and $R = (4, -2)$ in a Cartesian coordinate system.

b) Find the distance between P and R.

2. Sketch the graph of the function $y = \frac{1}{2}x^2 - 2$ by plotting a few points until you get the basic picture.

3. Use the conversion formula given in the text to find the temperatures in Celsius degrees corresponding to $95^0 F$ and $212^0 F$.

4.   If the temperature is $25^0 C$, what is the temperature in degrees Fahrenheit? More generally, write out explicitly a conversion formula from degrees Celsius to Fahrenheit.

5. Which of the functions $f(x) = 4x^2 + 2$ and $g(x) = 4x - 2$ is one-to-one on the domain $\mathbb{R}$? Explain!

6. Consider the following example of a tax function T. For incomes $x \leq \$10,000$, the tax $T(x) = 0$. If $\$10,000 \leq x \leq \$30,000$, the tax is 10% of the income above $\$10,000$, for $\$30,000 \leq x \leq \$50,000$, the tax is $\$2,000 + 20\%$ of the amount over $\$30,000$, and for incomes $x \geq 50,000$, the tax is $\$6,000 + 30\%$ of the amount above $\$50,000$.

a) Write out formulas for $T(x)$. There will be different expressions for the various income levels.

b) Sketch the graph of the function T.

7. Sketch the set of points $\{(x, y) : x = y^2\}$ in a Cartesian coordinate system. Is the curve so obtained the graph of a function? Explain.

8. Do Exercise 7 with the set $\{(x, y) : y \leq 0$ and $x = y^2\}$. Is this set the graph of a function? If so, write a formula for that function.

9. a) Is the function $h : (0, 1) \to \mathbb{R}$ defined by $h(x) = 1/x$ bounded? Explain!

b) Show that if $0 < c < 1$, then h defined in a) is bounded on the smaller domain $(c, 1)$.

c) With c as in b), is h bounded on the unbounded interval $(c, \infty) = \{x : x > c\}$?

I.3   Linear Functions and Slope

I.3.1   *The Slope of a Line*

We shall now discuss in detail linear functions f, which are given by a formula $f(x) = mx + b$. We begin by explaining the role of the two numbers

m and b, and of course we shall rely on the graph of f. Clearly the constant b is the value of f for $x = 0$, i.e., it identifies the point $(0, b)$ where the graph of f intersects the y-axis. To understand the geometric meaning of m, fix a point (x_1, y_1) on the graph and let $(x, y) = (x, mx + b)$ be any other point on the graph with $x \neq x_1$. Then

$$\frac{y - y_1}{x - x_1} = \frac{mx + b - (mx_1 + b)}{x - x_1} = \frac{m(x - x_1)}{x - x_1} = m.$$

This shows that the various right triangles that depend on $(x, mx + b)$, as shown in Fig. I.14, are all similar, since the ratios of the two legs are constant, that is, do not depend on the particular point $(x, mx + b)$. In particular, the angle α is independent of that point as well. Therefore, all such points lie on the unique line through (x_1, y_1) that forms an angle α with the horizontal x-axis.

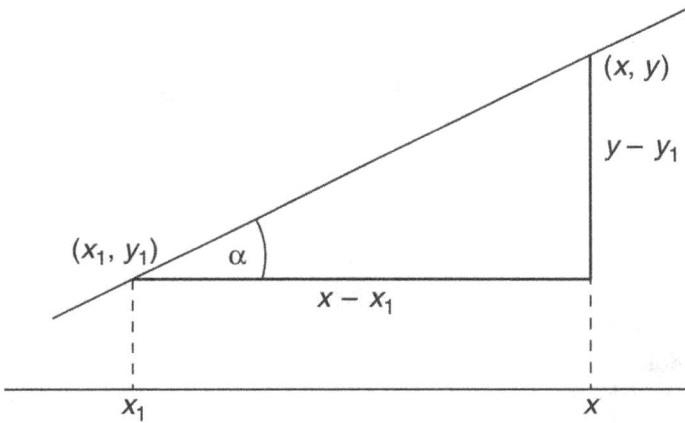

Fig. I.14 Constant $(y - y_1)/(x - x_1)$ implies constant angle α.

So we see that the graph of f is a line, justifying calling such a function *linear*.

The ratio $m = (y - y_1)/(x - x_1)$—sometimes also referred to as the "rise" over the "run"—is called the *slope* of the line. It measures the inclination, or steepness, of the line, and—for those familiar with basic trigonometry—it relates to the angle α by the formula $\tan \alpha = m$. Note that positive slopes correspond to lines that point upwards (as one moves along the positive direction given by the x-axis. The larger the slope, the steeper the line. If $m = 0$, the line is horizontal, i.e., parallel to the x-axis, while if m is negative, the line points downwards. The slope is really the central piece of information for a line. As we shall see, it is the most basic version of a rate of change (i.e., rise over run), which in the context of motion is known as *velocity*.

Given the geometric interpretation of m and b in the function $f(x) = mx + b$, this particular description of the line that is the graph of f is referred to as the *slope-intercept form* of that line, to distinguish it from other representations that we shall discuss shortly.

Example. Suppose we want to draw the graph of $f(x) = 2x - 3$. We start at the y-intercept, that is, at the point $(0, -3)$. We know that the graph is a line, so we just need another point to be able to draw the line. Let us choose the x-coordinate equal to 2. Then the corresponding y-coordinate equals $y = f(2) = 4 - 3 = 1$. Thus our second point is $(2, 1)$, and using a ruler, we draw the line going through $(0, -3)$ and $(2, 1)$, and we are done.

I.3.2   *Equations of a Line*

The whole process can be reversed, that is, given any line in $\mathbb{R}^2$ that satisfies the vertical line test, that is, any non-vertical line, we can readily determine the function whose graph equals this line. Simply identify the point where the line intersects the y-axis; that point is $(0, b)$ for some b that we read off on the y-axis. Pick any other different point, say (x_1, y_1), with $x_1 \neq 0$. Then the ratio $\frac{y_1 - b}{x_1 - 0}$ is the rise over the run, that is, it is the slope m. Our given line is thus the graph of $f(x) = mx + b$.

Going all the way back to Euclid, a line is most often identified by two distinct points P_1 and P_2, reflecting the fact that no reference frame, i.e., coordinate system was known in antiquity, so that slope and y-intercepts were not known. So it is useful to describe the function whose graph is the line through two points directly. For example, suppose the points are $(3, -1)$ and $(7, 2)$. Then the slope $m = \frac{2 - (-1)}{7 - 3} = \frac{3}{4}$, which is of course the same as $\frac{(-1) - 2}{3 - 7} = \frac{-3}{-4}$ if the order of the points is reversed. Now pick one of the two points, say $(3, -1)$ and let (x, y) be the coordinates of an arbitrary point on the line, with $x \neq 3$. Then

$$\frac{y - (-1)}{x - 3} \text{ must equal the slope } m = \frac{3}{4}.$$

So the points (x, y) on the line must satisfy the equation

$$y - (-1) = \frac{3}{4}(x - 3).$$

It follows that $y = \frac{3}{4}(x-3) + (-1) = \frac{3}{4}x - \frac{9}{4} - 1 = \frac{3}{4}x - \frac{13}{4}$, so that

$$f(x) = \frac{3}{4}x - \frac{13}{4}$$

is the slope-intercept form of the line. Note that $f(7) = \frac{21}{4} - \frac{13}{4} = \frac{8}{4} = 2$, which confirms that the other point $(7,2)$ is indeed on the graph of f.

What we have done for two specific points can of course be done for two distinct arbitrary points $P_1 = (x_1, y_1)$ and $P_2 = (x_2, y_2)$. We must assume $x_1 \neq x_2$, so that the line through the two points satisfies the vertical line test. Then the slope $m = \frac{y_2 - y_1}{x_2 - x_1}$. So the equation that must be satisfied by an arbitrary point $P = (x,y)$ with $x \neq x_1$ is

$$\frac{y - y_1}{x - x_1} = \frac{y_2 - y_1}{x_2 - x_1},$$

which we rewrite

$$y - y_1 = \frac{y_2 - y_1}{x_2 - x_1}(x - x_1).$$

This latter equation is valid also for $x = x_1$, and we call it the *two point form* of a line.

I.3.3   *The Point-Slope Form of a Line*

Finally, if the slope m is given (not necessarily derived from two distinct points), the equation of the line through $P_1 = (x_1, y_1)$ with slope m is given by

$$y - y_1 = m(x - x_1).$$

This last equation is known as the *point-slope form* of a line.

Note that we have described lines by equations which must be satisfied by the coordinates of the points on the line. All these points, i.e., the solutions of the equation, describe the graph of the corresponding function, whose explicit formula is obtained by solving the equation for y. For example, from the point-slope form above, we see that $y = m(x - x_1) + y_1$, so that the function is given by $f(x) = m(x - x_1) + y_1$. The corresponding slope-intercept form is then $f(x) = mx + (y_1 - mx_1)$.

Example. Let us find the equation of the line of slope -2 that passes through the point $(4,3)$. The point-slope form immediately gives the answer

$$y - 3 = (-2)(x - 4).$$

Solving for y gives

$$y = (-2)(x - 4) + 3, \text{ i.e.,}$$

$$y = -2x + 11.$$

While the last equation identifies the y intercept 11, the coordinates of the original point $(4, 3)$ have been lost in the process. As we shall see in the next chapter, when we discuss tangent lines to curves, the point-slope form is particularly suitable for those investigations. In contrast, the y-intercept usually plays an insignificant role.

I.3.4   *Some Geometric Properties*

Note that *horizontal* lines (i.e., those lines parallel to the x-axis), which are the graphs of constant functions, are precisely those lines that have slope 0. In contrast, observe that the *vertical* line (parallel to the y-axis) through the point (x_1, y_1) definitely is not the graph of a function f (the vertical line test fails); also, the formula for the slope becomes meaningless in this case, since $x = x_1$ for all points (x_1, y) on this line. Therefore the slope of vertical lines is NOT defined. On the other hand, the vertical line through (x_1, y_1) is described by the simple equation $x = x_1$, where it is understood that y is arbitrary.

The description of lines with respect to a coordinate system allows to translate geometric notions into algebraic properties. For example, two different lines are parallel if and only if they have the same slope. Furthermore, we can also give an algebraic proof of the basic result of Euclidean Geometry that two non-parallel lines in the plane intersect in exactly one point. In detail, if two non-parallel lines are given in the plane, introduce a coordinate system in the plane so that neither line is vertical. Hence we can view the two lines as the graphs of two functions $f(x) = m_1 x + b_1$ and $g(x) = m_2 x + b_2$. Since the lines are not parallel, we must have $m_1 \neq m_2$. A point of intersection (a, b) of the two lines must satisfy both $b = f(a)$ and $b = g(a)$, so that $f(a) = g(a)$. Thus the x-coordinate of such a point of intersection must be a solution of the equation

$$m_1 x + b_1 = m_2 x + b_2.$$

By using the familiar rules, this implies that $(m_1 - m_2)x = b_2 - b_1$, and since $(m_1 - m_2) \neq 0$, it follows that this equation has the unique solution

$$x = \frac{b_2 - b_1}{m_1 - m_2}.$$

So $a = \frac{b_2 - b_1}{m_1 - m_2}$ is the x-coordinate of the point of intersection, and the y-coordinate is $f(a) = g(a)$. You may practice your skills working with fractions by verifying that the latter equality is indeed correct.

I.3.5 *Perpendicular Lines*

Besides finding the coordinates of the point where two non-vertical lines intersect, it is often useful to be able to construct *perpendicular* lines. While there is a simple geometric process to accomplish this, we want to describe an algebraic process that involves the slopes of the lines.

Figure I.15 shows two perpendicular lines that intersect at the point P.

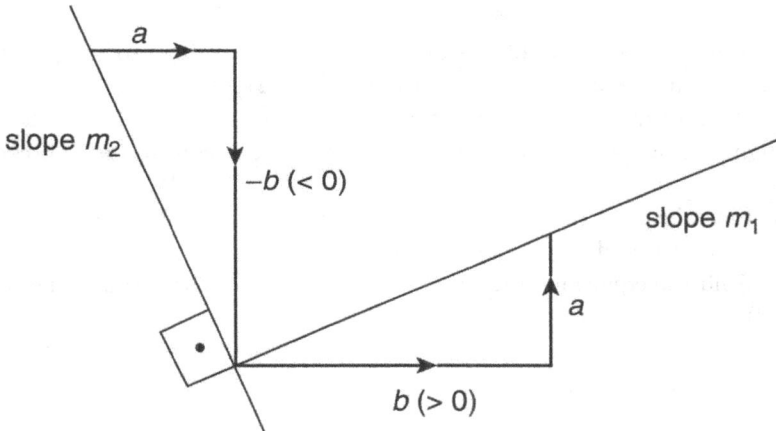

Fig. I.15 Slopes of two perpendicular lines.

Note that $m_1 = a/b$ is the slope of the line that is pointing upwards. If the right triangle with sides a and b is rotated clockwise by 90^0 around the point of intersection, so that its hypothenuse, which lies on the first line, ends up on the second line, we see that the second line has slope $m_2 = (-b)/a = -b/a$, the minus sign accounting for the fact that the "rise" is a "drop" in this case. It follows that

$$m_1 \cdot m_2 = -1.$$

The process can clearly be reversed. We have thus verified the following result.

Two (non-vertical) lines are perpendicular if and only if their slopes m_1 and m_2 satisfy $m_1 \cdot m_2 = -1$.

I.3.6 *Exercises*

1. a) Find the slope of the line that goes through the points $(3, 1)$ and $(6, -2)$.

b) Find the equation of the line in a).

2. Consider the line through the point $(-1, 3)$ with slope $m = -3/2$.

a) Find the equation of the line.

b) Does the point $(1, 2)$ lie on that line?

3. Consider the line given by $y = 3x + 2$. Find the coordinates of the points on the line at distance 4 from the point $(0, 2)$. (Hint: Make a sketch!)

4. Find the point of intersection of the two lines given by $f(x) = 3x - 5$ and $g(x) = -2x + 3$.

5. Find the equation of the line that is perpendicular to the graph of $y = \frac{1}{3}x + 4$ and that goes through the point $(1, 2)$.

6. A tangent through a point P on a circle with center C is the line through P that is perpendicular to the radial line connecting the center to P.

a) Find the equation of the tangent through the point $(4, 3)$ on the circle of radius 5 centered at the origin $(0, 0)$.

b) Find the equation of the tangent to the circle in a) through the point $(-5, 0)$.

Chapter II

Simple Algebra and Tangents

II.1 Quadratic Equations and Functions

II.1.1 *Quadratic Functions and Their Graphs*

The next class of functions that we shall study in detail involves quadratic functions, or polynomials of degree 2, given by

$$f(x) = ax^2 + bx + c,$$

where a, b, and c are fixed real numbers. Recall that for the time being we only consider rational numbers, so the reader should not worry about the definition of real numbers at this time. Note that if $a = 0$, we would just have the linear function $y = bx + c$ we studied in the previous chapter. We therefore shall assume that $a \neq 0$ to ensure that the quadratic term x^2 does indeed appear in the formula for the function.

We begin by visualizing the graph of the simple special case $s(x) = x^2$. We use the letter s to denote this function to remind us that given the input x, the output of this function is simply the square x^2 of x. Since $(-x)^2 = x^2$, we see that the graph is symmetric with respect to the y-axis, that is, if we take any point (x, x^2) on the graph and reflect it on the y-axis, we get the point $(-x, x^2)$, which is also on the graph by the preceding equality. Let us calculate a few values of s on positive numbers and place them in a table

x	$s(x) = x^2$
0.5	0.25
1	1
1.5	2.25
2	4
2.5	6.25
3	9
3.5	12.25

Table of some values for $s(x) = x^2$.

By plotting these points, as well as the reflected ones, in a coordinate system, we get a good idea of the general shape of the graph. Adding more points would lead to filling in the gaps with a curve as shown in Fig. II.1.

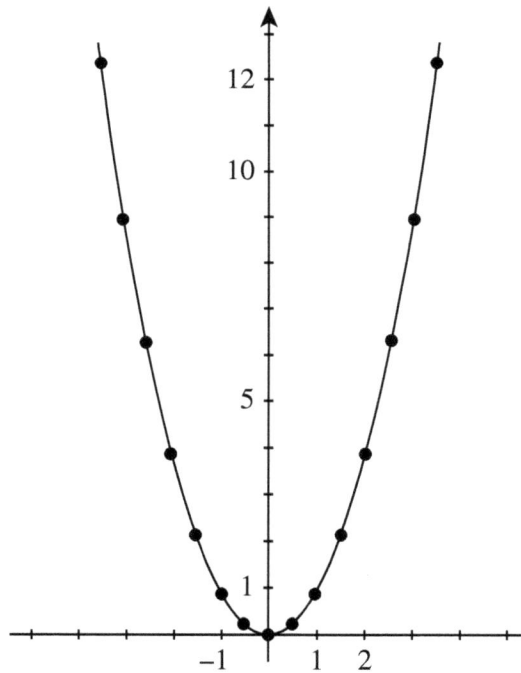

Fig. II.1   Graph of $s(x) = x^2$.

Perhaps you have seen a curve similar to this one before. Does *parabola* sound familiar? Indeed, we will now show that the graph of $s(x) = x^2$

is a parabola. In order to make sense of this, we must recall the precise definition of a parabola. Parabolas arose in antiquity as special cases of so-called conic sections, that is, those curves that are obtained by intersecting a circular cone with a plane. Depending on the angle between the plane and the axis of the cone, such curves are either ellipses (these include circles as special cases), parabolas, or hyperbolas. In particular, parabolas arise when the plane is parallel to the mantle of the cone. The great Greek philosopher and geometer Apollonius (3rd century B.C.) is credited with systematically recording the geometric definitions and known properties of the conic sections, and with discovering many additional properties. Other Greek geometers expanded on these investigations and introduced equivalent geometric definitions for the conic sections.

In particular, a "parabola" is defined geometrically by fixing a line L in the plane, the so called *directrix*, and by choosing a point F not on L, which is called the *focus* of the parabola. The parabola with focus F and directrix L is then defined to be the set of all points P in the plane that satisfy

$$dist(P, F) = dist(P, L),$$

where the distance from a point to the line L is the length of the perpendicular projection from P onto L. Figure II.2 illustrates the situation.

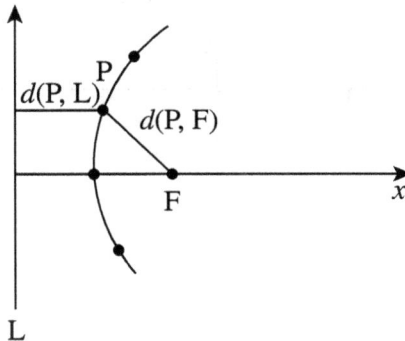

Fig. II.2 Geometric definition of a parabola.

The line through F that is perpendicular to the directrix is called the *axis* of the parabola. As the picture above shows, the parabola is symmetric about the axis.

The parabola enjoys the remarkable physical property that light rays that enter the parabola parallel to its axis are reflected on the parabola so that they then go through the focus F. This property has important applications in optics. For example, the 3-dimensional version obtained by rotating the parabola around its axis provides the theoretical foundation for today's *parabolic telescopes.* We shall verify this property later on in Section 2.5, after we have introduced the fundamental concept of a tangent, which is necessary in order to describe reflection mathematically.

In order to determine the equation satisfied by the points on the parabola we introduce a coordinate system, and we place the focus F on the y-axis at the point $(0, q)$, where $q > 0$, and we let the directrix be the horizontal line given by $y = -q$, as seen in the picture below. The origin O of the coordinate system is then on the parabola, since $dist(O, L) = |-q| = q = dist(O, F)$. We refer to this point that is in the middle between focus and directrix as the *vertex* of the parabola. We also note that the axis of the parabola coincides with the y-axis.

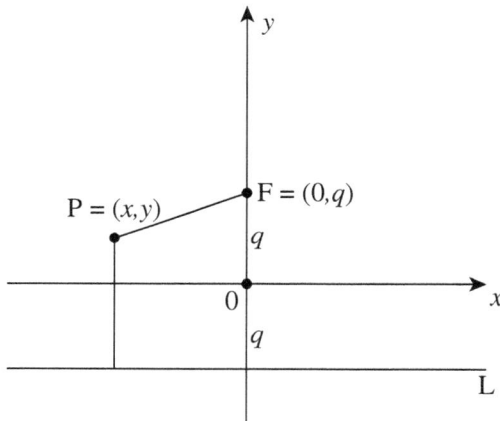

Fig. II.3   Parabola in coordinates.

With this setup, a point $P = (x, y)$ in the upper half plane $(y > 0)$ satisfies

$$dist(P, F)^2 = x^2 + (y - q)^2 \text{ and } dist(P, L) = y + q.$$

The defining property of the parabola, $dist(P, F) = dist(P, L)$ then implies

$$x^2 + (y - q)^2 = (y + q)^2,$$

so that

$$x^2 + y^2 - 2qy + q^2 = y^2 + 2qy + q^2,$$

which simplifies to

$$x^2 = 4qy.$$

Solving for y one obtains the function

$$y = \frac{1}{4q}x^2,$$

whose graph is the parabola with focus $F = (0, q)$. If we choose $q = 1/4$, we get exactly $y = s(x) = x^2$.

Since we can clearly reverse the process, we see that for any $a > 0$ the graph of $y = ax^2$ is always a parabola with vertex at O and focus $F = (0, \frac{1}{4a})$.

Replacing a by $-a$ just reflects the parabola given by $y = ax^2$ on the x-axis, that is, the point (x, ax^2) is replaced by the point $(x, -ax^2)$. The result is a parabola that opens to the bottom, with focus $(0, \frac{1}{4a})$, where $a < 0$ now.

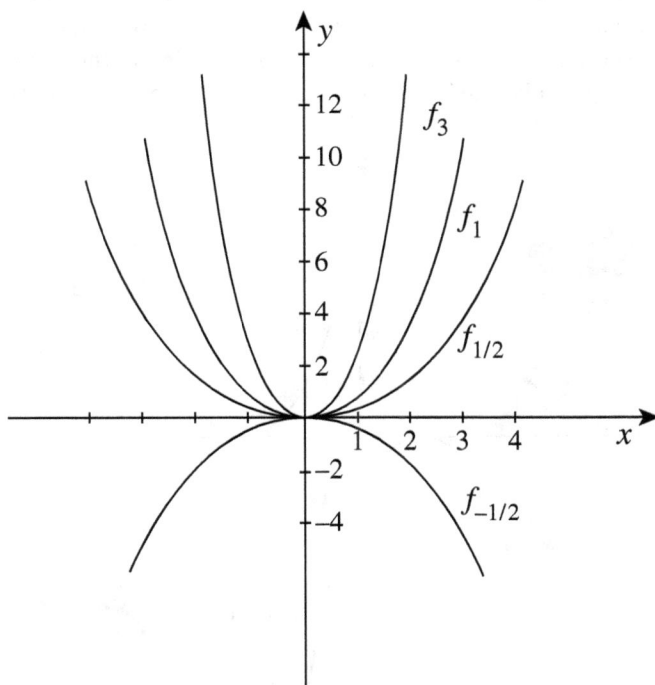

Fig. II.4

The visual effect of varying $a > 0$ is readily seen. As a gets larger, $q = \frac{1}{4a}$ gets smaller, that is, the focus and directrix move closer to the vertex, forcing the parabola to get narrower. Conversely, as a gets smaller, the focus $(0, q)$ moves higher on the y-axis, resulting in a parabola that opens up more. Figure II.4 shows the graphs of $y = f_a(x) = ax^2$ for different values a.

The effect on the graph of adding a constant c to the function, that is, to consider $y = ax^2 + c$, is pretty obvious: the graph of f_a is simply moved up c units if $c > 0$, and it is moved down $|c|$ units if $c < 0$.

Finally, in order to understand the effect of the term bx in the general quadratic function, we are guided by the fundamental equation

$$(x - p)^2 = x^2 - 2px + p^2,$$

which is an example a quadratic function. So how is the graph of $y = (x - p)^2$ related to the graph of f_1? Set $x' = x - p$. We then must consider the graph of $y = (x')^2$, which clearly is, geometrically, the same as the one of $y = x^2$, except that the vertex $(x', 0)$, with $x' = 0$, is not at the origin O, but at the point where $x' = x - p = 0$, that is, at the point $(p, 0)$. If $p > 0$, we simply take the graph of $y = x^2$ and move it p units to the right

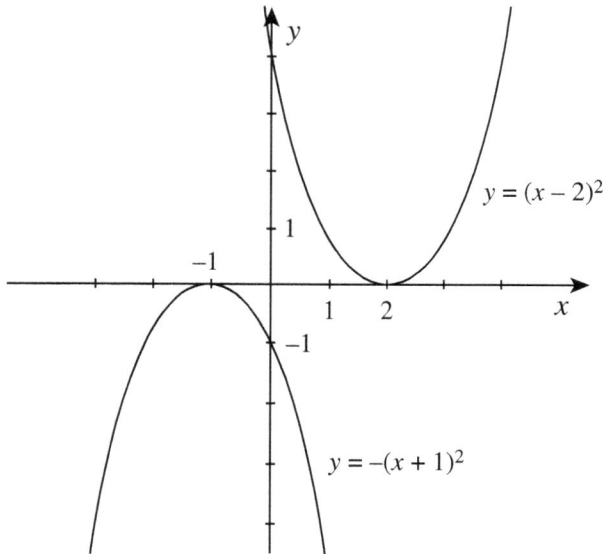

Fig. II.5   Translated parabolas.

to obtain the graph of $y = (x')^2 = (x - p)^2$, while if $p < 0$, we move to the left — see Fig. II.5 for some examples.

This discussion suggests that the graph of a general quadratic function is obtained from the basic graph of $s(x) = x^2$ by a combination of opening up or narrowing, reflections on the x-axis, and translations, both vertically and horizontally. In particular, the graph of any quadratic function is a parabola. So how can we make this more precise?

To simplify matters, let us first fix $a = 1$, and let us consider the function $f(x) = x^2 + bx + c$. We would like to write f in the form

$$f(x) = (x - p)^2 + d,$$

for suitable constants p and d. In other words, we must have

$$f(x) = x^2 + bx + c = (x - p)^2 + d$$
$$= x^2 - 2px + p^2 + d.$$

We see that the numbers p and d must satisfy

$$-2p = b \text{ and } p^2 + d = c.$$

The first equation implies that we must choose $p = -b/2$. Given this value, the second equation then tells us that we must choose $d = c - p^2 = c - b^2/4$. That's all there is to it! We thus have verified that

$$f(x) = x^2 + bx + c = \left(x + \frac{b}{2}\right)^2 + \left(c - \frac{b^2}{4}\right).$$

The technique we just employed is known as *Completing the Square*.

Example. Consider $g(x) = x^2 - 6x + 2$. The result above shows that g can be written as

$$g(x) = (x - 3)^2 + 2 - \frac{36}{4} = (x - 3)^2 - 7.$$

This latter form makes it easy to draw the graph of g. Just take the graph of f_1, move it 3 units to the right, and then 7 units down, to obtain the graph shown in Fig. II.6.

The form for g that we have obtained makes it easy to determine the points where the graph of g intersects the x-axis, that is, where $g(x) = 0$. Clearly such points must satisfy $(x-3)^2 = 7$; it then follows that $x - 3 = \sqrt{7}$ or $-\sqrt{7}$, which implies that the two points of intersection are $3 + \sqrt{7}$ and $3 - \sqrt{7}$, where we assume that $\sqrt{7}$ is a number r (not necessarily rational) that satisfies $r^2 = 7$. Using a calculator, we see that $\sqrt{7} \approx 2.6457$, so that the left point of intersection is approximately 0.3543, while the one on the right is about 5.6457.

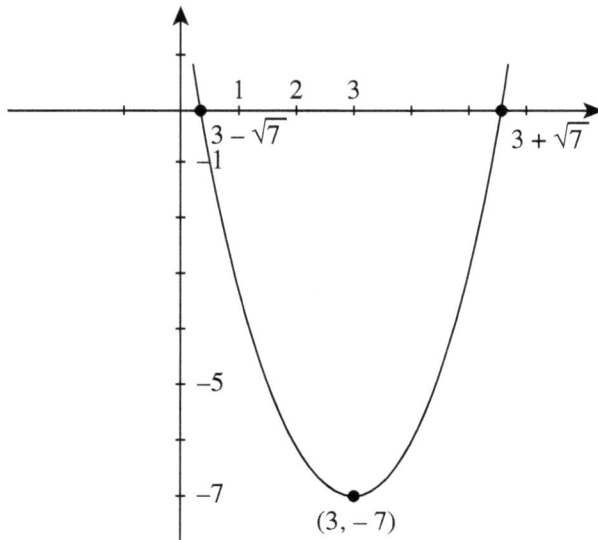

Fig. II.6   Graph of general quadratic function.

Finally we consider the general case $f(x) = ax^2 + bx + c$, where $a \neq 0$, by reducing it to the case where $a = 1$, as follows. Note that

$$ax^2 + bx + c = a\left[x^2 + \frac{b}{a}x + \frac{c}{a}\right] = a\left[\left(x + \frac{b}{2a}\right)^2 + \left(\frac{c}{a} - \frac{b^2}{4a^2}\right)\right]$$

$$= a\left(x + \frac{b}{2a}\right)^2 + \left(c - \frac{b^2}{4a}\right).$$

While the result looks a bit messy (don't try to remember it), what is important is to understand the procedure that we followed. The result shows that the graph of this general function is obtained from the graph of $s(x) = x^2$ by moving it $|b/2a|$ units to the right or left, depending on whether $b/2a$ is negative or positive, then opening or narrowing the graph according to the factor a, and, if $a < 0$, reflecting it across the x-axis, and finally, moving it up or down $|c - b^2/4a|$ units, depending on the sign of $c - b^2/4a$. In particular, this shows that the graph of f is still a parabola, with vertex at the point $(-b/2a, c - b^2/4a)$, and focus $q = |1/4a|$ units above or below the vertex, depending on the sign of a.

Again, we repeat that while this may look complicated, the important lesson is that we can identify all the relevant features of the graph of the general quadratic function by just using basic simple algebra.

II.1.2 *The Quadratic Equation and its Roots*

So far we have discussed the quadratic *function*. Closely related to it is the quadratic *equation*

$$ax^2 + bx + c = 0.$$

Stated this way, the main problem is to find its *roots*, i.e., those numbers x that satisfy the equation. In the previous section we essentially already solved this problem, when we identified the *zeroes* of the quadratic function, that is, those points where the graph of the function intersects the x-axis. At those points the function has the output 0, and the task was to find those inputs that get sent to 0. Given that this is the first significant nontrivial algebraic problem where the roots can be completely described, it is worthwhile to analyse the result and learn about some other features related to roots that are relevant for more general polynomials and, in particular, for the tangent problem.

As we saw, the crux of the matter is to write the above quadratic expression in the special form

$$a(x - p)^2 + d.$$

We showed that this can always be done, and we found explicit formulas for p and d that depend on the original coefficients a, b, c. This particular form makes it very easy to find the roots of $a(x - p)^2 + d = 0$. Since we assume $a \neq 0$, we can write this as

$$(x - p)^2 = -\frac{d}{a}.$$

We see that there are two distinct cases. For once, if $-d/a < 0$, then there is no solution at all, since $(x - p)^2 \geq 0$ for all (real) numbers, i.e., in this case the equation $a(x - p)^2 + d = 0$ has no *roots* at all. On the other hand, if $-d/a \geq 0$, then we can find the roots, assuming that we allow algebraic points on the number line, so that there exists a (square)-root $\sqrt{-d/a}$ of $-d/a$. (Perhaps this is the origin of the terminology "roots" for solutions of quadratic and more general equations.) In this latter case we see that the roots are

$$x_1 = p + \sqrt{-d/a} \text{ and } x_2 = p - \sqrt{-d/a}.$$

Note that in case $d = 0$, the two roots coincide. We shall investigate this special case in detail in the next section. For now, we just want to highlight that the problem of finding the roots is completely solved, subject to allowing appropriate algebraic (real) numbers in the discussion.

Our discussion has emphasized understanding the simplicity of the problem when the equation is written in a special form. In order to be able to use this efficiently in the case of the general equation $ax^2 + bx + c = 0$, we need to translate the preceding answer back to the original coefficients a, b, c. The process is straightforward based on our earlier work, but unfortunately the final answer is a bit messy, but that's the way it is.

Recall that we had determined that

$$p = -\frac{b}{2a} \text{ and } d = c - \frac{b^2}{4a}.$$

So $-d/a = \frac{b^2}{4a^2} - \frac{c}{a} = \frac{1}{4a^2}(b^2 - 4ac)$, and $-d/a \geq 0$ if and only if $b^2 - 4ac \geq 0$. This latter expression is known as the *discriminant* of the quadratic equation. Assuming that the discriminant is ≥ 0, it follows that $\sqrt{-d/a} = \frac{1}{2a}\sqrt{b^2 - 4ac}$ (if $a < 0$, we should take the negative of this, so that $\sqrt{-d/a} \geq 0$, but the final formula involves both roots, so it's not critical). In terms of the original coefficients, the two roots are then given as follows:

$$x_1 = -\frac{b}{2a} + \frac{1}{2a}\sqrt{b^2 - 4ac} \text{ and } x_2 = -\frac{b}{2a} - \frac{1}{2a}\sqrt{b^2 - 4ac},$$

which is commonly written as

$$x_{1,2} = \frac{-b \pm \sqrt{b^2 - 4ac}}{2a}, \tag{II.1}$$

where it is understood that the $+$ sign gives one solution, while the $-$ sign gives the other solution. Again, if the discriminant is *zero*, the two solutions coincide, and if the discriminant is negative, then there are NO (real) solutions.

This last formula (II.1) is known as the solution formula for the quadratic equation. It is one of the few formulas that should be memorized, although it is much more important to understand the process that led us to this answer.

Given a quadratic function f, the roots of the quadratic equation $f(x) = 0$ are also referred to as the roots of f.

Time to give some explicit numerical examples.

Example. Find the roots of $5x^2 + 5x - 60 = 0$. The discriminant is $5^2 - 4(5)(-60) = 25 + 1200 = 1225 > 0$. So the equation has two distinct solutions. Since $\sqrt{1225} = 35$ (check it out: $35^2 = 1225$), the solutions are

$$x_1 = \frac{-5 + 35}{2 \cdot 5} = \frac{30}{10} = 3 \text{ and } x_2 = \frac{-5 - 35}{10} = -4.$$

You may wish to check these answers by entering their values into the quadratic expression; the result should be 0.

Example. Find the roots of $4x^2 + 8x + 4 = 0$. The discriminant is $8^2 - 4 \cdot 4 \cdot 4 = 0$. So the two roots coincide. Their common value is

$$x_{1,2} = \frac{-8 \pm 0}{2 \cdot 4} = -1.$$

Of course, with some experience, one might just recognize that $4x^2 + 8x + 4 = 4(x^2 + 2x + 1) = 4(x + 1)^2$, so the roots must satisfy $x + 1 = 0$, that is $x = -1$.

Example. Consider the equation $-3x^2 + 12x - 17 = 0$. The discriminant $12^2 - 4(-3)(-17) = 144 - 204 = -60$ is negative, so this equation has no (real) roots.

Remark. In case the discriminant is *negative*, we have highlighted that the equation has no *real* roots. While this may appear redundant, since we only consider points on the number line (i.e., what we also call the real numbers), history has shown that one shouldn't give up so easily. Just as considering square roots forces us to go beyond the familiar rational numbers, mathematicians have found that it is most useful to enlarge the known real numbers by introducing so-called *complex* numbers to be able to consider square roots of negative numbers. Just like we introduced the symbol $\sqrt{2}$ to denote a number r that satisfies $r^2 = 2$, one introduces a symbol i (for *imaginary*) for a "quantity" that satisfies $i^2 = -1$, so that one may also write $i = \sqrt{-1}$. This leads us to consider "numbers" of the form $a + bi$, where $a, b \in \mathbb{R}$, and work with these new "complex" numbers by following all the familiar rules Q1-Q7, and adding the additional rule that $i \cdot i = -1$. For example,

$$
\begin{aligned}
(3 - 2i)(2 + 5i) &= 3 \cdot 2 + (3 \cdot 5)i + (-2)2i + (-2)5(i \cdot i) \\
&= 6 + (15 - 4)i + (-10)(-1) \\
&= 16 + 11i.
\end{aligned}
$$

More generally, if $b > 0$, a "square root" of the negative number $-b$ is given by $\sqrt{-b} = \sqrt{(-1)b} = \sqrt{-1}\sqrt{b} = \sqrt{b} \cdot i$. Indeed, $(\sqrt{b} \cdot i)^2 = (\sqrt{b})^2(i)^2 = b(-1) = -b$. Again, following the familiar rules, the "number" $-\sqrt{b} \cdot i$ also satisfies

$$(-\sqrt{b} \cdot i)^2 = (-\sqrt{b})^2 \cdot (i)^2 = b \cdot (-1) = -b,$$

so, unless $b = 0$, the number $-b$ has two distinct square roots.

This new number system is known as the *complex numbers*, and it is usually denoted by $\mathbb{C}$. In particular, a quadratic equation always has two roots in $\mathbb{C}$, which may happen to coincide. It looks like magic, but the complex numbers are perhaps the most important number system in mathematics, with wide application in the physical sciences.

II.1.3   *Factorization of Quadratic Polynomials*

We now want to discuss another important property of the roots of a quadratic equation. The key technical result is the following lemma that is just a special case of a result valid for arbitrary polynomials.

Lemma 5 Let f be the quadratic polynomial $ax^2 + bx + c$. For any $p \in \mathbb{R}$,

$$f(x) - f(p) = g(x)(x - p),$$

where g is a linear function.

Proof. We rearrange $f(x) - f(p) = ax^2 + bx + c - (ap^2 + bp + c) = a(x^2 - p^2) + b(x - p)$. Using the familiar factorization $x^2 - p^2 = (x + p)(x - p)$, we see that the last expression equals $[a(x + p) + b](x - p)$, which proves the result. ∎

In particular, we see that the linear factor g is given by $g(x) = a(x + p) + b$.

Corollary 6 If p is a root of f, then $f(x) = g(x)(x - p)$, that is, f has a factor $(x - p)$.

Note that it is obvious that the converse statement is true as well, that is, if f has a factor $(x - p)$, then p is a root of f.

Corollary 7 If p_1 and p_2 are the roots of f, then

$$f(x) = a(x - p_1)(x - p_2).$$

Proof. We apply the previous corollary with one on the roots of f, say $p = p_2$. Then the factor $g(x) = a(x + p_2) + b$ is a non-constant linear function, and thus it has a zero, which is then also a zero of f, and hence must be the other zero p_1 of f. Since $(p_1, 0)$ is a point on the graph of this linear function, the point-slope form for g shows that $g(x) = a(x - p_1)$, where a is the slope. Therefore, $f(x) = a(x - p_1)(x - p_2)$. ∎

In particular, this result shows that if p_1 and p_2 are the roots of $ax^2 + bx + c = 0$, then $p_1 + p_2 = -b/a$ and $p_1 p_2 = c/a$. Note that this result also follows from the explicit formulas for p_1 and p_2 in terms of a, b, c that we had found in the previous section.

In case $a = 1$, the factorization $f(x) = (x - p_1)(x - p_2)$ obviously implies

Corollary 8 The roots p_1 and p_2 of $f(x) = x^2 + bx + c$ satisfy $p_1 + p_2 = -b$ and $p_1 \cdot p_2 = c$.

This result is known as Viète's theorem, after the French mathematician F. Viète (1540–1603) who first proved it. (The Latin version Vieta of the name is also used.) It is commonly used to find (integer) roots by inspection. For example, given $x^2 + 5x + 6 = 0$, analysing the coefficients 5 and 6 with Viète's result in mind may readily lead to $-5 = (-2) + (-3)$ and $(-2)(-3) = 6$. This usually works quite well if the constant c has only very few prime factors.

It might be of interest to mention that Viète's theorem leads to an alternate method for finding the roots of a quadratic equation which is much easier to remember than the method we developed in the previous section that is based on completing the square. It apparently goes back to the Babylonians over 4,000 years ago, and it has recently been brought back to light by the mathematician Po-Shen Loh, as mentioned in an article in the New York Times.[1] Incidentally, this suggests that the basic idea of Viète's theorem was known long before Viète. This method is based on representing the two roots p_1 and p_2 as symmetrical about their midpoint m (in fact, m is the x-coordinate of the axis of the parabola), so that $p_1 = m - z$ and $p_2 = m + z$ for certain numbers m and z. By Viète's theorem one has

$$-b = p_1 + p_2 = 2m,$$

so that $m = -b/2$, and

$$c = p_1 \cdot p_2 = (m - z)(m + z) = m^2 - z^2.$$

It then follows that

$$z^2 = m^2 - c = \frac{b^2}{4} - c = \frac{1}{4}(b^2 - 4c),$$

so that $z = \pm\frac{1}{2}\sqrt{b^2 - 4c}$. The roots of $x^2 + bx + c = 0$ therefore are

$$p_{1,2} = m \pm z = \frac{-b \pm \sqrt{b^2 - 4c}}{2}.$$

Of course, in order to understand and apply this method, one must first learn about the relevant factorization of the quadratic equation and Viète's theorem, so it does not come for free. However, given that preparation, it does seem that this method is indeed easier to remember, just in case you have forgotten the formula for the roots of the quadratic equation. Furthermore, this alternate method suggests that factorization is perhaps the most

---

[1] *This Professor's "Amazing" Trick Makes Quadratic Equations Easier*, by Kenneth Chang and Jonathan Corum, NY Times, Feb. 5, 2020.

important idea to remember. Not only does it readily lead to the solution of the quadratic equation, but, as we shall see in Section 4, factorization generalizes to arbitrary polynomials, with most important consequences, while in general there is NO solution formula comparable to the one in the quadratic case.

II.1.4   *Exercises*

1. Find the coordinates of the vertex and of the focus of the parabola that is the graph of $g(x) = 3x^2 - 12x + 7$.

   2. Find the zeroes of the function $f(x) = 5x^2 - 6x + 1$.

   3. Here is an exercise to test your algebra skills. Verify that the formulas for the zeroes given by formula (II.1) above do indeed satisfy the equation $ax^2 + bx + c = 0$. Show that this also works if the discriminant d is negative by just proceeding formally and using the property $(\sqrt{d})^2 = d$.

   4. Find numbers p and q so that $\frac{1}{2}x^2 + 4x + 4 = \frac{1}{2}(x - p)(x - q)$.

   5. Find the vertex of the parabola given by $y = 2(x - 1/2)(x + 3/4)$.

   6. Find the zeroes of $x^2 + bx + c = 0$ by using the symmetry of the parabola about its axis, as follows. The zeroes are symmetric about a certain number q (actually the x-coordinate of its vertex), and therefore are equal to $q + \alpha$ and $q - \alpha$ for a certain number α. Use Viète's theorem to find q and α in terms of b and c.

   7. Use Viète's theorem to find the roots of the equation $x^2 + x - 12 = 0$. Do not use the general formula for the roots.

II.2   Double Roots and Tangents

II.2.1   *Multiplicity of Roots*

Given a quadratic function f, the factorization $f(x) = a(x - p_1)(x - p_2)$, where p_1, p_2 are the roots of f, reveals that, algebraically, there is nothing special about the case when $p_1 = p_2$, which arises when the discriminant equals zero. Each root accounts for one factor, and if the two roots coincide, we still have *two* factors which now happen to be equal. In any case, if the quadratic equation $f(x) = 0$ has any (real) roots at all, there always are two roots, each one accounting for one linear factor, where of course it may happen that the two roots are equal. In that case we say that the root is a *double root*, or that it has *multiplicity* 2, accounting for the fact that the one number p that describes the roots now appears in both linear factors, i.e., the corresponding factor $(x - p)$ appears 2 times, in the form

$(x-p)(x-p) = (x-p)^2$. In contrast, if the roots are different, i.e., $p_1 \neq p_2$, then each corresponding factor appears only *once*, and we say that in the case of different roots each root has multiplicity 1.

On the other hand, the *geometric* visualization reveals a significant difference that turns out to have most important consequences. We therefore want to *see* this difference and try to understand its deep implications in the simple case of the quadratic function, before studying the more general case of arbitrary polynomials in the next section.

Recall that the properties of the graph of a quadratic function f are most readily seen if the function is written in the special form

$$f(x) = a(x - p)^2 + d.$$

To simplify further, we shall assume that $a = 1$, so that the graph of f, a parabola with vertex (p, d), opens to the top. Here are two typical graphs.

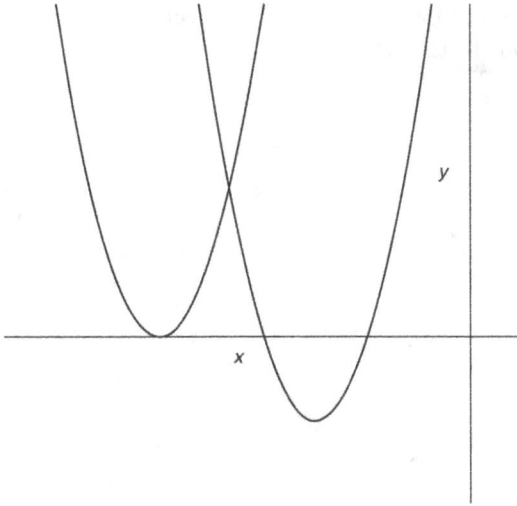

Fig. II.7 Graphs of two quadratic functions.

The graph on the left corresponds to the case where $d = 0$, while the graph on the right shows an example where $d < 0$. We do not show the case $d > 0$, as in this case the graph lies above the x-axis, that is, there are no zeroes of f in this case. Note that if we ignore the x-axis, the two graphs do not show any differences whatsoever — one is just a translation of the other. On the other hand, the points of the graph on the x-axis determine the roots of f. Clearly we have two distinct roots on the right, while on the left there is only one root, although we know that this is a double root of multiplicity 2. Note that if the graph on the left is moved ever so slightly downwards, the double root splits into two separate points, thus providing a visual justification to count the root of multiplicity 2 as two points, which just happen to lie one on top of the other.

II.2.2 *Touching Lines and Tangents*

Another feature that we notice is the very different relationship between the graphs and the x-axis in the two cases. Clearly on the right the x-axis "cuts" through the graph, while on the left the x-axis just "touches" the graph without cutting it. Perhaps you have seen this "touching" relationship before. A widely known example is the case where a line "touches" a circle, but does not cut it. Such lines are known as *tangents* to the circle. A concrete familiar version occurs with a wheel rolling on a flat road — the road is *tangential* to the wheel.

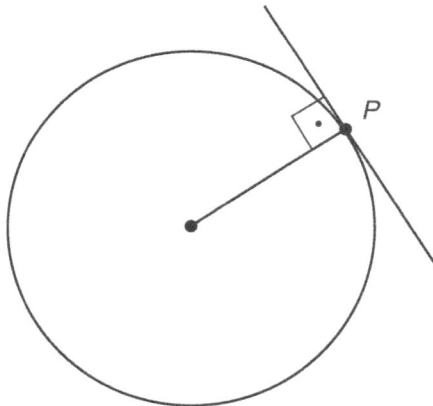

Fig. II.8   Circle with Tangent.

It is a well-known fact, which we will verify in the next section, that the tangent at the point P on the circle is perpendicular to the radius that connects the center C to P.

Tangents to circles and more general curves, including parabolas, ellipses, etc., have a long history. They were considered by Greek geometers as early as the fourth century B.C. In fact, the language we just used goes back to that time, when geometers described a tangent to a curve at the point P on the curve as a line that "touches" the curve, but does not "cut" it. While this terminology is somewhat vague, it surely accurately describes the situation with a tangent to a circle. And it also fits well with the x-axis as it touches the vertex of the parabola on the left in the figure shown in Section 1; it is consistent with this terminology to say that the x-axis is tangential to the curve on the left, while on the right it surely "cuts" the curve, so is NOT a tangent.

In the 17th century the problem of precisely describing the tangents to the graphs of curves became a very central area of investigation for philosopher and scientists, especially as its relationship to the concept of "instantaneous" velocity was recognized. This made the tangent problem most relevant for investigations of the physical world. In particular, René Descartes (1596–1650), one of the intellectual "giants" of that period, and whom we already mentioned in the context of the Cartesian coordinate systems, recognized the relationship between tangents and double points (or roots of multiplicity 2) and was able to find the tangents to an ellipse, and other curves, by a purely algebraic method. Note that a line that "touches" the circle at the point P or the parabola at its vertex really touches the curve at P in two points, that just happen to coincide. In fact, the "double point" splits into *two* separate points as soon as the line is rotated just so slightly about P.

The heart of Descartes' insight is clearly visible in the two parabolas shown in the figure in Section 1, which we may state as follows.

The x-axis is tangential to the parabola $f(x) = (x - p)^2 + d$
if and only if the equation $f(x) = 0$ has a double root.

II.2.3 *Tangents to a Circle*

To gain some appreciation of this insight, let us apply it in the familiar context of a circle. To keep matters simple we place the center of the circle at the point $(0, 0)$, i.e., at the origin of the coordinate system, and we choose the radius to be 1, so that the equation of the circle is $x^2 + y^2 = 1$. Any (non vertical) line through a fixed point $P = (a, b)$ on the circle can easily be described via its point-slope form $y - b = m(x - a)$, where m is the *slope* of the line, which measures the inclination or direction of the line. Since $P = (a, b)$ is on the circle, we must have $a^2 + b^2 = 1$. Figure II.9 shows several lines through P.

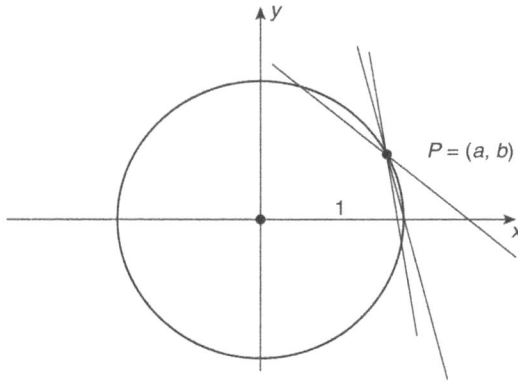

Fig. II.9   Circle of radius 1 with lines through the point P.

Based on Descartes' insight, the line is a tangent if it intersects the circle at P in a double point. So we need to determine the slope m so that the relevant equation that identifies the point P of intersection has a double root. This can be done by using simple familiar algebraic tools, as follows. The points of intersection (x, y) of the circle with such a line must satisfy the two equations

$$x^2 + y^2 = 1 \text{ and } y - b = m(x - a). \tag{II.2}$$

While the straightforward substitution of $y = b + m(x - a)$ into the first equation in (II.2) leads to a quadratic equation for x that can readily be solved by the techniques we discussed in Section 1.2, it is easier to take advantage of the fact that we already know that $x = a$ is one of the two solutions, and therefore, by Lemma 5, the resulting equation must have a factor $(x - a)$. We use the equation $a^2 + b^2 = 1$ (the point (a, b) lies on the circle!) and subtract it from the left equation in (II.2). One obtains

$$x^2 - a^2 + y^2 - b^2 = 0,$$

which can be factored into

$$(x + a)(x - a) + (y + b)(y - b) = 0.$$

Now substitute $m(x-a)$ for $y-b$ and $2b+m(x-a)$ for $y+b$, and rearrange, so that the equation for the x-coordinates of the point of intersection turns into the form

$$(x + a)(x - a) + [(2b + m(x - a))][m(x - a)]$$
$$= \{(x + a) + [(2b + m(x - a))]m\} \cdot (x - a) = 0.$$

This clearly shows—as expected—that $x = a$ is one of the solutions. Since we are looking for the slope m for which the point (a, b) is a *double* point of intersection, $x = a$ must be a zero of multiplicity 2 of this equation. Therefore, the factor in $\{...\}$ also must have a zero at $x = a$, that is, we must have

$$\{2a + [2b + 0]m\} = \{2a + 2bm\} = 0.$$

If $b \neq 0$, that is, if $(a, b) \neq (1, 0)$ or $(-1, 0)$, it follows that $m = -a/b$ is the slope of the unique line for which the point (a, b) of intersection with the circle is a *double* point. So $m = -a/b$ is the slope of the *tangent line* at the point (a, b). Note that this result confirms the classical geometric condition, namely that the tangent at P is perpendicular to the radial line from the center $(0, 0)$ to $P = (a, b)$. In fact, if we also assume that $a \neq 0$, the slope m_N of this radial line is b/a, and since $(-a/b)(b/a) = -1$, the tangent we determined algebraically using Descartes' insight is indeed perpendicular to the relevant radial line.

II.2.4 *Tangents to a Parabola*

Next we consider the tangent to a parabola at an arbitrary point on the curve. Let us first mention that in the third century B.C. the Greek geometer Apollonius, whom we already mentioned earlier in the context of conic sections, had discovered a geometric construction of the tangent at an arbitrary point on a parabola. The process is best described by Fig. II.10.

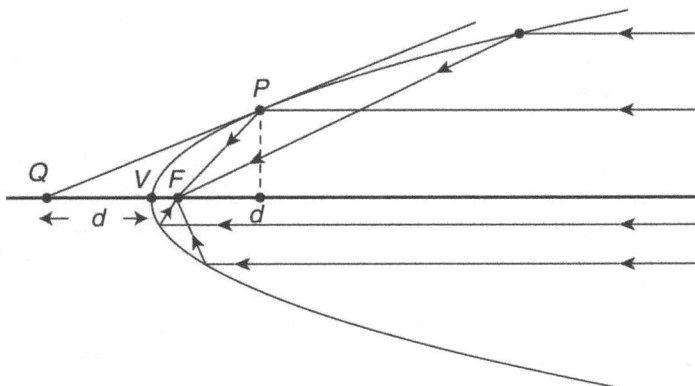

Fig. II.10 Apollonius' construction of tangents to the parabola.

As shown above, the (perpendicular) projection of P onto the axis of the parabola identifies a point at distance d from the vertex V. Consider the point Q on the extended axis that is at the same distance d from the vertex V on the opposite side. By a geometric argument involving the "touching" property of the tangent, Apollonius showed that the tangent to the parabola at P is that line through P that goes through the point Q. We shall verify this result a bit later, based on the algebraic construction of the tangent we shall give shortly. Knowledge of the tangent allows to describe the reflection of light rays entering the parabola parallel to the axis. As we mentioned earlier, all these light rays are reflected so that they then go through the focus F. We shall prove this property in the next section.

Let us now translate geometry into algebra and apply the double point method—which was so successful for a circle—to determine the tangents to a parabola. We place the vertex V at the point $(0,0)$ and choose the axis of the parabola along the positive y-axis. The equation of the parabola is then $y = \lambda x^2$ for some $\lambda > 0$ that depends on the distance between the vertex and the focus. Let us fix a point (a,b) on the parabola. As before, any (non-vertical) line through (a,b) has an equation $y = b + m(x - a)$, which is just a simple rearrangement of the point-slope form. Its points of intersection with the parabola are the solutions of

$$\lambda x^2 - b - m(x - a) = 0.$$

After replacing $b = \lambda a^2$ (the point (a,b) is on the parabola), this equation factors into

$$\lambda(x + a)(x - a) - m(x - a) = [\lambda(x + a) - m](x - a) = 0.$$

The two solutions are a and $m/\lambda - a$. Consequently (a,b) is a *double* point of intersection of the line with slope m precisely when these two solutions are equal, i.e., $a = m/\lambda - a$, or $m = 2\lambda a$.

Example. At the point $(-1,1)$ the slope of the tangent to the graph of $f(x) = x^2$ equals $2(-1) = -2$. Hence the equation of the tangent line at that point is $y = 1 + (-2)(x - (-1))$, or $y = 1 - 2(x + 1)$.

To complete the discussion, let us compare the algebraic result with the classical geometric construction of Apollonius. The projection of $(a, \lambda a^2)$ onto the axis of the parabola gives the point $(0, \lambda a^2)$ that is at distance $d = \lambda a^2$ from the vertex $V = (0,0)$. According to Apollonius, the tangent at $(a, \lambda a^2)$ goes through the point $(0, -\lambda a^2)$, and consequently that tangent has slope $m = [\lambda a^2 - (-\lambda a^2)]/(a - 0) = 2\lambda a^2/a = 2\lambda a$. As expected, this agrees with the result obtained by the double point method.

Finally, we want to show how this algebraic method works on a parabola that is the graph of an arbitrary quadratic function $f(x) = ax^2 + bx + c$.

Let us fix an arbitrary point $P = (p, f(p))$ on the parabola. As usual, the equation of an arbitrary line through P is given by $y = f(p) + m(x - p)$, and the x-coordinates of its points of intersection with the parabola are the solutions of the equation

$$f(x) = f(p) + m(x - p).$$

Clearly p is one of the roots (the line was constructed to go through P!). By Lemma 5, we know that $f(x) - f(p)$ can be factored into $g(x)(x - p)$, where g is a linear function. In the proof of that Lemma we saw that $g(x) = a(x + p) + b$. Therefore the equation for the points of intersection becomes

$$g(x)(x - p) = m(x - p),$$

which we rearrange in the form

$$[g(x) - m](x - p) = 0.$$

One root is clearly $x = p$, and it will have multiplicity 2 precisely when the factor $g(x) - m$ also has a root at p, that is, when $g(p) - m = 0$. Thus the slope m of the tangent line at the point $(p, f(p))$ must be equal to $g(p)$, and from the formula for g we obtain

$$m = g(p) = a(p + p) + b = 2ap + b.$$

Conclusion. *The tangent line to the graph of $f(x) = ax^2 + bx + c$ at the point $(p, f(p))$ has slope $2ap + b$.*

We note that if $b = 0$ this conclusion agrees with our earlier result, since the graph of $f(x) = ax^2 + c$ is just a vertical translation of the graph of $y = ax^2$.

Example. Find the equation of the tangent line to the graph of $f(x) = 3x^2 + 5x - 2$ at the point where $x = 3$. We note that the point is $(3, 10)$, and the desired slope is given by $m = 3(2 \cdot 3) + 5 = 23$. Therefore the point-slope form of the tangent line is

$$y - 10 = 23(x - 3).$$

Remark. The formula for the slope of the tangent line gives a very simple way to find the vertex of the parabola, as follows. Just note that the tangent line at the vertex is horizontal, i.e., has slope 0. For example, for the parabola we just considered, where $m = 6x + 5$ is the slope of the tangent at the point $(x, f(x))$, one sees that $m = 0$ exactly when $x = -5/6$. So the vertex is at $(-5/6, f(-5/6))$. Once we have identified the vertex of the parabala, and hence its axis, the coefficient a of x^2, that is, 3 in the example considered, determines the shape of the parabola, as we had discussed earlier. We also note that the vertex, identified as the point where the tangent has slope 0 is the lowest (or highest point, if $a < 0$) of the graph, so we see how determining the slope of the tangent provides a tool to identify such (local) minima or maxima of the graph.

II.2.5   *Reflection Property of the Parabola*

We now want to verify the important physical property of the parabola, namely that light rays that enter the parabola parallel to its axis are reflected on the parabola, so that they then go through the focus. Conversely, any light ray starting at the focus is reflected on the parabola, so that it is then parallel to the axis.

We first need to recall a basic fact about reflection of light on a flat surface, such as a mirror. The physical law of reflection states that the incoming and outgoing rays form the same angle α with the surface, or, equivalently, the same angle β with the line perpendicular to the surface at the point P where the ray hits the surface, as shown in Fig. II.11.

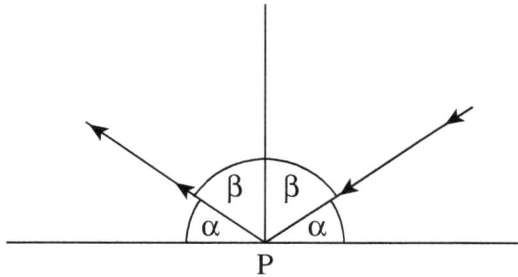

Fig. II.11   Law of reflection on a flat surface.

It is then verified experimentally that if the reflecting line is a curve, then reflection at the point P on the curve follows the law of reflection applied to the tangent line to the curve at P. So we see why it is important to have information about the tangents to a parabola.

Let us now consider the parabola with vertex at the origin O whose axis is the y-axis, and whose focus is at the point $F = (0, d)$. We had seen that this parabola is the graph of the function $y = \frac{1}{4d}x^2$. Let us follow a light ray that comes in from the top, parallel to the axis of the parabola, that is, to the y-axis. This light ray follows a line $x = a$ for some a which hits the parabola at the point $P = (a, \frac{1}{4d}a^2)$. We also draw the line from P to the focus F, and we want to show that this line is indeed the reflection of the (vertical) light ray at P according to the law of reflection applied to the tangent at P.

These details are visible in Fig. II.12, which also shows the tangent at P, whose slope $m = 2(\frac{1}{4d})a = \frac{a}{2d}$, where we of course have used the key result obtained in the previous section. Next, we draw a line through the focus F parallel to the tangent. Its equation is $y = d + \frac{a}{2d}x$. Let Q be the point of intersection of that line with the vertical line $x = a$. Then $Q = (a, d + \frac{a}{2d}a)$. We will now show that the triangle QPF in Fig. II.12 has two sides $\overline{QP}$ and $\overline{PF}$ of equal length. Note that

$$dist(Q, P) = \left| d + \frac{a^2}{2d} - \frac{a^2}{4d} \right| = d + \frac{a^2}{4d}.$$

This last number is precisely the distance from P to the directrix L of the parabola, i.e., $dist(P, L)$! By the defining geometric property of the parabola, $dist(P, L) = dist(P, F)$, and we have verified that $dist(Q, P) = dist(P, F)$. In Fig. II.12, we also show a piece of the line through P that is perpendicular to the tangent, and hence also to the line $\overline{FQ}$, and which meets that line at the point M. The two right triangles FMP and QMP share the same side $\overline{MP}$, and we just saw that their hypothenuses have

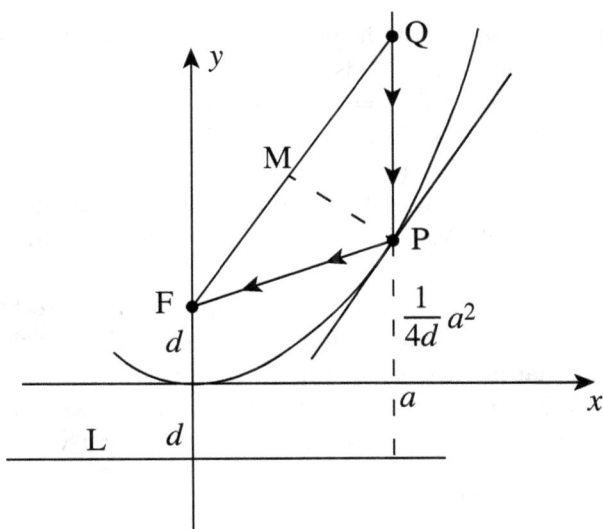

Fig. II.12 Light reflected into focus F.

equal length as well. So these two triangles are congruent, and therefore their angles at the point P must be equal. We have thus verified that the incoming ray from Q to P is reflected on the parabola at P, i.e., on its tangent at P, so that it will go exactly to the focus F.

We note how this argument involves both the geometric and algebraic description of a parabola, as well as information about the slope of the tangent. One could translate the algebraic components into geometric arguments, and obtain a purely geometric proof of the reflection property, and that is what Apollonius had done. However, that argument turns out to be quite a bit more complicated. The lesson to be remembered is that it is most useful to be open minded and to consider both algebra and geometry.

II.2.6   *Exercises*

1. Find the equation of the tangent line to the parabola given by $y = \frac{3}{2}x^2$ at the point $(2, 6)$.

2. Consider the point $(3, 9)$ on the parabola given by $y = x^2$,

   a) Determine the equation of the line through $(3, 9)$ with slope m,

   b) Substitute the equation in a) into the equation $y = x^2$ to obtain a quadratic equation in x of the form $x^2 + bx + c = 0$ for the x-coordinates of the points of intersection of the line with the parabola, where the coefficients b, c depend on m.

   c) Determine m so that the discriminant $b^2 - 4c = 0$. Determine the solution(s) of the equation for this value m.

3. Consider the parabola $y = x^2/4$. Use the construction of Apollonius to find the points of tangency on the parabola for the tangents to the parabola through the point $(0, -6)$.

4. Find the equations of each of the tangents to $y = x^2$ that go through the point $(3, 0)$. (Hint: Make a sketch of the situation before starting any computations.)

5. Consider the function $f(x) = x^3$. Apply the same techniques used for the parabola to find the slope m of the line through the point (a, a^3) on the graph of f that intersects the graph with multiplicity greater than one.

6. Work out Exercise 5 at the point $(0, 0)$. What do you notice about the multiplicity of the point of intersection of the graph of f with the x-axis?

II.3   Motion with Variable Speed

II.3.1   *Galileo's Discovery*

Before continuing with the study of tangents for other curves we first want to discuss the relationship of the tangent problem with a fundamental problem of motion. In fact, the search for a deeper understanding of motion and related phenomena in the physical world in the 17th century was arguably the major driving force that led to the development of calculus by Newton and Leibniz.

Experience shows that a stone that is dropped from the top of a building falls towards the ground at an *increasing* speed. The higher the building, the faster the stone will be falling just before impact. As we already mentioned in Section I.2.1, it was Galileo Galilei (1564–1642) who first analyzed the situation precisely in order to discover the underlying laws of motion.

Based on numerous observations of falling stones[2] and balls rolling down inclined planes, and trusting that the observed motion is governed by simple principles, Galileo recognized in 1604 that the motion of a freely falling body is *uniformly accelerated*, that is, the increase in speed over a time interval from t_1 to t_2 is a fixed multiple of the length $t_2 - t_1$ of that interval. In particular, if at time $t = 0$ the speed is zero, then at later times $t > 0$ the speed $v(t)$ equals $a \cdot t$ for a certain fixed number a, the so-called *acceleration*. Another relevant quantity—more easily measurable than speed—is the distance that an object has moved in a given time interval.[3] Indeed, Galileo proved by geometric arguments that in case of uniformly accelerated motion starting from rest, the ratio of the distances d_1 and d_2 traveled in corresponding times t_1 and t_2 equals the ratio of the *squares* of the times, i.e., $d_1 : d_2 = t_1^2 : t_2^2$. This translates into the formula $d(t) = ct^2$ for the distance $d(t)$ traveled in time t, where c is another constant. Galileo was able to confirm the validity of this latter formula in numerous experiments, thereby also obtaining a numerical value—which depends on the particular units chosen to measure distance and time—for the constant c. In the

---

[2]It is often reported that Galileo carried out such experiments by dropping stones from the Leaning Tower of Pisa.

[3]For example, one could envision a long ruler placed vertically on the side of the building, with its initial point 0 placed at the top. A stop watch is started at the moment the stone is dropped, and one reads off the position of the falling stone against the ruler after $1, 2, ...,$ seconds.

case of a freely falling object, and using today's standard units *meters* for distance and *seconds* for time, c is approximately 4.9 m/sec^2.[4]

In hindsight, Galileo's emphasis on using mathematical relationships to describe physical phenomena, rather than trying to explain the causes of phenomena by hidden actions of some mysterious higher being, turned out to be the breakthrough that—empowered by new mathematical tools—led to the amazing progress in mankind's understanding of the physical world since Galileo's days.

Incidentally, Galileo's formula for a freely falling stone provides a practical technique to estimate the height of buildings or rock walls, as follows.

Example. Suppose a stone is dropped from the top of a building of unknown height H meters. Its height $h(t)$ above ground after t seconds is then given by the formula

$$h(t) = H - 4.9t^2 \ m,$$

where the minus sign accounts for the fact that the distance $d(t)$ traveled by the stone needs to be subtracted from the initial height H in order to get the height after t seconds. The rock hits the ground when $h(t) = 0$. Suppose this happens after t_0 seconds. Then $H - 4.9t_0^2 = 0$ implies that $H = 4.9t_0^2$ m. This formula is sometimes applied by rock climbers who need to estimate their height above a ledge in order to judge whether their rope is long enough to rappel down. Suppose a climber drops a stone, and by using a watch (a stop watch would be nice) she determines that the stone hits the ledge after 3.5 seconds. By the preceding formula, the height above ground thus is approximately $5 \cdot (3.5)^2 \approx 61$ m, which is quite a bit more than her 50 m long rope. The climber thus decides not to rappel down at that location.

II.3.2   *From Distance to Velocity*

Returning to Galileo's result, the basic question that arises is how to derive a formula for the *velocity* $v(t)$[5] of the falling stone at time t from the formula for the distance $d(t) = ct^2$. In particular, one needs to give precise

---

[4]The unit m/sec^2 for the constant c is a consequence of the relationship *distance* $= c \times (time)^2$. If *feet* are used instead of meters, the numerical value for c is approximately 16 ft/sec^2.

[5]*Velocity* is the term generally used in science for what common language calls *speed*; velocity is allowed to be both positive and negative (or zero), with the sign accounting for the direction of motion along a line. More generally, when the motion is not constrained to a line, the velocity is represented by a so-called *vector*, a more complicated quantity that encodes, for example, the direction of the motion in space.

meaning to the concept of *velocity at a single moment in time*. As commonly understood, velocity is a measure of the rate of change of position over time, that is

$$velocity = \frac{distance}{time}.$$

More precisely, for two distinct moments in time t_1 and t_2, the *average* velocity over the time interval $I = [t_1, t_2]$ (assume $0 < t_1 < t_2$) is

$$v_I = \frac{d(t_2) - d(t_1)}{t_2 - t_1}, \tag{II.1}$$

where $d(t)$ is the distance traveled from the starting point $t = 0$, so that $d(0) = 0$.

If the *average* velocity of a motion is independent of the time interval I, we say that the motion has *constant* velocity $v = v_I$. In case of constant velocity, the velocity $v(t)$ at any moment t is always this same number v that equals the average velocity over any interval I, and it then follows easily that the distance $d(t)$ equals vt.

This is a good moment to highlight the relationship between velocity and slopes of lines. In both cases the relevant value is expressed as a ratio, which is interpreted as a *rate of change*: the slope measures the rate of change in height (measured along the y-axis) as we move along the x-axis, and velocity measures the rate of change of position as time moves on. As we just saw, in case of *constant* velocity v, the distance $d(t)$ traveled in t seconds is given by the *linear* function $d(t) = vt$. So in this case there is a perfect match: the velocity is exactly the slope of the graph of the (linear) distance function.

However, in the case of the falling stone the velocity is not constant, so how do we define the velocity $v(t)$ at any particular moment? Intuitively, we agree that at any moment the falling stone is moving with a certain velocity, which increases with time until the stone hits the ground. Similarly in modern times, when traveling in a car, we do experience the (variable) velocity (or speed) at any moment, and the speedometer even gives us a number that measures this speed. If we apply the brakes, the speedometer indicates a decreasing speed. So what exactly is the speedometer measuring? We shall now examine this problem more in detail, and eventually recognize its relationship with the tangent problem.

Notice that for a fixed moment t_0, while we agree that there is a velocity $v(t_0)$, surely we can not compute the average velocity over the interval $[t_0, t_0]$ by formula (II.1), since this formula now gives the meaningless expression $\frac{0}{0}$. However, if we rewrite the equation that defines velocity as the product *distance = velocity × time*, then the problem becomes more manageable. In fact, let us consider the simple case considered by Galileo, i.e., $d(t) = ct^2$. If we fix a particular time t_0, then $d(t) - d(t_0) = ct^2 - ct_0^2$, which factors into

$$d(t) - d(t_0) = c(t + t_0)(t - t_0). \tag{II.2}$$

Note that if $t > t_0$ the factor $c(t + t_0)$ in this last formula obviously equals the average velocity over the time interval from t_0 to t. (Just divide both sides of (II.2) by $t - t_0 \neq 0$.) This also holds if $t < t_0$, where the time interval now goes from t to t_0. (See Exercise 4.2.) Therefore, trusting in the consistency of the formula (II.2), we are led to define the velocity at t_0 by taking the value of this factor at $t = t_0$, that is, we define

$$v(t_0) = c(t_0 + t_0) = 2ct_0.$$

Just looking at the underlying algebra, note that the factor $c(t + t_0)$ in formula (II.2) agrees with the factor we had labelled $q(x)$ in the analogous factorization of the quadratic function $s(x) = x^2$, whose value at x_0 gave us the slope of the tangent line to the graph at the point (x_0, x_0^2). We thus realize that, when a single point t_0 is considered, we define the velocity $v(t_0)$ by the *slope of the tangent line* to the relevant graph.

Perhaps you have some doubts about the validity of this definition. After all, the basic formula *distance = velocity × time* reduces, in case $t = t_0$, to the equation $0 = c(t_0 + t_0) \cdot 0$, which surely is correct, but then any other number k also satisfies the equation $0 = k \cdot 0$. So you may ask why do we single out the particular number $c(t_0 + t_0)$ among all the other possible numbers k that satisfy the equation?

One justification surely comes from the fact that $c(t_0 + t_0)$ is exactly that number that arises when t is replaced by t_0 in the algebraic formula $d(t) - d(t_0) = c(t + t_0)(t - t_0)$. Since this formula does represent a "universal truth", the value of $c(t + t_0)$ at $t = t_0$ should have an interpretation that is analogous to that for all other values t, that is, it should represent a velocity. And since only one moment in time t_0 is involved, it is reasonable to think of $c(t_0 + t_0)$ as the velocity at t_0.

Another justification is based on the geometric interpretation involving tangents to parabolas that we discussed earlier in Section 2.4. As we showed then (just replace $x = t$ and $y = d(t) = ct^2$), and as we just recalled, the line through the point (t_0, ct_0^2) with slope $2ct_0$ is the tangent to the graph of the function $d(t) = ct^2$, i.e., it is that line that fits the graph in an "optimal" way. Rephrasing this in the context of motion we thus can say that at the moment $t = t_0$, the *constant* speed motion $l(t) = ct_0^2 + 2ct_0(t - t_0)$ with velocity $2ct_0$ (i.e., the equation that defines the tangent) provides an optimal description of the motion given by $d(t) = ct^2$ at that moment. More precisely, this constant speed motion matches the given motion described by $d(t)$ at the moment t_0 "with multiplicity *two*", that is, at two points in time that just happen to coincide. Alternatively, think of a vehicle starting from rest at $t = 0$ under the same uniform acceleration as a falling stone, so that—according to Galileo—the distance traveled at time $t > 0$ equals $d(t) = ct^2$. At time t_0 the driver takes off his foot from the accelerator. Neglecting minor factors such as friction, air resistance, and so on, the car would continue rolling with *constant* velocity equal to $2ct_0$. In any case, we see in these historical significant investigations of Galileo the connection to the geometric tangent problem.

Finally we can also consider a *dynamic* point of view, which perhaps reflects most closely the crux of motion with variable speed, as follows. As we saw, for $t \neq t_0$ the value $q(t) = c(t + t_0)$ gives the *average* velocity during the time interval $[t_0, t]$ (or $[t, t_0]$ if $t < t_0$). Surely we expect that the velocity at t_0, no matter how defined, should be very close to the average velocity over very short time intervals, i.e., when t is very close to t_0, and furthermore, this approximation should improve as the time interval gets shorter, i.e., the closer t gets to t_0. The chosen value $v(t_0) = q(t_0)$ fulfills this expectation perfectly, since

$$|q(t) - q(t_0)| = |c(t + t_0) - 2ct_0|$$
$$= |c|\,|t + t_0 - 2t_0| = |c|\,|t - t_0|. \tag{II.3}$$

Evidently formula (II.3) shows that when t is "very close" to t_0, then the average velocity $q(t)$ from t_0 to t is "very close" to $q(t_0)$ as well. For example, let us use meters and seconds, so that $c \approx 4.9\,\mathrm{m/sec^2}$. Suppose $t_0 = 5$ *sec* and $t = t_0 + 1/1000 = 5.001$ *sec*; then the average velocity $q(t)$ during the interval $[t_0, t]$ equals $4.9 \times 10.001\,\mathrm{m/sec}$, which differs from the velocity $q(5) = v(5) = 2 \times 4.9 \times 5\,\mathrm{m/sec}$ by $4.9 \times 1/1000 = 0.0049$ m/sec. Stated differently, formula (II.3) gives a precise meaning to the intuitive statement that as t approximates t_0 (we write $t \to t_0$), then $q(t) \to q(t_0)$ as well. As we shall see later, the property we just discussed and that we encode in the statement

$$\text{if } t \to t_0, \text{ then } q(t) \to q(t_0),$$

is an elementary example of a fundamental abstract property that is known as *continuity*.

Our discussion shows that the concept of *instantaneous velocity*, that is, velocity at a particular moment, is really just another version of the tangent problem. The techniques one develops in order to find the slope of tangents also allow us to define and calculate the velocity at a single moment in time. In particular, returning to Galileo's investigations of freely falling bodies, where $d(t) = ct^2$, we have determined that the velocity after t seconds is given by $v(t) = 2ct$. This confirms that the motion indeed is uniformly accelerated, with the acceleration a given by $2c \approx 9.8$ m/sec^2. Thus the distance formula under uniform acceleration takes the more informative form

$$d(t) = \frac{1}{2} \times acceleration \times t^2.$$

II.3.3 *Exercises*

1. Suppose a stone is pushed off a tower which is 60 m high. After how many seconds will the stone hit the ground?

   2. Explain why the formula (II.1) gives the same value regardless of whether $t_1 < t_2$ or $t_2 < t_1$.

   3. A coin dropped into a deep well hits water after 2.5 seconds. How deep is the well?

   4. Let $f(x) = x^2 + 4x$.

   a) Establish an estimate $|f(x) - f(a)| \leq c\,|x - a|$ for $|x - a| \leq 1$ and some constant $c > 0$. (Hint: Factor $f(x) - f(a)$.)

   b) Explain why this implies that $f(x) \to f(a)$ as $x \to a$.

II.4 Tangents to Graphs of Polynomials

II.4.1 *Polynomials*

So far we have studied in detail *linear* and *quadratic* functions (perhaps the latter should be called *parabolic* functions). The mathematical expressions used to define the "function machine" in these cases, that is, the instructions that tell us how to find the output $f(x_0)$ that corresponds to the input x_0 are the simplest examples of what is called a *polynomial*. In general a polynomial P is an expression

$$P = a_n x^n + a_{n-1} x^{n-1} + a_{n-2} x^{n-2} + \ldots + a_2 x^2 + a_1 x + a_0,$$

where n is a nonnegative integer, and $a_n, a_{n-1}, \ldots, a_2, a_1, a_0$ are certain fixed numbers that are called the *coefficients* of the polynomial. If the coefficient

a_n of the highest power x^n appearing in the expression that defines the polynomial P is not 0, one says that the polynomial P has *degree* n. In particular, the functions defined by polynomials of degree 1 are also called *linear* functions, and those defined by polynomials of degree 2 are also called *quadratic* (maybe better *parabolic*) functions. Similarly, we may use the names linear and quadratic polynomials. Note that *nonzero* numbers are polynomials of degree 0. The degree of the 0 polynomial, where all coefficients are 0, is NOT defined.

For the time being, we mainly consider coefficients that are rational numbers, although occasionally we will need to also consider certain algebraic numbers, such as $\sqrt{2}, \sqrt{3}$, etc. Eventually, we will consider coefficients that can be any real number, that is, any point on the number line, or more generally, one may even consider polynomials with complex numbers as coefficients, or any collection of elements for which an addition and a multiplication is defined that satisfies the rules Q1–Q6 that we have singled out in the study of rational numbers. The symbol x stands for an unspecified quantity that may represent any of the numbers that are under consideration. Since x may be thought of as a *variable* number, the familiar operations on numbers are meaningful for x as well, such as $x + x = 2x$, $x \cdot x = x^2$, and so on. This is particularly evident when we interpret a polynomial P as a function with domain $\mathbb{Q}$ (or $\mathbb{R}$, or $\mathbb{C}$) as follows:

Given a number q (the input), the output $P(q)$ is defined by

$$P(q) = a_n q^n + a_{n-1} q^{n-1} + a_{n-2} q^{n-2} + \dots + a_2 q^2 + a_1 q + a_0.$$

Now the expression has a precise meaning according to the rules that are satisfied by numbers, and in particular $P(q)$ is again a number. It therefore is meaningful to apply these same rules when a specific number is replaced by the symbol x. In particular, two polynomials can be added and multiplied by formally following the usual rules, as if the symbol x were a number. We already applied this process without much thought, when for example, we considered a product

$$(2x + 3)(4x - 2) = 2x \cdot 4x + 2x(-2) + 3 \cdot 4x + 3(-2)$$
$$= 8x \cdot x + (-4)x + 12x + (-6)$$
$$= 8x^2 + 8x - 6.$$

Lemma 9 Suppose P is a polynomial of degree n and G is a polynomial of degree m. Then PG is a polynomial of degree $n + m$.

Proof. If $a_n x^n$ and $b_m x^m$ are the terms with the highest powers in P, resp. G, where a_n, $b_m \neq 0$, then $a_n x^n b_m x^m = a_n b_m x^{n+m}$ is the term with the highest power in the product PG, and since $a_n b_m \neq 0$, the degree of PG is $n + m$. ∎

Corollary 10 If P, G are nonzero polynomials, then $PG \neq 0$. Stated differently, the product of two polynomials is 0 if and only if at least one of the factors is 0.

The set of all polynomials with rational coefficients is denoted by $\mathbb{Q}[x]$, and it is called the *ring of polynomials* (or *polynomial ring*) over $\mathbb{Q}$. The name "ring" is a technical mathematical term that refers to a (possibly very abstract and general) set of elements, for which two operations labelled addition and multiplication are defined, and so that the rules Q1–Q6 are satisfied. We have not gone through the detailed verification that these rules apply to polynomials. We shall just assume that they are valid for polynomials, just as they are for numbers, and this justifies calling $\mathbb{Q}[x]$ a ring. Note that at this point we have not introduced multiplicative inverses of polynomials. In general, a *ring* does not require the existence of a multiplicative inverse, hence rule Q7 does not apply. So the polynomial ring $\mathbb{Q}[x]$, even though it involves rational numbers as coefficients, is formally comparable to the ring of integers $\mathbb{Z}$, at least as far as the rules that apply to it. By restricting the coefficients to just $\mathbb{Z}$, one obtains the ring $\mathbb{Z}[x]$ of polynomials with integer coefficients, or, in case of more general coefficients, one obtains the rings $\mathbb{R}[x]$ and $\mathbb{C}[x]$.

As we shall see in the next section, basic algebraic concepts and results that we studied in the special case of linear and quadratic polynomials will apply to arbitrary polynomials. However things get much more complicated when we consider graphs. Note that the geometric structure of the graphs in the linear case remains the same since they always are (non-vertical) lines, and in case of quadratic polynomials (i.e., parabolas) it essentially remains the same, except for translations, reflections, and rescaling. But as soon as one considers polynomials of degree 3, there is a much greater variety of curves.

For example, consider the graph of $C(x) = x^3$ in Fig. II.13 shown below.

Note the special role of the origin O, which is a point on the graph. Moving along the graph from the left, we see that the graph is curved towards the right until we get to O, at which point the graph begins to turn to the left. Think of a car travelling along the graph. Right at the origin the steering wheel is straight, since it is changing from a right turn to a left turn. Such a point on the graph is called a *point of inflection*. We

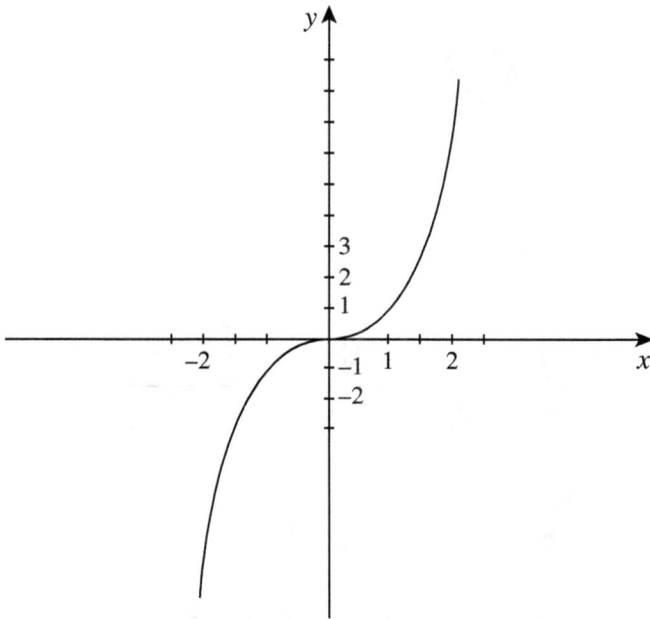

Fig. II.13   Graph of $y = x^3$.

shall see eventually that the tools of calculus can be used to characterize such points, as well as other geometric properties of graphs of functions.

A slight change leads us to the graph of $y = (x - p)^3$. Just as we saw in the case of the parabola, this graph is obtained from the graph of C by translating it to the right by p units if $p > 0$, or to the left if $p < 0$.

On the other hand, as one sees in Fig. II.14 (most easily generated by a graphing calculator), the graph of a typical general third degree polynomial $P(x) = 2x^3 + 3x^2 - 12x + 1$ looks quite a bit more complicated than the graph of C.

Again, we see that there is a point of inflection at the point on the graph where $x = -1/2$. Furthermore, there are two points that seem to be comparable to the vertex of a parabola, the first one is a "local peak" where the x-coordinate is -2, the second one is "local valley", where the x-coordinate is 1.

As you can imagine, things will get even more complicated as one considers polynomials of degree 4 or higher. The reader should use a graphing calculator to visualize several more examples.

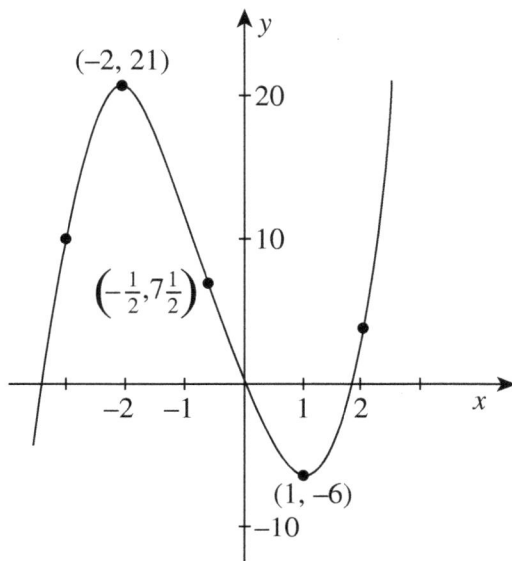

Fig. II.14   Graph of general cubic polynomial.

II.4.2   *Roots and Factorization of Polynomials*

We shall now generalize certain basic results we had obtained for quadratic polynomials to arbitrary polynomials. First of all, given a polynomial P, a number p is called a *root* or *zero* of P if $P(p) = 0$. If we consider the graph of P, then the roots identify those points on the x-axis where the graph intersects the x-axis.

Finding the zero of a linear function just involves solving a linear equation, and that is a very simple problem indeed. In Section 1.2, we obtained the formula for the zeroes of a quadratic polynomial. That formula is somewhat more complicated, and in particular it involved the so-called discriminant, whose properties ($> 0, = 0$, or < 0) determine whether the equation has 2, 1, or NO zeroes in $\mathbb{R}$. In any case, the problem of the zeroes of a quadratic polynomial is completely solved.

Moving to polynomials of degree 3 or even higher, and looking at the graph of the polynomial of degree 3 in the figures above, we may expect that matters become quite a bit more complicated. In fact, as far as *finding* the roots of polynomials, the situation is really bad. In the 16th century mathematicians struggled to solve polynomials of degree 3, even setting up competitions with prizes. Eventually, Gerolamo Cardano (1501–1576) published a solution in 1545 that he attributed to Scipione del Ferro (1465–1526), although it is commonly referred to as *Cardano's Formula*. While today only specialists in Algebra might be interested to study this formula, there is a detail that is worth mentioning. If the coefficients of the polynomial satisfy certain conditions, Cardano's formula involves explicitly square roots of *negative* numbers, which of course did not make any sense at all at that time. However, working with these quantities by following the usual rules, it turned out that the final result of the formula does lead back to (real) numbers which are indeed roots of the polynomial under consideration, even in case there are 3 distinct roots, as we saw in the last figure. That was quite a remarkable discovery, and it made people wonder if square roots of negative numbers do "exist" and could be useful. Perhaps this marks the beginnings of complex numbers, but it took many years until mathematicians became comfortable working with such mysterious quantities. Major progress was achieved only in 1806, when Jean-Robert Argand (1768–1822) introduced a geometric visualization of complex numbers in a plane, the so-called Argand Plane. This became widely accepted once Carl Friedrich Gauss (1777—1855), generally recognized as one of the greatest mathematicians of all times, used this idea extensively, so that the Argand Plane became also known as the Gaussian Plane.

With much effort mathematicians eventually found a formula also for the roots of a polynomial of degree 4, but in spite of many attempts, nothing could be done for general polynomials of degree 5. The matter was finally clarified in 1824 when Niels Henrik Abel (1802–1829) published a complete proof that it is NOT possible to find a formula to solve the general polynomial equation of degree 5. Eventually, this was verified for any degree ≥ 5. In spite of his brief life, Abel made major contributions to mathematics, and in 2002 the Norwegian Academy of Sciences established the *Abel Prize in Mathematics*, awarded each year since then. The prestige and financial rewards associated with this prize make it the equivalent of the (non-existing) Nobel Prize in Mathematics.

Fortunately, we do not need solution *formulas* for our investigations of tangents to graphs of polynomials. However, the important results involving factorization and multiplicity of roots that we learned about with the quadratic polynomial do carry over to arbitrary polynomials without any significant complications.

Theorem 11 *Let P be a polynomial of degree $n \geq 1$. If p is a zero of P,* *then there exists a (unique) polynomial q of degree $n-1$, so that*

$$P(x) = q(x)(x-p).$$

Corollary 12 *Given P as above, if p is any number, there exists a polynomial q (that depends on p) of degree $n-1$, so that*

$$P(x) - P(p) = q(x)(x-p).$$

Proof. Clearly the Corollary is an immediate consequence of the Theorem. To prove the Theorem, let us first consider the special case $p=0$. If $P(x) = a_n x^n + a_{n-1} x^{n-1} + a_{n-2} x^{n-2} + \dots + a_2 x^2 + a_1 x + a_0$, then $P(0) = 0$ clearly implies that $a_0 = 0$, so every summand in P has a factor x. Hence we immediately obtain a factorization $P(x) = q(x) \cdot x$. More generally, if P has a zero at the number p, consider the polynomial P^* defined by $P^*(x) = P(x+p)$. Then $P^*(0) = 0$, so by the special case just considered we have a factorization $P^*(x) = q^*(x) \cdot x$. Finally, observe that $P(x) = P^*(x-p) = q^*(x-p) \cdot (x-p)$, and setting $q(x) = q^*(x-p)$, we are done. As for uniqueness, suppose we have $P(x) = q_1(x)(x-p)$ and $P(x) = q_2(x)(x-p)$ for two polynomials q_1 and q_2. After subtracting one equation from the other, we get

$$0 = q_1(x)(x-p) - q_2(x)(x-p) = [q_1(x) - q_2(x)](x-p).$$

Since $x-p$ is clearly not the 0 polynomial, it follows that $q_1 - q_2$ must be 0, so that $q_1 = q_2$. ∎

II.4.3   *Division of Polynomials with Remainder*

Note that in the proof above we were not concerned about the precise expression for q. While in principle we could carefully keep track of all the details in the above proof to end up with a precise formula for the factor q, this would certainly be messy, even if one starts with an explicit polynomial with given numerical coefficients. However, there is an alternate proof of the Theorem that takes a detour through an important result that is the polynomial version of the familiar (long) division of integers with remainder. If needed, this result does yield a precise formula for the factor q in a systematic way.

On the other hand, the subsequent discussion does not use this more general result, so you may feel free to skip the details of the somewhat complicated proof.

Theorem 13 *Suppose P is a polynomial of degree $n \geq 1$ and G_l is a polynomial of degree l, with $1 \leq l \leq n$. Then there are polynomials q of degree $n - l$ and R_{l-1} of degree $\leq l - 1$ or $R_{l-1} = 0$ (the remainder), so that*

$$P = qG_l + R_{l-1}.$$

Note that the Theorem in the previous section is an immediate consequence of this Theorem. In fact, if p is a zero of P, apply the Theorem to P and $G_1 = (x - p)$. Then R_0 has degree 0 or is 0, so in any case R is constant. Evaluation of $P(x) = q(x) \cdot (x - p) + R_0$ at $x = p$ shows that $R_0 = 0$.

Proof. The idea is to subtract an appropriate multiple of $G = G_l$ from P to eliminate the highest degree term in P, leaving a polynomial of degree $\leq n - 1$, and then to repeat this procedure a total of at most $n - l + 1$ times, lowering the degree by 1 at each step, so that we end up with a polynomial of degree $\leq l - 1$, which will be the remainder. In detail, suppose a_n is the leading coefficient of P, and let $b_l \neq 0$ be the leading coefficient of G. Note that $a_n x^n = \frac{a_n}{b_l} x^{n-l} (b_l x^l)$. Consequently, by splitting up $G = b_l x^l + (G - b_l x^l)$, we obtain

$$P - \frac{a_n}{b_l} x^{n-l} G = [P - a_n x^n] - \left[\frac{a_n}{b_l} x^{n-l} (G - b_l x^l) \right] = R_{n-1},$$

where each of the terms inside the square brackets has degree $\leq n - 1$ (check it out carefully!), and therefore the preliminary remainder R_{n-1} is a polynomial with leading term $c_{n-1} x^{n-1}$, where the coefficient could be 0. In any case, if $n - 1 \geq l$, we can repeat this process formally with R_{n-1}, resulting in

$$R_{n-1} - \frac{c_{n-1}}{b_l} x^{n-1-l} G = [R_{n-1} - c_{n-1} x^{n-1}] - \left[\frac{c_{n-1}}{b_l} x^{n-1-l} (G - b_l x^l) \right]$$

$$= R_{n-2},$$

where R_{n-2} is a polynomial of degree $\leq n - 2$. Note that if $c_{n-1} = 0$, the above reduces to $R_{n-1} = R_{n-2}$. We repeat this process as long as the remainder has degree $\geq l$, so that after the last step we end up with a polynomial R_{l-1} of degree $\leq l - 1$, and then we must stop. By combining the terms by which we multiply G in each step we get a polynomial q whose leading term is $\frac{a_n}{b_l} x^{n-l}$. Since $a_n \neq 0$, q indeed has degree $n - l$. The result is $P - qG = R_{l-1}$ so that $P = qG + R_{l-1}$. ∎

There is a convenient detailed procedure for carrying out these computations explicitly that is modelled after the analogous division process for integers. This is best described by considering an explicit example. Let us take

$$P = 6x^4 - x^3 + 2x^2 \text{ and } G = 2x^2 + x + 1.$$

We write P under a horizontal line, and place G to its left, separated by a vertical bar, as follows:

$$(2x^2 + x + 1)\lceil \overline{} 6x^4 - x^3 + 2x^2$$

In the next step we write the factor $3x^2$, which is obtained by "dividing" $6x^4$ by the leading term of G, i.e., $2x^2$, above the line, and below we subtract the product $3x^2 G$, showing the result in the very bottom line, which is our preliminary remainder R_3.

$$
\begin{array}{l}
 \underline{3x^2} \\
(2x^2 + x + 1)\lceil\ 6x^4 - x^3 + 2x^2 \\
 \underline{-(6x^4 + 3x^3 + 3x^2)} \\
 0 \quad - 4x^3 - x^2
\end{array}
$$

We now repeat this process with R_3, resulting in

$$
\begin{array}{l}
 3x^2 - 2x \\
(2x^2 + x + 1)\lceil\ 6x^4 - x^3 + 2x^2 \\
 \underline{-(6x^4 + 3x^3 + 3x^2)} \\
 0 \quad - 4x^3 - x^2 \\
 \underline{- (-4x^3 - 2x^2 - 2x)} \\
 0 \quad + x^2 + 2x
\end{array}
$$

The bottom line now shows the next preliminary remainder R_2. So we repeat the process one more time with R_2, resulting in

$$
\begin{array}{l}
 3x^2 - 2x + \tfrac{1}{2} \\
(2x^2 + x + 1)\lceil\ 6x^4 - x^3 + 2x^2 \\
 \underline{-(6x^4 + 3x^3 + 3x^2)} \\
 0 \quad - 4x^3 - x^2 \\
 \underline{- (-4x^3 - 2x^2 - 2x)} \\
 0 + \quad x^2 + 2x \\
 \underline{-(x^2 + \tfrac{1}{2}x + \tfrac{1}{2})} \\
 0 + \tfrac{3}{2}x - \tfrac{1}{2}
\end{array}
$$

The bottom line now shows the remainder R_1 of degree $1 < 2$. The final result then is

$$(6x^4 - x^3 + 2x^2) = \left(3x^2 - 2x + \frac{1}{2}\right)(2x^2 + x + 1) + \left(\frac{3}{2}x - \frac{1}{2}\right).$$

By multiplying the first two factors on the right, you can verify that this equation is indeed correct.

II.4.4 *Multiplicities of Roots*

Returning to the key factorization of the polynomial P of degree n, with a root at p, given by

$$P = q(x)(x - p),$$

we can check whether P has more roots, that is, look for a root of q. To keep track of the different roots, let us write $q = q_1$ and $p = p_1$. Suppose indeed that there is p_2 with $q_1(p_2) = 0$. By the factorization theorem, there exists a polynomial q_2 of degree $n - 2$, so that

$$q_1 = q_2(x)(x - p_2),$$

and hence,

$$P = q_2(x)(x - p_2)(x - p_1).$$

We can continue this process as long as there are additional roots for P, but in any case, there can be at most n roots, since after n steps, the remaining factor q_n will have degree 0, so is a constant, and there is no more room for additional roots.

As we carry out this process, it is of course possible that the same root occurs several times, that is, we have repeated roots, just like in the case of the quadratic polynomial. For example, consider the polynomial

$$(x^2 + 2)(x + 3)^4,$$

which has degree 6 and the one root -3 that occurs 4 times.

Definition. A root p of the polynomial P has *multiplicity* $m \in \mathbb{N}$ if $P = q(x)(x - p)^m$, where p is NOT a root of q.

By keeping track of the multiplicities of the different roots, we conclude that given a polynomial P of degree n, if $p_1, p_2, ..., p_l$ are all the distinct roots (that is, $p_j \neq p_k$ if $j \neq k$), and p_j has multiplicity m_j for $j = 1, ..., l$, then

$$P = q(x)(x - p_l)^{m_l}...(x - p_2)^{m_2}(x - p_1)^{m_1},$$

where q has no more roots. Note that we must have $m_1 + m_2 + ... + m_l \leq n$.

We thus have established the following important property.

A polynomial of degree n has at most n roots, counting multiplicities.

We already mentioned that complex numbers include the imaginary number i, which satisfies $i^2 = -1$. Of course, $-i$ also satisfies $(-i)^2 = -1$. So i and $-i$ are the (complex) roots of the polynomial $x^2 + 1$, and one has the factorization

$$x^2 + 1 = (x - i)(x + i).$$

What is most amazing and provides evidence for the importance of the complex numbers $\mathbb{C}$, is the following

Fundamental Theorem of Algebra. *Every polynomial of degree $n \geq 1$ over the complex numbers $\mathbb{C}$ has a root in $\mathbb{C}$.*

This result was sort of suspected since the 17th century, and various incomplete proofs were published since then. The first complete proof was given by Argand in 1806, and while this is referred to as a theorem of Algebra, there is no known proof that only uses purely algebraic tools.

By using this result in the process of factorization discussed above, one obtains the following

Corollary 14 A polynomial $P \in \mathbb{C}[x]$ of degree n has exactly n roots in $\mathbb{C}$, counting multiplicities.

II.4.5   *Tangents and Double Points*

We saw that in the cases of a circle or a parabola the problem of finding the tangents is completely solved by simple algebra, involving the concept of a double point, or, more precisely, of a root of multiplicity 2 of the relevant quadratic equation. Before considering more general cases, we want to get a better understanding of the geometric idea of a tangent to a curve. We shall use the language from over 2000 years ago, when a tangent was defined as a line that "touches" the curve at a point P on the curve, but does not "cut" it.

In order to illustrate the idea of a double point more in detail, consider Fig. II.15.

It shows how in a typical situation the tangent touches the curve in just one point P, which however splits up into *two* points as soon as the tangent is moved just so slightly while staying connected with the curve. The single point of tangency thus really accounts for two points that become clearly visible if the tangent is suitably moved. That is why we call such a point

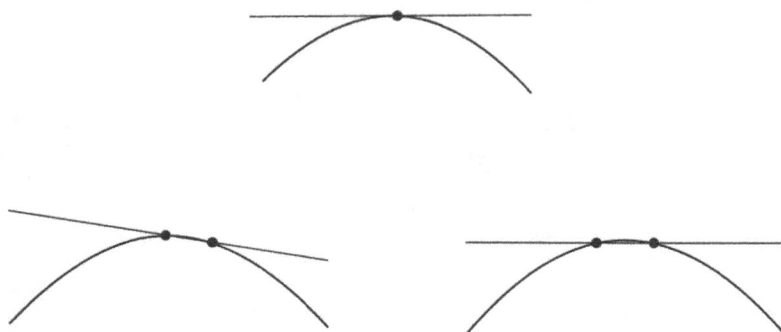

Fig. II.15   The "touching" point splits into two points.

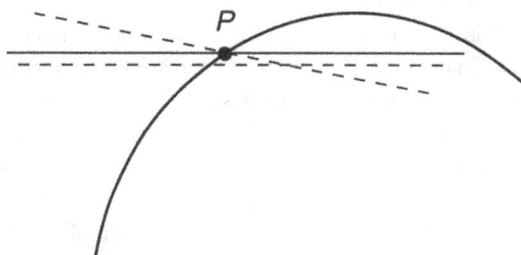

Fig. II.16   A cutting line intersects at a single point.

of intersection a *double point.* Note that, in contrast, a line that obviously "cuts" the curve at P does not have this property, as shown in Fig. II.16.

Any small change of the line still intersects the curve only at *one* point, at least in a neighborhood of P.

We thus see that the ancient geometric definition of a tangent as a line that touches the curve but does not cut it has been made more precise by introducing the notion of a double point:

*A line through the point P on a curve is a **tangent** to that curve precisely when the point P of contact is a double point!*

Descartes' insight then is to identify double points by finding double roots (or zeroes of multiplicity 2) of appropriate algebraic equations.

In certain situations we need to generalize the concept of a tangent. For example, let us consider the graph of $C(x) = x^3$, which we had already considered at the beginning of this section, at the point $(0, 0)$:

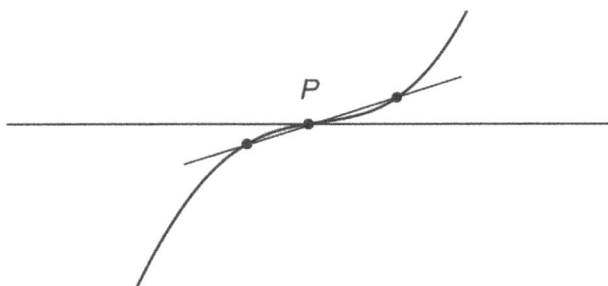

Fig. II.17   A triple point of tangency.

Here the x-axis "touches" the graph at $(0,0)$ in a special way, since the curve now meets both sides of the line, so this line also "cuts" the curve. But the situation here is very different from the obvious cut considered earlier. In fact, note that if the x-axis is rotated just so slightly counter-clockwise, as shown in Fig. II.17, it suddenly intersects the curve in 3 points! So the single point of contact $(0,0)$ between the curve and the x-axis really covers 3 points that become visible as the axis is rotated just so slightly in an appropriate way. Correspondingly, we say that the point of intersection of the curve and the x-axis is a *triple* point, or a point of *multiplicity* 3. Note that the equation that identifies the x-coordinate of the point of intersection is simply $x^3 = 0$, which has the 3 equal solutions $x = 0$, so 0 is a root of multiplicity 3.

You may wonder why a small *clockwise* rotation of the x-axis only shows **one** point of intersection with the curve. Such a line is given by $y = -\varepsilon x$ for a small positive $\varepsilon > 0$; the points of intersection of this line with the curve are given by the solutions of the equation

$$-\varepsilon x = x^3 \text{ or } x^3 + \varepsilon x = (x^2 + \varepsilon)x = 0.$$

Since $\varepsilon > 0$, the equation $x^2 + \varepsilon = 0$ has no (real) zeroes, so that is why we only see the one point of intersection corresponding to $x = 0$. However, if one allows complex numbers, then this equation has the zeroes $i\sqrt{\varepsilon}$ and $-i\sqrt{\varepsilon}$, so in the complex world there still are 3 points of intersection.

Similarly, the curve given by $y = x^4$ intersects the x-axis at $(0,0)$ with multiplicity 4, since the equation $x^4 = 0$ has 4 equal solutions. However, in this case we can not visualize the separation into 4 distinct points as the x-axis is rotated just so slightly. In fact, by looking at the relevant picture, only 2 points become visible. More precisely, looking at the situation algebraically, the slightly rotated x-axis is the graph of $y = mx$ for some small number m, so that the x-coordinates of the points of intersection of this line with the graph of $y = x^4$ are the roots of the equation

$$x^4 - mx = 0.$$

This equation factors into $x(x^3 - m) = 0$, whose roots are 0 and $\sqrt[3]{m}$, that is, that unique point on the number line that satisfies $(\sqrt[3]{m})^3 = m$. On the other hand, if we allow *complex* numbers, then, by the Fundamental Theorem of Algebra, the polynomial $x^3 - m$ has 3 zeroes in $\mathbb{C}$, counting multiplicities. In fact, for this simple polynomial of degree 3 one can easily show (see Exercise 5) that if $m \neq 0$, then these zeroes are all distinct, resulting in a total of 4 distinct points of intersection of the curve with the line $y = mx$ in the *complex* world.

Based on these examples we therefore make the following *geometric* definition of a tangent.

Definition 15 *A tangent to a curve at the point P on the curve is a line that touches the curve at P with multiplicity 2 (or possibly higher), that is, in a double (or triple, or higher) point. Geometrically this means that some suitable arbitrarily small rotation of the line will split the point P in 2 (or possibly more) distinct points of intersection.*

This geometric definition is an attempt to give more precise meaning to the ancient, somewhat vague definition that we have recalled earlier. We saw that in the case of a parabola, i.e., a quadratic polynomial, the geometric property captured above is made precise by considering the multiplicity of roots of the quadratic equation. In the next section we shall see that this simple method generalizes most naturally to arbitrary polynomials.

II.4.6 *Tangents to Polynomial Curves*

First of all, recall from Section 4.1 that a polynomial is a formal algebraic expression P given by

$$P = a_n x^n + a_{n-1} x^{n-1} + \ldots + a_1 x + a_0,$$

where n is an integer ≥ 0, and $a_0, \ldots, a_n$ are fixed numbers. If $a_n \neq 0$, we say that the polynomial P has *degree* n. Every polynomial defines a *polynomial function* $P(x)$, which to each input $r \in \mathbb{Q}$ (or more generally in $\mathbb{R}$ or $\mathbb{C}$) assigns the output $P(r)$ that is calculated by the expression $a_n r^n + a_{n-1} r^{n-1} + \ldots + a_1 r + a_0$. In common language the distinction is usually not expressed clearly, and a polynomial is often identified with the related function. But it is important to view a polynomial as just a symbolic expression, which however can naturally be used to define a function. Thus we may talk about the graph of a polynomial, while what we really mean is the graph of the associated polynomial function $y = P(x)$. We also saw that as soon as the degree is ≥ 3, the graphs are no longer just described by a simple geometric feature, but can become quite complicated indeed. If you have not yet done so, we encourage you again to use your graphing calculator and enter various explicit polynomials to see the variety.

What is amazing is that, no matter how complicated the graph of a polynomial may be, there is a very simple procedure to identify the tangents, which is a straightforward generalization of what we did in the case of a parabola.

We fix a point $(a, P(a))$ on the graph of P, and want to determine the tangent to the curve at this point, that is, that line that intersects the curve in a double (or higher) point. Note that the tangent problem is trivial for polynomials of degree 1, whose graphs are lines; a line can intersect another line at two distinct points only if the two lines are identical, so a line can "touch" another line only if the two lines coincide. Therefore we may assume that our polynomial has degree ≥ 2. As before, any line through $(a, P(a))$ is described by an equation $y = P(a) + m(x - a)$, where m is the slope, and the points of intersection of such a line with our curve are given by the solutions of the equation

$$P(x) = P(a) + m(x - a).$$

As we mentioned earlier, in contrast to the simple quadratic equation, in general there is no way to find the zeroes of such a polynomial equation. But do not worry! We only are interested in the ONE obvious zero, namely $x = a$, and we want to determine when this is a zero of multiplicity greater than 1.

We rearrange the above equation in the form

$$P(x) - P(a) - m(x - a) = 0.$$

In the simple case of the parabola we used the familiar factorization $x^2 - a^2 = (x + a)(x - a)$. In the general case we use the corresponding factorization given by Corollary 12, namely

$$P(x) - P(a) = q(x)(x - a),$$

where q is a certain polynomial of degree $n - 1$. Recall also that this factor q is uniquely determined by P and the point a.

By using this result, we can now factor our equation $P(x) - P(a) - m(x - a) = 0$ and rearrange it in the form

$$[q(x) - m](x - a) = 0.$$

Again, this clearly shows that a is a solution of the equation—after all we only consider lines that intersect the curve at $(a, P(a))$! Most significantly, this factorization shows that a is a zero of multiplicity greater than one if and only if the factor $[q(x) - m]$ also has a zero at a, and this occurs precisely when the slope $m = q(a)$.

Let us formalize our result in the following theorem that completely solves the tangent problem for any polynomial.

Theorem 16 *Let P be a polynomial of degree $n \geq 2$ and let $(a, P(a))$ be a point on its graph. Then there exists a unique line through $(a, P(a))$ that intersects the graph at that point with multiplicity greater than 1. The slope m of that line is given by $q(a)$, where q is the (unique) polynomial of degree $n - 1$ in the factorization $P(x) - P(a) = q(x)(x - a)$.*

Of course we call this unique line given by the theorem the *tangent to the graph of P at the point* $(a, P(a))$.

II.4.7   *Exercises*

1. Use a graphing calculator to visualize the graph of $P(x) = 2x^4 - 4x^3 - 6x^2 + 5x + 1$.

  2. Find all roots of the equation $(x^2 - 1)(x^3 + 1)(x^2 - 2x - 3) = 0$, including their multiplicities. Hint: Look at each factor separately.

  3. a) Use long division of polynomials to find polynomials q and R, with degree of $R \leq 1$, so that

$$2x^3 - 3x^2 + x - 5 = q(x)(x^2 + 1) + R(x).$$

  b) Verify your answer by explicitly calculating the right side of the above equation and checking that it agrees with the left side.

  4. Let $n \in \mathbb{N}$. Find the polynomial q so that $x^n - a^n = q(x)(x - a)$.

  5. Consider the equation $x^3 - m = 0$, where $m \neq 0$. It has exactly one root $a = \sqrt[3]{m}$ on the (real) number line.

  a) Use the result in Exercise 4 to determine *explicitly* (involving a) the polynomial q of degree 2 that satisfies $x^3 - a^3 = q(x)(x - a)$.

  b) Using the explicit form of q found in a), verify that the discriminant of q is a non-zero *negative* number.

c) Explain why b) implies that if we allow complex numbers, then q has two distinct roots in $\mathbb{C}$.

6. Suppose f and g are polynomials of degree m and n respectively. Prove that $\deg(f + g) \leq \max\{m, n\}$. Give an explicit example where the inequality is strict.

7. The polynomial $P(x) = 2x^4 + x^3 + 2x^2 - 3$ has a zero at $a = -1$.

a) Find the polynomial q so that $P(x) = q(x)(x + 1)$.

b) Find the equation of the tangent line to the graph of P at the point $(-1, 0)$.

8. Find the equation of the tangent line to the graph of $y = 3x^4$ at the point $(1, 3)$.

II.5   Simple Differentiation Rules for Polynomials

II.5.1   *Derivatives of Polynomials*

In the last section we discovered a simple procedure for finding the slope, and hence the equation, for the tangent line at an arbitrary point $(a, P(a))$ on the graph of a polynomial P. This slope $m = q(a)$ is called the *derivative of P at the point a*, and it is denoted by $D[P](a)$ or also by $P'(a)$. Note that this process of finding the derivative $D[P]$ of P defines a new function, whose value at the point x is the derivative $D[P](x) = P'(x)$. We shall see shortly that $D[P]$ is also a polynomial. This process is also known as the *differentiation* of P.

Remarks on Notation. The symbol D refers to **D**erivative, and it is used to indicate that it is an operation applied to the polynomial P that results in a new function $D[P]$, the *derivative of P*. The symbol P' is often used for the sake of brevity. Historically, the derivative has also been denoted by the "differential quotient" dP/dx, a *formal* quotient of "differentials" dP and dx that were used to denote the vague concept of *infinitesimals*, or *infinitely small* quantities. This notation reminds us of the relationship of the derivative to the average rates of change $\Delta P/\Delta x$ that we encountered, for example, in the context of average velocity in Section 2.3. Since the approach chosen in this book emphasizes the factorization formula as a product, and since we avoid quotients that lead to $0/0$, we shall limit the use of the notation dP/dx mainly to applications, when we want to highlight the relevant variables under consideration and the interpretation of derivatives as rates of change.

As we discussed earlier, the factor q in the critical factorization, whose value $q(a) = D[P](a)$ is the derivative at a (i.e., the slope of the tangent at $(a, P(a))$), can be calculated in principle by using the division algorithm for polynomials. A much easier way to find the derivative of any polynomial is based on the application of some simple rules, as follows.

Example. *We already discussed the simple quadratic case* $P(x) = x^2$. *Let us now look at the general power function* $P(x) = cx^n$, *where c is a nonzero constant and $n \geq 2$.*
We need to factor $cx^n - ca^n = q(x)(x - a)$. One easily verifies that

$$q(x) = c[x^{n-1} + x^{n-2}a + x^{n-3}a^2 + \ldots + xa^{n-2} + a^{n-1}]$$

is the desired factor, and consequently, $D[cx^n](a) = q(a) = cna^{n-1}$. Note that this result holds also for $n = 1$ (the line $y = cx$ has slope c) and $n = 0$ (the horizontal line $y = c$ has slope 0).

Since a polynomial is simply a sum of power functions, we now can handle any polynomial by applying the following almost trivial rule.

If P and Q are polynomials, then $(P \pm Q)'(a) = P'(a) \pm Q'(a)$.

The proof of this rule is straightforward, as follows. Let

$$P(x) - P(a) = q_P(x)(x - a) \text{ and } Q(x) - Q(a) = q_Q(x)(x - a)$$

be the relevant factorizations of P and Q. To simplify, let us just consider $+$; so add the two equations and rearrange to get

$$(P + Q)(x) - (P + Q)(a) = (q_P + q_Q)(x) \cdot (x - a).$$

Clearly this is the relevant (unique) factorization for $P + Q$. Therefore

$$(P + Q)'(a) = (q_P + q_Q)(a) = q_P(a) + q_Q(a) = P'(a) + Q'(a).$$

It follows that for

$$P = a_n x^n + a_{n-1} x^{n-1} + \ldots + a_1 x + a_0,$$

its derivative at the arbitrary point x is given by

$$P'(x) = [a_n x^n]' + [a_{n-1} x^{n-1}]' + \ldots + [a_1 x]' + [a_0]'$$
$$= na_n x^{n-1} + (n-1)a_{n-1} x^{n-2} + \ldots + a_1 + 0.$$

In particular, we see that $D[P]$ is a polynomial of degree $n - 1$ if P has degree n.

Example. *Suppose* $P(x) = 7x^4 - 4x^3 + 3/2x^2 - 5x + 3$. *Then* $D[P](x) = 4 \cdot 7x^3 - 3 \cdot 4x^2 + 2(3/2)x^1 - 5 = 28x^3 - 12x^2 + 3x - 5$.

There is another most elementary rule for derivatives, namely $(c \cdot P)'(x) = c \cdot P'(x)$ for any number c. In fact, if $P(x) - P(a) = q(x)(x - a)$ is the relevant factorization for P, then multiplying with the constant c,

$$cP(x) - cP(a) = [cq(x)](x - a).$$

Thus $(cP)'(a) = cq(a) = cP'(a)$.

Let us point out right now that there is no simple formula for the derivative of a product that is comparable to the case of the derivative of a sum. Just note that $(x^2)' = 2x$, while $x^2 = x \cdot x$, and $x' \cdot x' = 1 \cdot 1 = 1$. So there certainly cannot be a rule $(P \cdot Q)' = P' \cdot Q'$. We will study the appropriate product rule in the next chapter in the context of the larger class of *rational* functions.

II.5.2   *Compositions of Functions and the Chain Rule*

In contrast to what we just mentioned at the end of the previous section, there is another very simple, yet most important formula for derivatives that applies to the most natural operation on functions that we already discussed in Section I.2.2, namely the so-called *composition*. Recall that in order to consider addition or multiplication of functions, one needs the corresponding operations on the numbers that are used for the output. However, if the codomains of the functions are some abstract set without any multiplication, it does not make sense to consider products of functions. On the other hand, given two functions $f, g : \mathbb{Q} \to \mathbb{Q}$, it is possible to apply one of them after the other, that is, to consider the function machine

$$x \to g(x) \to f(g(x)).$$

This clearly defines a new function that is called the composition of g with f, and that is labelled $f \circ g$. Note that nothing of the algebraic structure is involved, only the concept of a function needs to be understood, and therefore composition can be introduced for functions $f, g : \Omega \to \Omega$, where Ω is an arbitrary set.

In particular, one can consider the composition of any two polynomial (functions); it is easily verified that the result is again a polynomial.

Example. Suppose $f(x) = x^4$ and $g(x) = 3x^2 - 5x + 1$. Then $f \circ g$ is the function defined by the rule

$$(f \circ g)(x) = (3x^2 - 5x + 1)^4.$$

By expanding the right side, we see that $f \circ g$ is again a polynomial. Note that the order matters. The function $g \circ f$ (that is, first apply f, then g) is

given by the formula
$$(g \circ f)(x) = 3(x^4)^2 - 5x^4 + 1,$$
and clearly $f \circ g \neq g \circ f$. In fact the leading term of $g \circ f$ is $3x^8$, while the one of $f \circ g$ is $(3x^2)^4 = 81x^8$.

While composition is not commutative, even if both $f \circ g$ and $g \circ f$ are defined, whenever successive compositions of 3 or more functions are defined, the associative rule holds, that is, suppose f, g, h are three functions with domain and codomain $\mathbb{Q}$, then
$$h \circ (f \circ g) = (h \circ f) \circ g.$$
Just note that for an input $x \in \mathbb{Q}$, either side results in $h(f(g(x)))$. Again, as we had already mentioned in Section I.2.2, this remains true in the most general setting, as long as the domains and codomains satisfy the appropriate conditions, so that the compositions are defined.

We want to emphasize that composition is a most natural idea that is closely associated with the concept of a function (machine), and that does not require any other structural properties of domain and codomain, except for the obvious requirement that the codomain (or even just the range) of the first function is contained in the domain of the second function. Given this natural operation, it is perhaps not surprising that the rule to handle the derivative of compositions of polynomial functions is as simple as it gets, as we shall now describe it.

So let P and Q be two polynomials, and we shall consider $P \circ Q$. Fix an input $a \in \mathbb{Q}$ (or $\mathbb{R}$, or $\mathbb{C}$), and let $b = Q(a)$. Consider the relevant factorizations for Q at a and for P at b:
$$Q(x) - Q(a) = q_Q(x)(x - a),$$
and
$$P(y) - P(b) = q_P(y)(y - b),$$
where q_Q and q_P are the polynomials given by the basic factorization theorem 6. When we consider $P \circ Q$ we must replace the input y in the second equation by $Q(x)$, and the number b by $Q(a)$, resulting in
$$P(Q(x)) - P(Q(a)) = q_P(Q(x)) \cdot [Q(x) - Q(a)].$$
Here we can now substitute the factor on the right by the relevant factorization given above, so that
$$(P \circ Q)(x) - (P \circ Q)(a) = P(Q(x)) - P(Q(a)) = q_P(Q(x)) \cdot [q_Q(x)(x - a)]$$
$$= [q_P(Q(x)) \cdot q_Q(x)](x - a).$$

Clearly this is the relevant (unique) factorization for $P \circ Q$ at the point a, so that

$$D[P \circ Q](a) = q_P(Q(a)) \cdot q_Q(a) = D[P](Q(a)) \cdot D[Q](a).$$

We have thus obtained the following important Theorem, which is known as the *Chain Rule for Derivatives*.

Theorem 17 If P and Q are two (polynomial) functions, then

$$(P \circ Q)' = (P' \circ Q) \cdot Q'.$$

Example. Recall the functions f and g from the example considered earlier, so that $(f \circ g) = (3x^2 - 5x + 1)^4$. To calculate its derivative, we could of course evaluate the power on the right side in detail and obtain its standard polynomial form, and then use the basic rule for derivatives of polynomials that we established in the preceding subsection. However, proceeding in this way can get messy, and furthermore the original structure of $f \circ g$ is completely obscured, and the roles of f and g are lost. Using the Chain Rule is much easier, and it does retain the information about the structure. In detail, since $f'(x) = 4x^3$ and $g'(x) = 6x - 5$, we immediately obtain

$$(f \circ g)'(x) = f'(g(x)) \cdot g'(x) = 4(3x^2 - 5x + 1)^3(6x - 5).$$

While so far we have only considered derivatives of (polynomial) functions, we do want to mention that the Chain Rule, as well as the other rules discussed in this section remain valid for much more general functions, in fact the most general class of *differentiable functions* that we will introduce much later. Most remarkable is the fact that the proof we just discussed in the case of polynomials will carry over to the general case, enhanced by the relevant details that relate to the more general class of functions considered. So we encourage the reader to work diligently through the Exercises to practice these rules and become familiar with them, since they will apply throughout this book.

II.5.3 *Exercises*

1. Find the derivatives of

$$f(x) = 6x^7 - 4x^6 - 2x^5 + 5x^4 + 7x^3 + 8x - 10$$

and

$$g(x) = 10x^{99} - 2x^{57} + 3x^{19} - 4x^7.$$

2. Find the equation of the tangent line to the graph of $k(x) = 5x^8 - 3x^5$ at the point $(2, k(2))$.

3. Find the points on the graph of $P(x) = 2x^3 + 3x^2 - 36x - 10$ where the tangents are horizontal.

4. a) Find the points, if any, on the graph of $y = x^3 + 2x$ where the tangents are horizontal.

   b) Confirm your answer by displaying the graph in a graphing calculator.

5. Find the derivative of $G(x) = 3(x^3 - 4x^2 + 1)^7 - 5(x^4 + 10)^4$.

6. Find the derivative of $H(x) = [2(x^5 - x^3 + 4x)^5 + 3x^2]^{10}$. Hint: Be careful while applying the chain rule more than once.

7. Suppose f, g, h are polynomials. Generalize the chain rule to find a formula for the derivative of $f \circ g \circ h$ at the point a in terms of the derivatives of f, g, h at the appropriate points.

8. Find the equation of the tangent line to the graph of $P(x) = (x^3 - 3x^2 + 4)^4$ at the point $(1, P(1))$.

Chapter III

The Differential Calculus of Rational Functions

While polynomial functions are useful to describe properties of a parabola, or to analyze the motion of a freely falling stone, there are numerous important applications where other functions are needed. For example, later on we will describe problems related to growth and decay, or to periodic motions. The relevant functions cannot be described by simple algebra, and are usually called *transcendental*. At this point however, while we are still studying algebraic concepts and constructions, we shall introduce so-called *rational functions*, a generalization of polynomials. Their construction is comparable to the way rational numbers are built up from the integer numbers.

One familiar example where such functions arise is the description of the gravitational force between two objects in space, say between the earth and the moon. It was Isaac Newton (1642–1727) who discovered the law that governs this process. Precisely, if we denote by $d = d(O_1, O_2)$ the distance between the two objects, then the gravitational force F between the two objects is given by

$$F(d) = \gamma \frac{M_1 M_2}{d^2},$$

where M_1 and M_2 are the masses of the two objects, and γ is a universal constant, known as the gravitational constant, whose numerical value depends on the units chosen to measure the masses and the distance. We note that $F(d)$ is certainly not a polynomial, but it does seem to be closely related to a polynomial, something like a "fraction" that involves a polynomial in the denominator. In this chapter we shall discuss such functions and solve the problem of tangents in this more general setting, guided by what we have learned in the case of polynomial functions.

III.1   Rational Expressions and Functions

III.1.1   *Multiplicative Inverses for Polynomials*

We shall now apply the basic principles that we used when extending the set $\mathbb{Z}$ of integers to the rational numbers $\mathbb{Q}$ to the set of polynomials $\mathbb{Q}[x]$. We remind the reader that this latter set satisfies all the rules Q1–Q6, which justifies calling it a *commutative ring with identity*. We also emphasize that these rules are the guiding principle in the construction of a larger set, and that we need to ensure that these rules continue to hold. The critical new item that we want to add is a *multiplicative inverse* for polynomials, in complete analogy to what was done to enlarge the integers to the rational numbers.

More precisely, given a polynomial $Q \neq 0$ of degree ≥ 0, we simply define a formal expression Q^{-1} to denote an object that satisfies $Q \cdot Q^{-1} = 1$ and add it to the set $\mathbb{Q}[x]$. Consistent with the commutative rule Q1, we also require that $Q^{-1} \cdot Q = 1$. We call Q^{-1} the multiplicative inverse of Q. If Q has degree 0, then Q equals a non-zero rational number c, and $Q^{-1} = c^{-1}$ is just the familiar rational number that is the multiplicative inverse of c. In case $c = m \in \mathbb{Z}$, we had introduced fractional notation and we write $m^{-1} = \frac{1}{m}$. In analogy, we introduce fractional notation for the multiplicative inverse Q^{-1}, that is, we write

$$Q^{-1} = \frac{1}{Q}$$

for an arbitrary non-zero polynomial Q. For example, $(x^2 + 1)^{-1} = \frac{1}{x^2+1}$, or $(2x^3 - 5x)^{-1} = \frac{1}{2x^3-5x}$.

We want to emphasize that this new concept, as well as the notation chosen, is motivated by the construction of the rational numbers and is fully consistent with it. In fact, when we consider the associated function, discussed in detail in the next subsection, this becomes clearly visible, as follows. Given a (rational) number p so that $Q(p) \neq 0$, the *function* Q^{-1} is defined for this input p by

$$Q^{-1}(p) = [Q(p)]^{-1} = \frac{1}{Q(p)}.$$

Since $Q(p)$ is a rational number, its multiplicative inverse $[Q(p)]^{-1}$ is a well defined number in $\mathbb{Q}$, and of course, by the definition in $\mathbb{Q}$, $[Q(p)]^{-1} \cdot Q(p) = 1$. As this holds for arbitrary p, as long as $Q(p) \neq 0$, surely the product $Q^{-1} \cdot Q$ of the functions equals the constant function 1, that is, the multiplicative identity in $\mathbb{Q}[x]$.

III.1.2 *Polynomial Fractions and Rational Expressions*

Next, we have to introduce the (formal) product $P \cdot Q^{-1}$ of a polynomial P with the multiplicative inverse Q^{-1} of a non-zero polynomial. Again, in order to satisfy the commutative rule Q1, we must have $P \cdot Q^{-1} = Q^{-1} \cdot P$, and just like we did for rational numbers, we write

$$P\frac{1}{Q} = \frac{1}{Q}P = \frac{P}{Q},$$

where again we call the expression on top the numerator, and the one at the bottom the denominator.

The "polynomial fractions" obtained by this process are called *rational expressions*. The collection of all such rational expressions is denoted by $\mathbb{Q}(x)$ — careful, the only difference to $\mathbb{Q}[x]$ are the round brackets (), instead of the square brackets []. It is quite common to just refer to this as the collection of *rational functions* over $\mathbb{Q}$, thereby blurring the distinction between the formal expression and the associated function, just as the notation $P(x)$ blurs the distinction between polynomials and polynomial functions.

Since we know how to add and multiply polynomials, we can now formally use the same process that we used for $\mathbb{Q}$ to define sum and product of rational expressions. In particular, we say that two polynomial fractions $\frac{P}{Q}$ and $\frac{F}{G}$ are equivalent, i.e., they represent the same rational expression, if and only if $PG = QF$. Just as in case of rational numbers, it then follows that if $G \neq 0$, then

$$\frac{P}{Q} = \frac{PG}{QG},$$

that is, a rational expression is not changed if common factors in numerator and denominator are either introduced or deleted. For example,

$$\frac{3x^2 + 2x}{4x^3 - 6x^2} = \frac{x(3x + 2)}{x(4x^2 - 6x^1)} = \frac{3x + 2}{4x^2 - 6x^1}.$$

Formally, adding and multiplying rational expressions looks identical to adding and multiplying rational numbers, and therefore the same rules Q1–Q7 hold. We are just defining certain operations with symbols that we write as fractions (really quite a level of abstraction), the only difference between $\mathbb{Q}$ and $\mathbb{Q}(x)$ is the meaning behind the symbols, and consequently the definitions of sums and products of the symbols in numerators and denominators depend on whether we deal with $\mathbb{Q}$ (and hence they are integers) or with $\mathbb{Q}(x)$, where they are polynomials. For example, compare

$$\frac{1}{3} + \frac{5}{7} = \frac{1 \cdot 7}{3 \cdot 7} + \frac{3 \cdot 5}{3 \cdot 7} = \frac{1 \cdot 7 + 3 \cdot 5}{3 \cdot 7} = \frac{22}{21}$$

with

$$\frac{1}{x} + \frac{5x}{x^2 - 1} = \frac{1 \cdot (x^2 - 1)}{x \cdot (x^2 - 1)} + \frac{x \cdot 5x}{x \cdot (x^2 - 1)}$$

$$= \frac{1 \cdot (x^2 - 1) + x \cdot 5x}{x \cdot (x^2 - 1)} = \frac{x^2 - 1 + 5x^2}{x^3 - x} = \frac{6x^2 - 1}{x^3 - x}.$$

Try to look at this carefully to fully recognize that in the two additions of fractions exactly the same formal rules are followed, except in the second case addition and multiplication of the underlying symbols in numerator and denominator (that is, polynomials in this case) are of course different than in the first addition. While the second addition surely looks quite a bit more complicated, it is important to understand that we are just following the same rules that we are familiar with from the rational numbers.

We also point out that while $x^{-1} = \frac{1}{x}$ is the multiplicative inverse of x, similarly the symbol x^{-n} for a natural number n turns out to be the multiplicative inverse of x^n. In fact, by following the rules for powers with integer exponents, one has

$$x^{-n} = x^{(-1)n} = (x^{-1})^n = (x^{-1}) \cdot ... \cdot (x^{-1}) \ (n \ \text{factors}),$$

so that

$$x^n \cdot x^{-n} = (x \cdot ... \cdot x)(x^{-1}) \cdot ... \cdot (x^{-1}) = (x \cdot x^{-1}) \cdot ... \cdot (x \cdot x^{-1}) = 1 \cdot 1 ... \cdot 1 = 1.$$

More generally, for any two integers n and m one has $x^{n+m} = x^n \cdot x^m$ and $(x^n)^m = x^{n \cdot m}$. Furthermore, these rules remain valid if the polynomial x is replaced by an arbitrary $non-zero$ polynomial P. In particular P^{-n} is the multiplicative inverse of P^n, so we obtain

$$P^{-n} = \frac{1}{P^n}.$$

For example, it follows that

$$(x^3 - 2x)(3x^2 - 5x + 1)^{-3} = \frac{x^3 - 2x}{(3x^2 - 5x + 1)^3}.$$

Finally, multiplication of two rational expressions is defined in complete analogy to multiplication of two rational numbers, that is,

$$\frac{P}{Q} \cdot \frac{F}{G} = \frac{P \cdot F}{Q \cdot G} = \frac{F \cdot P}{G \cdot Q} = \frac{F}{G} \cdot \frac{P}{Q},$$

where we have used the commutative rule Q1 in the numerator and denominator in the middle. In particular, this shows that rule Q1 holds for rational expressions.

To summarize, we have constructed the set $\mathbb{Q}(x)$ of rational expressions, whose elements can be written as fractions of two polynomials (the denominator non-zero), subject to the "equivalence" property mentioned above that we are familiar with from the usual rational numbers. We also note that any polynomial P can be written as a fraction, as follows: $P = P \cdot 1 = P \cdot 1^{-1} = \frac{P}{1}$. Therefore, the set of polynomials is a subset of the set of rational expressions. Finally, all the rules Q1–Q7 hold in $\mathbb{Q}(x)$. We emphasize again the fundamental importance of these rules. While the construction in this section has reached a more complicated level of abstraction, it should give comfort that no new rules have to be learned!

III.1.3   *Domains of Rational Functions*

Note that when considering rational expressions, the only restriction is that the polynomial in the denominator is not the *zero* polynomial. Just as for rational numbers, the only polynomial that does not have a multiplicative inverse is the additive identity 0. However, when we consider the *function* associated with a rational expression, we must be more careful, since a non-zero polynomial in the denominator may very well have a root p, and evaluation at that point would introduce the number 0 in the denominator, something that is not allowed.

More precisely, suppose P/Q is a rational expression, where $Q \neq 0$. Let $\Omega_Q = \{x \in \mathbb{Q} : Q(x) \neq 0\}$. We now define a function

$$f_{P/Q} : \Omega_Q \to \mathbb{Q}$$

by the "machine"

$$f_{P/Q}(a) = \frac{P(a)}{Q(a)} \text{ for every } a \in \Omega_Q.$$

As is commonly done for functions, we denote $f_{P/Q}$ also by $f_{P/Q}(x)$. However, in common usage this notation may also be used for the rational expression $\frac{P(x)}{Q(x)}$. While in case of a polynomial function P this ambiguity of the notation $P(x)$ is not crucial, when dealing with rational functions/expressions, one needs to make sure about what is meant, since when one considers the function, special attention must be given to the domain. In case of possible confusion, it is best to include the name *rational function* or *rational expression*. Also, unless something else is specified, given a rational function $f_{P/Q}$, it is always understood that its domain is Ω_Q, as defined above.

When one considers statements involving several rational functions, such as addition, multiplication, or equality, particular care must be given to the domains involved, as these will typically differ. Again, unless something else is explicitly specified, it is understood that such a statement is valid on the intersection of the domains of the functions that are involved. For example, while $f(x) = \frac{1}{x}$ has domain $\Omega_f = \{\, x \in \mathbb{Q} : x \neq 0\,\}$, and $g(x) = \frac{x}{x-3}$ has domain $\Omega_g = \{x \in \mathbb{Q} : x \neq 3\}$, the sum $f + g$ of the two functions has domain $\Omega_f \cap \Omega_g$. Therefore, while the statement about rational expressions

$$\frac{1}{x} + \frac{x}{x-3} = \frac{x-3+x^2}{x(x-3)}$$

is correct as it stands, the analogous statement about the corresponding functions, i.e.,

$$(f+g)(x) = \frac{x-3+x^2}{x(x-3)}$$

has to be made precise by adding that this equality only holds on the domain $\Omega_f \cap \Omega_g$. While this may not be explicitly stated in practice, it is important that the student firmly keep in mind the restriction on the domain. Matters get even more tricky when one considers multiplication of f and g, as follows. The equation

$$(f \cdot g)(x) = \frac{1}{x} \cdot \frac{x}{x-3} = \frac{1 \cdot x}{x(x-3)} = \frac{1}{x-3}$$

would tempt one to believe that the domain of $f \cdot g$ is all $x \neq 3$. However, the product of the two functions is only defined on the domain $\Omega_f \cap \Omega_g$, which does not include 0. The problem arises because of the possible equivalence of two polynomial fractions to describe a rational expression. In the case above, the statement

$$\frac{1 \cdot x}{x(x-3)} = \frac{1}{x-3}$$

about rational expressions is correct, since one can cancel the common nonzero factor x in numerator and denominator. However, if we view it as a statement about functions, then it is not quite correct, since the function on the left is not defined for $x = 0$, so the functions on the left and right have different domains, and hence are not equal. When viewed as a statement about functions, the above is correct only when the functions are restricted to the common domain $\Omega_f \cap \Omega_g$.

The key lesson to be learned is that equality of rational expressions does not imply equality of the corresponding functions, unless the domain is suitably restricted.

The preceding discussion was limited to rational numbers as inputs for functions. Later in this book, when we have formally introduced the real numbers $\mathbb{R}$ and investigated their key properties, the functions related to polynomials and rational expressions will have domains $\mathbb{R}$, or appropriate subsets of $\mathbb{R}$ that exclude the roots of the denominators. The same attention must be given to the relevant domains as in the cases we discussed just now. Furthermore, while the function $h(x) = \frac{1}{x^2-2}$ has domain $\mathbb{Q}$ in the rational world (recall that no rational number r satisfies $r^2 = 2$), when this expression is used to define a function over $\mathbb{R}$, its domain will be $\{x \in \mathbb{R} : x \neq \sqrt{2} \text{ or } -\sqrt{2}\}$.

These complications are part of the story, and we must learn to read statements very carefully and be certain that we understand what the particular symbols that are used refer to. Just like most people have a harder time working with fractions (i.e., rational numbers) than with integers, it certainly is more complicated working with rational expressions than just with polynomials.

III.1.4 *Graphs of Rational Functions*

The graphs of rational functions are typically quite a bit more complicated than those of polynomial functions. In particular, new phenomena arise near the points where the denominator has a zero. This already appears in the simplest case of $f(x) = 1/x$, as one can see from Fig. III.1.

One easily can get the general picture by calculating and plotting a few points of the graph, or, even better, by using a graphing calculator. As $x > 0$ approaches 0, the values of f get increasingly larger, without any bound. This is shown clearly in Fig. III.1. A similar phenomenon appears as one approaches 0 from the negative side. The graph of f approaches the y-axis closer and closer, without ever touching it—note that $x = 0$ is not in the domain of f. Because of this property, the y-axis is called a *vertical asymptote*. Correspondingly, as we see from Fig. III.1, the x-axis is a *horizontal asymptote*, that is, the graph of f gets closer and closer to the x-axis as $|x|$ gets larger, without ever touching it, since $f(x) = 1/x \neq 0$ for all x.

Our next example is the function $g(x) = \frac{x}{x^2-4}$, with domain $\Omega = \{x : x \neq -2, 2\}$. Its graph shown in Fig. III.2 below shows two vertical asymptotes, the lines given by $x = -2$ and $x = 2$. Again, the x-axis is a horizontal asymptote.

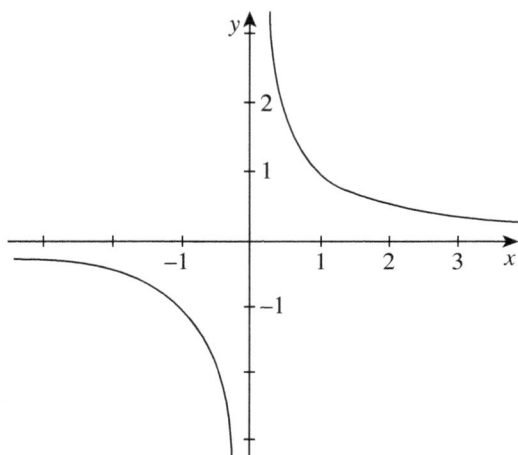

Fig. III.1   Graph of $f(x) = \frac{1}{x}$.

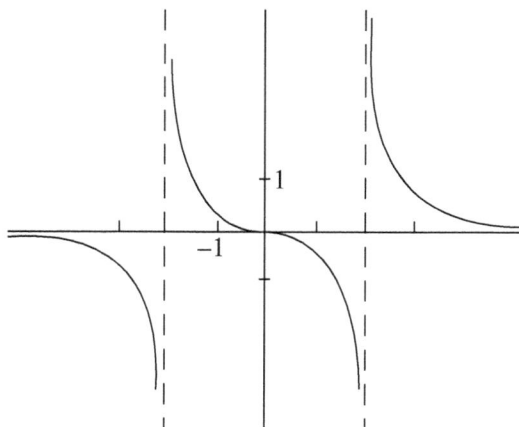

Fig. III.2   Graph of $g(x) = \frac{x}{x^2-4}$.

Another special feature is visible at the origin. Coming in from the left on the graph, the graph turns gently to the left, but after it goes through the origin, it begins a right turn. The point $(0,0)$ on the graph is an example of a *point of inflection*. Note that we had seen such a feature already with the graph of $y = x^3$.

Other lines may also exhibit the property of an asymptote. For example, consider the function $h(x) = \frac{x^2+1}{x}$. Its graph is shown below in Fig. III.3. Again, the y-axis is a vertical asymptote.

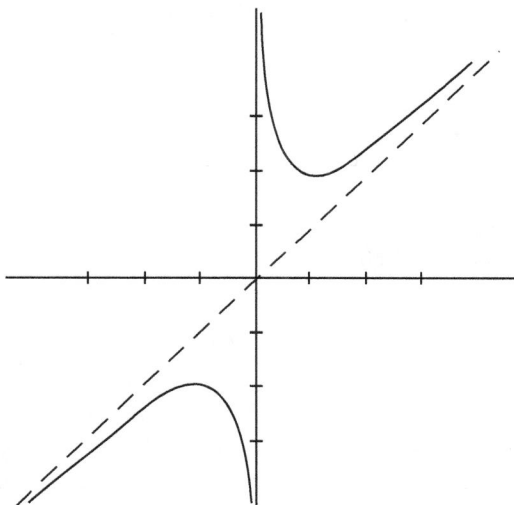

Fig. III.3   Graph of $h(x) = \frac{x^2+1}{x}$.

Furthermore, by rewriting h as $h(x) = x + \frac{1}{x}$, one can tell that the graph of h will get closer and closer to the line $y = x$ as $|x|$ gets larger and larger, without ever touching it. So this line is also an asymptote for h.

These few simple examples shall suffice to exhibit some of the new features that we may encounter. Graphing calculators allow us to easily visualize the graph of any function. Years ago, before such high tech tools were readily available, the concepts of calculus were the main resource available in order to study the graphs of functions. Even today calculus may be used in order to identify those sections of the graphs that show key features. These applications of calculus are typically discussed in the context of the general concept of a *differentiable function* defined for (subsets of) real numbers, which will be introduced later in Chapter V. For an introduction to these applications we refer the reader, for example, to Chapter III of our book *What is Calculus? From Simple Algebra to Deep Analysis*. That chapter is the natural sequel to the present book.

III.1.5   *Exercises*

1. Find the multiplicative inverse of a) x^{-5} and b) $\frac{x^3-2x^2+1}{x^2+3x-4}$ in $\mathbb{Q}(x)$.

2. Suppose $P = \frac{x^4-4x^2+3}{x-1}$ and $R = \frac{2x^3-x^2+4x-5}{(x+2)(x^2+2x)}$. Find formulas for P^{-1}, $P \cdot R$, and $P + R$.

3. Simplify $\frac{x^2-4}{x-1} \cdot \left(\frac{x+2}{x^3-1}\right)^{-1}$ as much as possible.

4. What is the domain of the function $G(x) = \frac{3x^3+1}{x(2x^2+2x-12)}$?

5. Carefully discuss whether $\frac{x+1}{x^2-1}$ and $(x-1)^{-1}$ are equal or not. Be sure to state relevant assumptions.

6. Evaluate $g(x) = \frac{1}{x^2+1}$ at sufficiently many points so as to be able to sketch its graph.

7. Use a graphing calculator to visualize the graph of $R(x) = \frac{3x^2+x}{(x+1)(x-3)}$. Identify all asymptotes of the graph.

III.2   Tangents and Simple Differentiation Rules

III.2.1   *Roots and Factorization*

Once we understand the basics about rational expressions and functions, the results we have established for polynomials carry over easily to this more general setting. Suppose $R = P/Q$ is a rational function with domain Ω_Q. A number $p \in \Omega_Q$ is a root (or zero) of R if $R(p) = 0$. Clearly p is a root of R if and only if p is a root of the numerator P, that is, if $P(p) = 0$. Note that we only consider numbers in Ω_Q; a root of the numerator P that is also a root of the denominator is NOT a root of P/Q!

The critical factorization theorem for polynomials extends immediately to a *rational* function $R = P/Q$ at any point $a \in \Omega_Q$ as follows. Let us assume first that $R(a) = 0$, then one must have $P(a) = 0$ as well, and hence $P(x) = q_P(x)(x - a)$ for some polynomial q_P. Consequently, $R(x) = q(x)(x - a)$, where $q = q_P/Q$ is a rational function with the same domain as R. If $R(a) \neq 0$, we apply this result to the rational function $R - R(a)$, which does have a zero at a. Consequently, one obtains a factorization

$$R(x) - R(a) = q_R(x)(x - a) \tag{P.7}$$

with another *rational* function q_R with the same domain Ω_Q as R.

Once we have established this factorization, we can define multiplicities of zeroes as in the case of polynomials. Precisely, we say that a rational function R has a *zero at a of multiplicity* $\geq m$, where m is a positive integer, if $R(x) = k_m(x)(x - a)^m$ for some rational function $k_m(x)$ with the same domain as R. We say that a is a zero of multiplicity m, if, in addition, $k_m(a) \neq 0$.

III.2.2 *Tangents and Derivatives*

Once we have established the basic factorization theorem, the following theorem is an immediate consequence.

Theorem 18 Suppose R is a rational function and a is a number in its domain. Then there exists a unique line through $(a, R(a))$ that intersects the graph of R at this point with multiplicity ≥ 2. The slope m of this line is given by $m = q_R(a)$, where q_R is the rational function with same domain as R in the factorization $R(x) - R(a) = q_R(x)(x - a)$.

The proof is identical to the one in the polynomial case, where now of course the critical factor is a rational function.

Of course we call the line given by the theorem the *tangent line* to the graph of R at the point $(a, R(a))$, and its slope $m = q_R(a)$ is called the derivative of R at a, denoted by $D[R](a)$ or $R'(a)$, just as in case of polynomials.

Example. Let us consider the simple function $f(x) = 1/x$. Fix $a \neq 0$, and note that

$$\frac{1}{x} - \frac{1}{a} = \frac{a - x}{x \cdot a} = \left[-\frac{1}{x \cdot a} \right] (x - a)$$

is the relevant factorization. So $D[\frac{1}{x}](a)$ equals the value of the rational function in square brackets at a, that is, $-\frac{1}{a^2}$. We thus established

$$\left(\frac{1}{x} \right)' = -\frac{1}{x^2} \text{ for all } x \neq 0.$$

More generally, we can consider $f_n(x) = \frac{1}{x^n}$ for any natural number n by essentially the same process. Note that

$$\frac{1}{x^n} - \frac{1}{a^n} = \frac{a^n - x^n}{x^n a^n} = -\frac{x^n - a^n}{x^n a^n} = -\frac{\sum_{j=0}^{n-1} x^{n-1-j} a^j}{x^n a^n}(x - a),$$

where we have used the familiar factorization of $x^n - a^n$. Now evaluate the rational factor at a, resulting in the rational number

$$-\frac{n \cdot a^{n-1}}{a^{2n}} = (-n)\frac{1}{a^{n+1}}$$

after cancellation of the common nonzero factor a^{n-1} in numerator and denominator. We have thus proved

$$D\left[\frac{1}{x^n}\right] = (-n)\frac{1}{x^{n+1}} \text{ for all } x \neq 0 \text{ and any } n \in \mathbb{N}.$$

Recall that we had established that $\frac{1}{x^n}$ is the multiplicative inverse $(x^n)^{-1} = x^{-n}$ of x^n, so we can write the above result in the form

$$D[x^{-n}] = (-n)x^{-n-1} \text{ for all } x \neq 0.$$

Does this look familiar? Recall that we had proved that $D(x^n) = nx^{n-1}$ for any integer $n \geq 0$. The result we just proved shows that this differentiation formula holds for negative integers as well. So we have established that

$$D[x^m] = mx^{m-1}$$

is correct for all integers m, provided we assume $x \neq 0$ in case $m < 0$.

The result we just proved is a special case of the so-called quotient rule, that in particular tells us how to differentiate rational functions. We will discuss this rule a bit later.

III.2.3 *Simple Differentiation Rules*

The following rules had been established for polynomials. They do carry over to rational functions, subject to appropriate restrictions on the domains, with essentially the same proofs.

Rule 0 (Power Rule). If m is an integer, then $(x^m)' = mx^{m-1}$, where we must assume $x \neq 0$ if $m < 0$.

This is the rule we established for the case $m < 0$ at the end of the previous subsection.

Rule I (Linearity). If f and g are rational functions, then

(1) $D[cf] = cD[f]$ for any constant c;
(2) $D[f \pm g] = D[f] \pm D[g]$.

Recall that these rules allow us to easily find the derivative of any polynomial.

Examples.

i) $(3x^2 - 5x^4)' = (3x^2)' - (5x^4)' = 3x - 5 \cdot 4x^3$.

ii) $(5x^7 - 3x^6 + 2x^4 - 5x^2 + 7x - 4)' = 35x^6 - 18x^5 + 8x^3 - 10x + 7$.

iii) $D\left[3x^{-2} + \dfrac{5}{x^3}\right] = 3D[x^{-2}] + 5D[x^{-3}] = -6x^{-3} - 15x^{-4}$.

Rule II (Chain Rule). Recall that for two functions f and g, the composition $f \circ g$ of f and g is defined by evaluating first g and then inserting the output into f, i.e., $(f \circ g)(x) = f(g(x))$. This of course applies in particular to rational functions, although one must be careful to limit the input x to values a for which g is defined, and moreover, so that f is defined at $b = g(a)$. (If both f and g are *polynomials*, there is no restriction on x.) The chain rule then states that

$$D[f \circ g](a) = D[f](b) \cdot D[g](a), \text{ where } b = g(a), \text{ or}$$
$$(f \circ g)'(a) = f'(g(a)) \cdot g'(a).$$

By using functional notation, the chain rule can be written

$$D[f \circ g] = (D[f] \circ g) \cdot D[g].$$

The crux of the matter is that the *derivative of a composition is the product of the derivatives*.

We had discussed the very simple and natural proof in the case of polynomials. That proof carries over verbatim by just replacing polynomials with more general rational functions, while giving appropriate attention to the inputs.

Here is a simple example that exhibits how various rules fit together consistently. Let m and n be two positive integers. Since $(x^m)^n$ is a composition of two power functions, we can apply the chain rule to obtain

$$[(x^m)^n]' = n(x^m)^{n-1}(mx^{m-1})$$
$$= nmx^{m(n-1)}x^{m-1}$$
$$= mnx^{mn-1},$$

where we have used the standard rules $(x^m)^n = x^{mn}$ and $x^s x^t = x^{s+t}$ for powers with natural numbers as exponents. The final answer agrees, of course, with the direct application of the power rule to x^{mn}.

We also notice that the chain rule allows us to easily find the derivative of certain rational functions. For example, note that we can view the rational function $R(x) = \frac{1}{x^2+3x-2}$ as the composition $f \circ g$ of $g(x) = x^2 + 3x - 2$ and $f(x) = 1/x$. Therefore,

$$D\left[\frac{1}{x^2 + 3x - 2}\right] = D[f \circ g](x) = D[f](g(x)) \cdot D[g](x)$$
$$= -\frac{1}{[g(x)]^2} \cdot g'(x) = -\frac{1}{(x^2 + 3x - 2)^2} \cdot (2x + 3)$$
$$= -\frac{2x + 3}{(x^2 + 3x - 2)^2}.$$

This result can also be obtained as a special case of the Quotient Rule that will be discussed in Section 3.2.

III.2.4   *A First Look at the Inverse Function Rule*

Next to composition, there is another natural operation that applies to functions in the most general settings. This is the *inverse* of a function, a concept that we had discussed already in Section I.2.2. Recall that if $S : \Omega \to S(\Omega)$ is a function that is one-to-one, that is, if $S(a) \neq S(b)$ whenever $a \neq b$, then one can "reverse" the function machine as follows. Start with an element $p \in S(\Omega)$. Then $p = S(a)$ for a *unique* $a \in \Omega$ (S is one-to-one!), and we can define the inverse $R : S(\Omega) \to \Omega$ of S by setting $R(p) = a$. Then $R \circ S = id_\Omega$, where id_Ω is the identity function on Ω that is defined by $id_\Omega(a) = a$. Similarly, we have $S \circ R = id_{S(\Omega)}$. We had discussed in Section I.2.2 that the notation S^{-1} for the inverse of S is quite appropriate, although one must be careful not to confuse it with the standard multiplicate inverse $1/S$, which is also denoted by S^{-1}.

We now want to discuss how to find the derivative of the inverse S^{-1} given the derivative of S. A very simple example involves linear functions $f(x) = mx + b$, with non-zero slope m. In fact, given $y \in f(\mathbb{Q})$, solving $y = mx + b$ for x gives the unique solution $x = \frac{1}{m}(y - b)$, so f is indeed injective. Its inverse f^{-1} is then defined by $f^{-1}(y) = y/m - b/m$, and we notice that it is again a linear function. Again, we must be careful not to confuse this with the multiplicative inverse $f^{-1}(x) = \frac{1}{mx+b}$, which certainly is very different from a linear function. We note that in this example one has $D[f^{-1}] = \frac{1}{m} = \frac{1}{D[f]}$. Since for linear functions the derivatives are constant, i.e., just the slope, there is no need to specify input values. In the next example we will see explicitly how the appropriate input values need to be chosen.

We shall now consider the squaring function $S : \mathbb{Q} \to S(\mathbb{Q})$ defined by $S(x) = x^2$. Since $S(-x) = S(x)$, S is not one-to-one. However, if we restrict S to the smaller domain $\mathbb{Q}^+ = \{x \in \mathbb{Q} : x \geq 0\}$, then $S^+ = S|_{\mathbb{Q}^+}$ is clearly one-to-one, and hence we can consider its inverse

$$R = (S^+)^{-1} : S^+(\mathbb{Q}^+) \to \mathbb{Q}^+.$$

Since $x = (S^+ \circ R)(x) = [R(x)]^2$, R is the familiar square root function, i.e., $R(x) = \sqrt{x}$, with the domain $S^+(\mathbb{Q}^+)$. Note that by restricting $\sqrt{x}$ to this particular domain, we avoid all problems about the existence of $\sqrt{x}$. Since R surely is not a rational function, we need to check if and how tangent lines are defined for points on the graph of R. So let $(b, \sqrt{b})$ be a point on the graph, where $b \in S^+(\mathbb{Q}^+)$. As usual, the tangent at this point should be a line that intersects the graph in a double point, so, if the algebra makes sense, the relevant equation should have a zero of multiplicity ≥ 2. As we know from our experience with polynomials and rational functions, the key is to obtain the appropriate factorization for $R(x) - R(b) = \sqrt{x} - \sqrt{b}$. Note that

$$\sqrt{x} - \sqrt{b} = (\sqrt{x} - \sqrt{b})\frac{\sqrt{x} + \sqrt{b}}{\sqrt{x} + \sqrt{b}} = \frac{1}{\sqrt{x} + \sqrt{b}}(x - b),$$

where we have used standard algebraic rules. Given that we are dealing with functions defined for certain rational numbers, whose values are again rational numbers, we can view each of the symbols in the preceding formula as a rational number, thereby justifying the particular process that is applied. So consider a line $y = \sqrt{b} + m(x - b)$ through the point $(b, \sqrt{b})$ with slope m. Its points of intersection with the graph of R are the solutions of the equation

$$\sqrt{x} = \sqrt{b} + m(x - b),$$

that is,

$$\sqrt{x} - \sqrt{b} - m(x - b) = 0.$$

After introducing the previous factorization and rearranging, we obtain

$$\left(\frac{1}{\sqrt{x} + \sqrt{b}} - m\right)(x - b) = 0.$$

Clearly the solution b has multiplicity greater than 1 if and only if the factor on the left also has a zero at $x = b$, and this occurs precisely if and only if $m = \frac{1}{2\sqrt{b}}$. This of course only makes sense if $b > 0$. So, assuming $b > 0$, you may well wonder if by choosing $m = \frac{1}{2\sqrt{b}}$, the preceding equation really has a root of multiplicity 2, that is, can we split off another factor $(x - b)$? The answer is indeed yes, although the verification is just a bit messy — after all, we are dealing with a new kind of function here. So consider

$$\frac{1}{\sqrt{x} + \sqrt{b}} - \frac{1}{2\sqrt{b}} = \frac{2\sqrt{b} - (\sqrt{x} + \sqrt{b})}{2\sqrt{b}(\sqrt{x} + \sqrt{b})} = -\frac{\sqrt{x} - \sqrt{b}}{2\sqrt{b}(\sqrt{x} + \sqrt{b})},$$

where we have used the familiar rules for rational numbers, given that x and b are in $S^+(\mathbb{Q}^+)$, so both $\sqrt{x}$ and $\sqrt{b}$ are indeed in $\mathbb{Q}^+$. But for the numerator we can now use the factorization we had obtained earlier, resulting in

$$\frac{1}{\sqrt{x} + \sqrt{b}} - \frac{1}{2\sqrt{b}} = -\frac{\sqrt{x} - \sqrt{b}}{2\sqrt{b}(\sqrt{x} + \sqrt{b})} = -\frac{1}{2\sqrt{b}(\sqrt{x} + \sqrt{b})}\frac{1}{\sqrt{x} + \sqrt{b}}(x - b).$$

So we see that the term $(\frac{1}{\sqrt{x}+\sqrt{b}} - \frac{1}{2\sqrt{b}})$ can be factored so as to exhibit a factor $(x - b)$, never mind the messy other factor.

To summarize, we have verified that indeed there is a tangent to the graph at the point $(b, \sqrt{b})$, defined in complete analogy to the case of rational functions, whose slope is given by the derivative

$$D[R](b) = \frac{1}{2\sqrt{b}}.$$

Since we know that $b = S^+(a)$ for a unique $a \in \mathbb{Q}^x$, $a > 0$, we can replace $\sqrt{b} = \sqrt{a^2} = a$, so that

$$D[R](b) = \frac{1}{2a} = \frac{1}{D[S^+](a)} = \frac{1}{D[S^+](R(b))}.$$

You should study this relationship between the derivatives of S^+ and of its inverse $(S^+)^{-1} = R$, including the inputs at which the two sides are evaluated, very carefully, making sure that you understand it well. The heart of the matter is of course the fact that the derivative of the inverse function is the (multiplicative) inverse of the derivative of the original function, just as we had seen in the case of a linear function.

We also want to point out that once the existence of a derivative, that is, the slope of the tangent, has been verified, the relationship between the derivatives of S^+ and of its inverse R is an immediate consequence of the chain rule. (To be precise, we should first prove the chain rule for the more general class of functions that we now consider, but we shall skip these details until we shall verify this for general differentiable functions in Section V.5.3.) In fact, since

$$(S^+ \circ R)(x) = x,$$

taking derivatives at $x = b$ while using the chain rule on the left side, gives

$$D[S^+](R(b)) \cdot D[R](b) = D[x](b) = 1.$$

So, while skipping some of the technical details to be discussed later, we have established the following

Rule III (Inverse Function Rule). The derivative $D(R)(b)$ at the point $b = S(a)$ of the inverse function R of an injective function S is the multiplicative inverse of the derivative $D(S)(a)$ at the point $a = R(b)$, that is,

$$\text{If } R = S^{-1}, \text{ then } D[R](b) = \frac{1}{D[S](R(b))}.$$

Note that for this to make sense it must be assumed that the relevant derivatives are different from 0.

As we can see from the discussion of the derivative of the inverse of $S^+(x) = x^2$, the details get sort of messy because of deficiencies of the rational numbers, and of course also because we are dealing with a new type of function that is no longer rational. So we will postpone the detailed proof of the general formula for derivatives of inverse functions until we have expanded the domains of functions to appropriate subsets of the *real* numbers and introduced differentiation of sufficiently general classes of functions. However, our discussion shows that the derivative of the inverse has a most natural relationship to the derivative of the original function, a relationship that is an obvious consequence of the Chain Rule.

III.2.5   *Exercises*

1. Show that if the rational function R is defined at the point $x = a$, then $D[cR](a) = cD[R](a)$ for any constant c.

2. Consider the rational function R defined by $R(x) = 1/x^2$ for all $x \neq 0$.

a) If $a \neq 0$, show that $\frac{1}{x^2} - \frac{1}{a^2} = -\frac{x+a}{a^2 x^2}(x - a)$.

b) Use the result in a) to find the derivative of R at the point a.

3. Find the equation of the tangent line to the graph of $f(x) = 3x^{-4}$ at the point $(1, 3)$.

4. Find the equation of the tangent line to the hyperbola described by $1 = xy$ at the arbitrary point $(a, 1/a)$ on the graph, where $a \neq 0$, by the following two methods.

a) Write the equation of an arbitrary line through $(a, 1/a)$ with slope m. Substitute this into $1 = xy$ and determine m, so that the point of intersection $(a, 1/a)$ of the line with the hyperbola is a *double* point.

b) Write the equation as $y = 1/x$ and find the slope of the tangent directly by differentiation.

5. Use the Chain Rule to find the derivative of $G(x) = \frac{1}{(x^2 + 3x + 2)^2}$.

6. Find the derivative of $R(x) = 3 \cdot x^{-3} + 5(x^3 + 2x + 1)^4 - 6(x^2 + 1)^{-5}$.

7. Consider the function $C(x) = x^3$ defined on $\mathbb{Q}$.

a) Show that C is one-to-one, and hence it has an inverse $F : C(\mathbb{Q}) \to \mathbb{Q}$. Note that $F(x)^3 = x$, so $F(x) = \sqrt[3]{x}$, which is a well defined number for $x \in C(\mathbb{Q})$.

b) Modify the technique used earlier for the squaring function S to find a factorization $F(x) - F(b) = q(x)(x - b)$ for $b \neq 0$, where q is a particular function involving $F(x)$.

c) Use b) to find the derivative $D[F](b)$ at points $b \neq 0$, and verify that it equals $\frac{1}{D[C](F(b))}$.

III.3   Product and Quotient Rule

We already mentioned that products, and also quotients of functions, are somewhat more complex operations, as they require the corresponding algebraic operations on the codomain of functions. Still, these operations are typically introduced to students before the composition of functions. Perhaps this is a consequence of the fact that the general concept of a function, something like a machine, is not discussed very much at the beginning. Instead, functions are introduced by looking at simple algebraic expressions, like linear functions or quadratic functions. So the underlying addition and multiplication are taken for granted, making it look like these are the most "natural" operations on functions. Instead, we have tried to highlight from the beginning the general concept of a function, independent of any algebraic properties of domain and codomain. Consequently, the natural operations in this context are compositions and taking inverses of functions, that is, reversing the machine. Of course, in order to consider tangents and derivatives, algebraic notions must be included, since we talk of lines, points of intersection, zeroes of polynomials, and so on. In fact, even the Chain Rule for compositions of functions requires multiplication, since the derivative of a composition is the *product* of the derivatives.

We shall now discuss the differentiation rules for products and quotients of functions. As we shall see, these rules turn out to be quite a bit more complicated, and it is difficult to view them as something "natural", something that one could just have guessed.

III.3.1   *Product Rule*

From the perspective of algebra, the product $f \cdot g$ of two functions defined by $(f \cdot g)(x) = f(x)g(x)$ might appear more natural and simpler than the composition $f \circ g$. However, for *derivatives*, the opposite is the case. Since by the chain rule the derivative of a composition is the product of the derivatives, we *cannot* expect the simple formula $D[f \cdot g] = D[f] \cdot D[g]$ for the product of two functions, because the right side is already "reserved". In fact, the rule for finding the derivative of a product is more complicated, as follows.

Rule IV.

Product Rule: $D[f \cdot g] = D[f] \cdot g + f \cdot D[g].$

Notice that rule **I.1** is a special case of the product rule: if c is a constant, then $(cf)' = c'f + cf' = cf'$, since $c' = 0$.

We shall prove this rule for two rational functions, and of course it is assumed that the relevant inputs are in the common domain of the two functions. So suppose the two rational functions f and g are defined at the point $x = a$, and rearrange the standard factorizations in the form

$$f(x) = f(a) + q_f(x)(x - a) \text{ and } g(x) = g(a) + q_g(x)(x - a).$$

This choice is motivated by the need to consider the product of f and g. In fact, we now obtain by standard algebra rules that

$$f(x)g(x) = f(a)g(a) + q_f(x)(x-a)g(a) + f(a)q_g(x)(x-a) + q_f(x)q_g(x)(x-a)^2.$$

It follows that the relevant factorization for $f \cdot g$ is given by

$$(fg)(x) - (fg)(a) = [q_f(x)g(a) + f(a)q_g(x) + q_f(x)q_g(x)(x - a)] \, (x - a)$$
$$= q(x)(x - a),$$

where q denotes the rational function in the edged bracket [...] in the preceding line. Therefore

$$(f \cdot g)'(a) = q(a) = q_f(a) \cdot g(a) + f(a) \cdot q_g(a) + 0 = f'(a) \cdot g(a) + f(a) \cdot g'(a).$$

∎

Example. Let us take $f(x) = g(x) = x$. Then $(fg)(x) = x^2$, and hence $(fg)'(x) = 2x$. Since $f'(x) = g'(x) = 1$, clearly $f'(x)g'(x) = 1 \neq (fg)'(x)$. On the other hand, the product rule

$$(f \cdot g)'(x) = 1 \cdot g(x) + f(x) \cdot 1 = 1 \cdot x + x \cdot 1 = 2 \cdot x$$

gives the correct derivative of $f \cdot g$. More generally, if $f(x) = x^n$ and $g(x) = x^m$ for two positive integers n and m, then, by the product and power rules,

$$(f \cdot g)'(x) = (x^n)' \cdot x^m + x^n \cdot (x^m)'$$
$$= n \cdot x^{n-1} \cdot x^m + x^n \cdot m \cdot x^{m-1}$$
$$= (n + m) \cdot x^{n+m-1}.$$

The answer agrees, as it should, with the direct application of the power rule to $x^{n+m} = x^n x^m$.

Example. Use the product rule to find the derivative of $f(x) = (x^3 - 4x + 1)(4x^5 + 2x^4 - x^3 + 20x)$.

Solution:

$$D[f] = D[x^3 - 4x + 1] \cdot (4x^5 + 2x^4 - x^3 + 20x)$$
$$+ (x^3 - 4x + 1) \cdot D[4x^5 + 2x^4 - x^3 + 20x]$$
$$= (3x^2 - 4) \cdot (4x^5 + 2x^4 - x^3 + 20x)$$
$$+ (x^3 - 4x + 1) \cdot (20x^4 + 8x^3 - 3x^2 + 20).$$

Do not simplify the answer any further. The form above, while some-what complicated, does exhibit the structure of the product rule, and hence makes it easy to check whether the rule has been applied correctly.

III.3.2 *Quotient Rule*

The rule for differentiating the *quotient* f/g (that is, f multiplied with the multiplicative inverse of g, denoted by g^{-1} or $\frac{1}{g}$) of two functions is even more complicated then the product rule. Let us first consider the simpler case of the *reciprocal* $1/g$ of a rational function g at the point a, where $g(a) \neq 0$. With q the appropriate rational factor that satisfies $g(x) - g(a) = q(x)(x - a)$, so that $q(a) = g'(a)$, it follows that

$$\frac{1}{g(x)} - \frac{1}{g(a)} = \frac{g(a) - g(x)}{g(x)g(a)} = \frac{-q(x)(x - a)}{g(x)g(a)}$$
$$= -\frac{q(x)}{g(x)g(a)}(x - a).$$

Clearly the first factor is a rational function with the same domain as $1/g$, so we have established the (unique) relevant factorization for $\frac{1}{g(x)} - \frac{1}{g(a)}$. Consequently, this factorization leads to the **reciprocal rule**

$$\left(\frac{1}{g}\right)'(a) = -\frac{q(a)}{(g(a))^2} = -\frac{g'(a)}{(g(a))^2},$$

or

$$D\left[\frac{1}{g}\right] = -\frac{D[g]}{g^2}.$$

Examples. i) Let us apply the reciprocal rule to find the derivative of $y = 1/x$ at $x \neq 0$. It follows that

$$\left(\frac{1}{x}\right)' = -\frac{x'}{x^2} = -\frac{1}{x^2}.$$

Of course, this result matches what we had already found earlier directly.

ii) More generally, let m be any positive integer. Then

$$\left(\frac{1}{x^m}\right)' = -\frac{mx^{m-1}}{x^{2m}} = (-m)\frac{1}{x^{m+1}}.$$

As we had seen already earlier, this translates into

$$(x^{-m})' = (-m)x^{-m-1},$$

that is, the power rule **0** holds for any integer exponent m, provided one assumes $x \neq 0$ if $m < 0$.

iii) By the reciprocal rule, one obtains

$$D\left[\frac{1}{x^3+1}\right] = -\frac{3x^2}{(x^3+1)^2} \text{ for all } x \neq -1.$$

Remark. The reciprocal rule can also be obtained directly from the product rule. (See Exercise 3.) Also, example i) above, which we had proved directly earlier, combined with the chain rule, gives an even simpler proof of the reciprocal rule, as follows. Note that $1/g$ can be viewed as the composition of g with $f(x) = 1/x$. Thus,

$$D\left[\frac{1}{g}\right](x) = D[f \circ g](x) = D[f](g(x)) \cdot D[g](x).$$

By i), $D[f](g(x)) = -\frac{1}{[g(x)]^2}$, and the result $D(\frac{1}{g}) = -\frac{D(g)}{g^2}$ follows.

Finally, the general case of a quotient f/g of rational functions follows by combining the product rule **IV** with the reciprocal rule, as follows.

$$D\left[\frac{f}{g}\right](a) = D[f \cdot \left(\frac{1}{g}\right)](a) = D[f](a) \cdot \frac{1}{g(a)} + f(a) \cdot D\left[\frac{1}{g}\right](a)$$

$$= D[f](a) \cdot \frac{1}{g(a)} + f(a) \cdot \left(-\frac{D[g](a)}{g(a)^2}\right).$$

Adding the two fractions gives the following formula.

Rule V. Quotient Rule:

$$D\left[\frac{f}{g}\right](a) = \frac{D[f](a) \cdot g(a) - f(a) \cdot D[g](a)}{g(a)^2}.$$

The expression in the numerator is very similar to the result of the product rule, except for the minus sign. It is thus very important to keep the order straight, i.e., to remember that differentiation begins with the numerator. Symbolically, if Num is the Numerator and Den is the Denominator, then

$$Quotient\ Rule: \quad \left[\frac{Num}{Den}\right]' = \frac{Num' \cdot Den - Num \cdot Den'}{Den^2}.$$

III.3.3   *Some Examples*

It is the nature of product and quotient rules that the result of differentiation can turn out to look very complicated. However, it is important to keep track of the structure of the rules, and not get overwhelmed by the increasing number of expressions that may appear. Let us consider some examples.

$$\left[\frac{x^3 - 4x^2 + 3x - 1}{x^2 - 9}\right]'$$

$$= \frac{(x^3 - 4x^2 + 3x - 1)' \cdot (x^2 - 9) - (x^3 - 4x^2 + 3x - 1) \cdot (x^2 - 9)'}{(x^2 - 9)^2}$$

$$= \frac{(3x^2 - 8x + 3) \cdot (x^2 - 9) - (x^3 - 4x^2 + 3x - 1) \cdot 2x}{(x^2 - 9)^2} \quad \text{for all } x \neq \pm 3.$$

It is best to leave the answer in this last form which reflects the structure of the quotient rule, rather than to attempt any algebraic "simplification".

Remark. The quotient rule implies that the derivative $D[R](x)$ of a rational function $R(x)$ is again a rational function that is defined wherever $R(x)$ is defined. Therefore one can define its derivative $D[R'] = R''$, which is again a rational function, so the process can be continued. Taking derivatives n times, the resulting nth derivative $R^{(n)} = D^n[R]$ is again rational with the same domain as R. We note however, that these higher order derivatives will get more and more complicated. In contrast, if we start with a polynomial P of degree n, every differentiation lowers the degree by 1, so that after n differentiations the degree is 0, i.e., $P^{(n)}$ is constant. Therefore $P^{(n+1)} = 0$. This shows, once again, that polynomials are so much simpler than general rational functions.

Next we consider an example that combines several rules of differentiation. It is best to proceed with one rule at a time, as appropriate, until all differentiations have been carried out. Moreover, do not attempt any "simplifications", neither during the calculations nor at the end.

$$\left(\frac{(x^3 + 2x)^6 \sqrt{x^2 + 1}}{4x + 5}\right)'$$

$$= \frac{D[(x^3 + 2x)^6 \sqrt{x^2 + 1}] \cdot (4x + 5) - [(x^3 + 2x)^6 \sqrt{x^2 + 1}] \cdot D[4x + 5]}{(4x + 5)^2}$$

$$= (I)$$

by Rule V, where of course we must assume $x \neq -5/4$. Next,

$$(I) = (\{D[(x^3 + 2x)^6] \cdot \sqrt{x^2 + 1} + (x^3 + 2x)^6 \cdot D[\sqrt{x^2 + 1}]\} \cdot (4x + 5)$$
$$- \{(x^3 + 2x)^6 \sqrt{x^2 + 1}\} \cdot D[4x + 5])/(4x + 5)^2$$
$$= (II),$$

where we have used the Product Rule IV. Finally, by using Rules 0-III, as well as the rule for the derivative of the square root function, one obtains

$$(II) = \left(\left\{6(x^3 + 2x)^5 \cdot (3x^2 + 2) \cdot \sqrt{x^2 + 1} + (x^3 + 2x)^6 \cdot \left(\frac{1}{2}\frac{1}{\sqrt{x^2 + 1}}2x\right)\right\}\right.$$

$$\left.(4x + 5) - \{(x^3 + 2x)^6 \sqrt{x^2 + 1}\}4\right)/(4x + 5)^2.$$

III.3.4 *Exercises*

1. Find the derivatives of the following functions:
    a) $P(x) = 4x^5 - 6x^4 - \frac{1}{5}x^3 + 3x^2 + 2$.
    b) $f(x) = 5x^3 + 7x^{1/2} - 3(x^2 + 1)^7$ for $x > 0$.
    c) $g(x) = 1/(3x^4 + 7x^2 + 2)^5$. (Hint: Use $1/(b^5) = b^{-5}$.)
    d) $k(x) = \frac{5x^3 - 2x}{4x^2 + x - 1}$.
    e) $h(x) = 4/x^3 + 5x^{1/5} - 2\sqrt{x^4 + 2}$ for $x \neq 0$.
2. Find the derivative of $G(x) = (x^3 - 2x^2 + 4x)\sqrt{3x^2 + 1}$.
3. Find the derivative of $Q(x) = \frac{(x+1)\sqrt{x^2 + 4}}{x^3(x+1)}$.
4. Note that the reciprocal $1/g$ of a function g with $g(a) \neq 0$ can be written as the composition $1/g = f \circ g$, where $f(y) = 1/y$ for $y \neq 0$. Use the chain rule and power rule to prove the reciprocal rule for the derivative $D[1/g](a)$.
5. a) Derive the reciprocal rule for differentiation directly from the product rule by differentiating both sides of the equation $g \cdot (1/g) = 1$. Note that the reciprocal (i.e., multiplicative inverse) $1/g$ of a rational function g is rational as well.
    b) Apply the analogous method to $g \cdot (f/g) = f$ to find the derivative of f/g.
6. Let n be a positive integer. The function $y = R(x) = x^n$ is one-to-one on $\mathbb{Q}^+ = \{x \in \mathbb{Q} : x > 0\}$. Use the inverse function rule **III** to find the derivative of the inverse $x = S(y) = y^{1/n}$ on $R(\mathbb{Q}^+)$. Verify that $D[y^{1/n}] = \frac{1}{n}y^{\frac{1}{n}-1}$. (Remark: Given the choice of domain for the inverse S, there is no problem whatsoever with the existence of nth roots.)

7. Prove that the power rule holds for arbitrary rational exponents $r = m/n$, $n > 0$. (Hint: Note that $f(x) = x^{m/n} = (x^{1/n})^m$ and apply the chain rule and the power rule for exponents $m \in \mathbb{Z}$ and $1/n$. Again, the inputs are restricted to values for which the functions are readily defined.)

8. Do Exercise 7 by reversing the order in the composition, i.e., write $f(x) = (x^m)^{1/n}$.

Chapter IV

Continuity and Approximation of Derivatives

IV.1 Local Boundedness and Continuity

IV.1.1 *An Estimate for Polynomials and Continuity*

Given a polynomial f, we have established the basic factorization

$$f(x) - f(a) = q(x)(x - a),$$

where a is an arbitrary point in $\mathbb{Q}$ (or $\mathbb{R}$), and q is another polynomial that of course depends on the choice of a. Taking absolute values on both sides one obtains

$$|f(x) - f(a)| = |q(x)| \cdot |x - a|.$$

Now, as we will see shortly, there is a constant $K > 0$, so that one has $|q(x)| \le K$ for all x with $|x - a| \le 1$. It then follows that

$$|f(x) - f(a)| \le K \cdot |x - a| \text{ for all } x \text{ with } |x - a| \le 1.$$

We had already seen this sort of estimate when we discussed Galileo's investigations about velocity. In fact, in that context, this estimate provides the justification for what seems to be obvious, namely, that the *instantaneous* velocity at one point in time t_0 is approximated by the *average* velocities over shorter and shorter time intervals with one end point at t_0. We now want to discuss this phenomenon more in detail.

Lemma 19 *Suppose f is a polynomial. Given any bounded interval I, there exists a constant $K = K(I)$, so that*

$$|f(x)| \le K \text{ for all } x \in I.$$

Proof. Suppose $f(x) = \sum_{j=0}^{n} c_j x^j$. Since the interval I is bounded, we may assume that I is contained in an interval $[-M, M]$ for some integer M. Since for $x \in I$ one then has $|x| \leq M$, standard estimations imply that

$$|f(x)| \leq \left| \sum_{j=0}^{n} c_j x^j \right| \leq \sum_{j=0}^{n} |c_j| \, |x|^j \leq \sum_{j=0}^{n} |c_j| \, |M|^j = K \text{ for } x \in I.$$

In the key estimate at the beginning of this section, since the interval $[a - 1.a + 1]$ is bounded, we apply the Lemma to the polynomial q in the factorization of f, and the desired result follows.

Even though the estimate is very elementary, its implications are far reaching. We note that as x gets closer and closer to a, $|x - a|$ gets smaller and smaller. The estimate then shows that $|f(x) - f(a)|$ also becomes smaller and smaller, the factor K only changing the degree of smallness. For example, suppose we want $|f(x) - f(a)| < 0.0001$. The estimate then shows that this will be the case for all x that satisfy $|x - a| < 0.0001/K$.

We shall use the notation $x \to a$ to indicate that x approaches a, which means that $|x - a|$ gets smaller and smaller, that is, it approaches 0. Similarly, we write $f(x) \to f(a)$ to indicate that $|f(x) - f(a)|$ approaches 0. Using this notation, the estimate then implies that

$$f(x) \to f(a) \text{ as } x \to a.$$

Geometrically this means that as x approaches the value a, the point $(x, f(x))$ on the graph of f moves towards the point $(a, f(a))$. This property is intuitively evident for the graph of any familiar polynomial, say a line or a parabola, or the graph of $y = x^3$. Using a graphing calculator to display the graph of your favorite polynomial, this property again becomes visually evident.

Here is the graph of a function g which evidently does not have this property at the point $x = 1$. See Fig. IV.1.

While $g(x) \to 2 = g(1)$ as x approaches 1 from the right side, i.e., as $x > 1$, the picture clearly shows that as x approaches 1 from the left, i.e., as $x < 1$, $g(x)$ surely does not get closer and closer to $g(1) = 2$. In fact, Fig. IV.1 suggests that $g(x)$ approaches 1 as $x \to 1$ from the left.

The property that we just identified is known as *continuity*. More precisely, we have verified that a polynomial is continuous at the point a. Let us formulate this new concept as follows.

Definition 20 *The function f is continuous at the point a in its domain if*
$$x \to a \text{ implies } f(x) \to f(a).$$
A more formal notation is $\lim_{x \to a} f(x) = f(a)$.

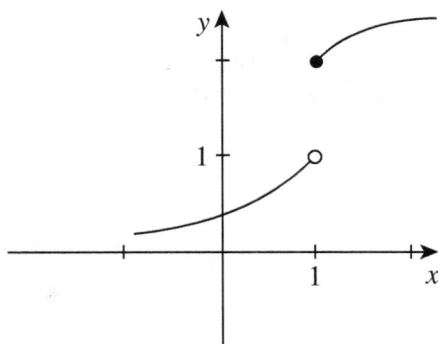

Fig. IV.1   Graph of a discontinuous function.

The letters lim stand for *limit*. The estimate that motivated the idea gives a precise meaning to this statement. Later, when we consider more general functions, we shall formulate the idea of limit precisely, in a more general form that does not use the basic estimate we identified for polynomials.

For the time being, the intuitive formulation given in the definition above is quite sufficient.

Our discussion has led to the conclusion that every polynomial is continuous at every point $a \in \mathbb{R}$. The function g whose graph is shown in Fig. IV.1 is NOT continuous at the point $x = 1$, although the graph suggests that it is continuous at every point $a \neq 1$ in its domain.

Continuity is a property of many natural phenomena, and it had been known intuitively for a long time. In fact, the ancient Latin statement

Natura non facit saltus,

that is, "Nature does not make jumps", is just another way of saying that the functions that describe natural phenomena are continuous. Modern science of course has identified processes that are not really continuous, such as in Quantum Mechanics. Also, in evolutionary biology, the phenomenon of "mutation" suggests a jump, that is, a discontinuity. Given that the basic structure of DNA is discrete, nature is now widely understood to make jumps at the biological level, if only on a very small scale.[1]

---

[1] Our reference for these facts is Wikipedia.

IV.1.2 *Continuity of Rational Functions*

We proved in Section III.2.1 that rational functions satisfy a factorization result that is completely analogous to the factorization of polynomials, the only difference being that the critical factor q is now a rational function. Clearly we can not expect that rational functions are bounded on bounded intervals, just look at the function $y = 1/x$ on the interval $(0, 1)$. However, there is a local version that will be sufficient to establish the continuity of a rational function at every point in its domain. To illustrate the idea, let us consider $y = 1/x$, and fix a point a in its domain, i.e., we must have $a \neq 0$. Hence $|a| > 0$. Then $\delta = |a|/2$ is positive as well. We claim that $1/x$ is bounded on the interval $I_\delta(a) = (a - \delta, a + \delta)$. In fact, if $x \in I_\delta(a)$, then $|x - a| \leq \delta$. Hence, by a version of the triangle inequality,

$$|x| = |a + (x - a)| \geq |a| - |x - a| \geq |a| - \delta = |a|/2.$$

It then follows that

$$\left|\frac{1}{x}\right| \leq \frac{2}{|a|} \text{ for all } x \in I_\delta(a),$$

which proves the claim.

For a general rational function $R = \frac{P}{Q}$, where P and Q are polynomials, we use the same strategy, with a little twist. Since the numerator, being a polynomial, is bounded on any bounded interval, the crux is to deal with the denominator. So, assume a is in the domain of R. Then $Q(a) \neq 0$, and hence, $|Q(a)| > 0$. Let $\delta = |Q(a)|/2 > 0$. Just as in the case where $Q(x) = x$ that we just considered, we want to estimate $|Q(x)|$ from below for x sufficiently close to a. Again, we have

$$|Q(x)| = |Q(a) + (Q(x) - Q(a))| \geq |Q(a)| - |Q(x) - Q(a)|.$$

Now Q, being a polynomial, is continuous at a. More precisely, there is a constant $K > 0$, so that we have an estimate

$$|Q(x) - Q(a)| \leq K \cdot |x - a| \text{ for all } x \in I_\delta(a).$$

This estimate remains correct if K is replaced by a larger number, so we can surely assume that $K \geq 1$. Now, in order to have $|Q(x) - Q(a)| \leq \delta$, it is sufficient to have $|x - a| \leq \delta/K$, which is $\leq \delta$. It follows that for all such x,

$$|Q(x)| \geq |Q(a)| - |Q(x) - Q(a)| \geq |Q(a)| - \delta = |Q(a)|/2.$$

Therefore, for all such x one has $1/|Q(x)| \leq 2/|Q(a)|$. Finally, since the numerator P is a polynomial, there is a constant L, such that $|P(x)| \leq L$

for all x with $|x - a| \leq \delta/K$. Putting the last two estimates together, one obtains

$$|R(x)| = \frac{|P(x)|}{|Q(x)|} \leq \frac{L}{|Q(a)|/2} \text{ for all } x \text{ with } |x - a| \leq \delta/K.$$

We have thus proved the following result.

Lemma 21 *Suppose R is a rational function. Then R is **locally** bounded near each point a in its domain, that is, there exists an interval $I(a)$ centered at a in the domain of R, so that R is bounded on $I(a)$.*

Corollary 22 *Suppose R is a rational function. Then R is continuous at every point a in its domain.*

Proof. As we had verified in Section III.2.1, there is a factorization $R(x) = q(x)(x - a)$, where q is a rational function with the same domain as R. By the last Lemma, q is locally bounded at a. We are thus in the same situation as in the polynomial case, and the result follows as before. ∎

IV.1.3   *Exercises*

1. Determine the limits

$$a) \ \lim_{x \to 2} (x^3 - 2x^2 + 3x), \ b) \ \lim_{x \to -1} \frac{3x}{x^2 + 3x - 1}.$$

(Hint: Use the continuity of the functions.)

2. The function $R(x) = \sqrt{x}$ defined on $\mathbb{Q}^+ = \{x \in \mathbb{Q} : x \geq 0\}$ is not rational. Show that it is continuous at each point $a \in \mathbb{Q}^+$. (Hint: If $a > 0$, use the relevant factorization obtained in Section 2.4. For $a = 0$ produce an appropriate estimate directly.)

3. Determine the limits

$$a) \ \lim_{x \to 25} \frac{x - 25}{\sqrt{x} - 5}, \qquad b) \ \lim_{r \to 2} \frac{4 - r^2}{r^3 - 8}.$$

(Hint: Apply algebra to simplify the expressions so as to eliminate the zero in the denominator.)

4. Let f be a rational function defined at $x = 2$, and define $G(x) =$
$$\begin{cases} \frac{f(x) - f(2)}{x - 2} & \text{for } x \neq 2 \\ f'(2) & \text{for } x = 2 \end{cases}.$$
Is G continuous at $x = 2$? Explain!

5. Define $g(x) = \begin{cases} \frac{x-1}{x+3} & \text{for } x \neq -3 \\ -4 & \text{for } x = -3 \end{cases}$. At which points $a \in \mathbb{R}$ is g continuous? Explain!

6. Define $P(t) = \begin{cases} 2t + 3\,t^2 & \text{for } t \le 1 \\ (t+2)^2 - 4t & \text{for } t > 1 \end{cases}$. Is the function P continuous at $t = 1$? Explain!

IV.2   Rates of Change

IV.2.1   *Average Rates of Change*

As we saw in the previous chapter, given a polynomial f, or more generally a rational function, and a point a in its domain, the elementary algebraic factorization

$$f(x) - f(a) = q(x)(x - a),$$

where q is a uniquely determined rational function, provides the critical information to solve the tangent problem for the curve that is the graph of f. In fact, the value $q(a)$ is the slope of that unique line through the point $P = (a, f(a))$ that intersects the graph of f with multiplicity 2 or higher. Recall that this is the special property that singles out the *tangent* line to the graph at P, and that the value $q(a)$, i.e., the slope, is called the derivative $D[f](a) = f'(a)$ of f at a.

While the calculation of an explicit formula for the factor q in concrete cases typically involves lengthy computations, we showed in Chapter III that its value at a, that is, the derivative $D(f)(a)$, can readily be found for all rational functions by a routine application of specific rules. We shall now consider in detail the values of q at inputs x that are *different* from a. Regardless of the nature of the function f, or whether a particular explicit expression for the factor q is available, the value $q(x)$ for $x \ne a$ is uniquely determined by

$$q(x) = \frac{f(x) - f(a)}{x - a} \quad \text{for } x \ne a. \tag{IV.1}$$

The quotient on the right side, which is NOT defined for $x = a$ (plugging in $x = a$ leads to the expression $0/0$, which is meaningless), contains important information that is most useful in applications. Even more, this quotient is critical in order to solve the tangent problem for more general *non-algebraic* functions, such as the exponential function that we shall study in the next section, where the value $q(a)$, i.e., the derivative, is not accessible by any elementary methods.

In order to illustrate the significance of the quotient (IV.1), we shall now consider several concrete situations.

Average Velocity Let us begin with the concept of "average velocity" that we had already considered in Section II.3.2 when we discussed Galileo's investigations. We are all familiar with the basic idea as it arises, for example, with a moving automobile. A car that travels a distance of 15 km between 12:10 p.m. and 12:22 p.m. is said to have travelled with a velocity of $\frac{15}{12}$ km/min over that time period. Note that this number does not take into account any changes that may occur during the time interval considered, such as slowing down to avoid an obstacle, stopping for a traffic light, or accelerating to pass another car. Instead, what has been measured is the *average* velocity between 12:10 p.m. and 12:22 p.m. Formally, for two distinct moments in time $t_1 < t_2$, one defines

$$\textbf{average velocity } \textit{between } t_1 \textit{ and } t_2 = \frac{\textit{distance traveled between } t_1 \textit{ and } t_2}{t_2 - t_1},$$

or, more briefly,

$$\textit{average velocity} = \frac{\textit{distance}}{\textit{time}}.$$

Note that the numerical value of the average velocity depends on the units chosen to measure distance and time. For example, since 12 minutes are 0.2 hours, the velocity of $\frac{15}{12}$ km/min corresponds to a velocity of $\frac{15}{0.2} = 75$ km/hour. Converting 15 km into 9.32 miles results in an average velocity of $\frac{9.32}{0.2} = 46.61$ miles/hour. Commonly used units for velocity are *m/sec = meters/second, ft/sec = feet/second, km/h = kilometer/hour, and mi/h = miles/hour.*

The distance traveled between two points in time t_1 and t_2 relates to the change in position of the automobile. Suppose the car moves along a highway, and let $s(t)$ measure its position at time t as given, for example, by the km-markers along the road. Between the times t_1 and t_2 the car will have traveled a distance $\Delta s = s(t_2) - s(t_1)$, as shown below.

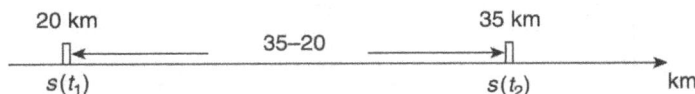

Fig. IV.2 Distance traveled on a road.

The average velocity between t_1 and t_2 is thus given by the quotient

$$\frac{\Delta s}{\Delta t} = \frac{s(t_2) - s(t_1)}{t_2 - t_1},$$

where the symbol Δ (= delta = capital Greek "D") is generally used to indicate a difference of relevant quantities. Note that this latter expression has exactly the structure of the general quotient (IV.1), with the function f replaced by the position function s.

A car is said to move with *constant* velocity (during a particular time period) if the average velocity between any two points in time during that period is always the same number. In that case the position of the car in dependence of time can easily be described precisely. Suppose that the car travels with *constant* (average) velocity v. Then

$$\frac{s(t_2) - s(t_1)}{t_2 - t_1} = v \text{ for any } t_1 \neq t_2.$$

Therefore, if we fix the initial time t_1, and let $t_2 = t$ be arbitrary, one obtains

$$\frac{s(t) - s(t_1)}{t - t_1} = v \text{ for any } t \neq t_1.$$

(This formula holds both when $t > t_1$ and when $t < t_1$.) This equation can be solved for $s(t)$, resulting in

$$s(t) = v(t - t_1) + s(t_1).$$

Note that the latter formula is valid also for $t = t_1$. We see that constant velocity implies that the position $s(t)$ is described by a *linear* function of time t, that is, by a polynomial of degree 1.

Lines and Slopes   An analogous discussion applies when one considers the inclination or steepness of a highway, as shown in Fig. IV.3.

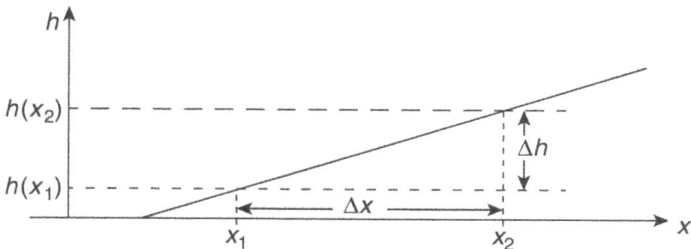

Fig. IV.3   Inclination of a straight line.

A measure of the "steepness" is given by the change in height $\Delta h = h(x_2) - h(x_1)$ that occurs over the (horizontal) distance $\Delta x = x_2 - x_1 > 0$.

What matters is not the value of Δh itself, but rather the value of the ratio
$$\frac{\Delta h}{\Delta x} = \frac{h(x_2) - h(x_1)}{x_2 - x_1}.$$
Note that if $x_1 < x_2$, then $\Delta h/\Delta x < 0$ is equivalent to $\Delta h < 0$, so a negative quotient indicates that the height is decreasing, i.e., that the road goes downhill. In case the road follows an inclined *straight* line, the quotient $\frac{\Delta h}{\Delta x}$ is a constant m independent of x_1 and x_2, since the corresponding right triangles shown in Fig. IV.4 are similar, and hence the ratios of corresponding sides are equal.

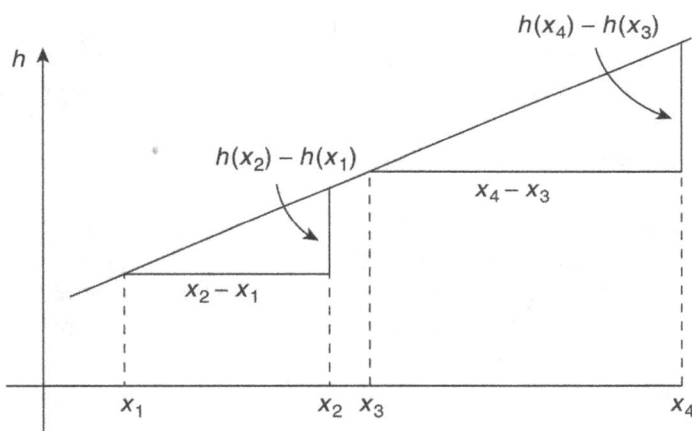

Fig. IV.4   Measure of steepness is independent of position.

This constant number
$$m = \frac{\Delta h}{\Delta x} = \frac{\text{``rise''}}{\text{``run''}}$$
which depends on the units chosen to measure length, is called the *slope* of the line. For example, if the road rises by 40 meters over a horizontal distance of 1 kilometer, $m = 40/1 = 40 \ m/km$. Using the same units for the height and the distance results in a slope of $40/1000 = 0.04 \ m/m$, which is a quantity without any "dimension." One often then writes the result as a percentage, i.e., in the example just considered one says that the road rises with a slope of 4% or at a rate of 4%.

Continuing with this example, let us fix x_1 and consider $x_2 = x \neq x_1$ as variable. We then solve the equation
$$m = \frac{h(x) - h(x_1)}{x - x_1}$$

for $h(x)$, resulting in

$$h(x) = m(x - x_1) + h(x_1) = mx + (h(x_1) - mx_1).$$

Again, the relevant function—the height h in this case—is given by a linear function.

Average Rate of Climb   More generally, suppose the steepness of the road changes along the way, perhaps alternating between climbing and descending. Again, denote by x the position in the horizontal direction and let $h(x)$ be the height of the road above sea level, measured in meters, at the position x. In this case the quotient $\Delta h/\Delta x = [h(x_2) - h(x_1)]/(x_2 - x_1)$ depends on the particular locations x_1 and x_2 chosen, i.e., it is not constant along the way. This quotient thus describes the *average* rate of change in altitude between the two points x_1 and x_2, or also the *average rate of climb* of the road between the points identified by x_1 and x_2.

Average Rates of Change   Returning to the case of a general function f, for a fixed value $x_1 \neq a$, the numerator $f(x_1) - f(a)$ of the quotient in (IV.1) simply measures the difference Δf, or change in the values of f, between the two input values a and x_1. As in the examples we just considered, division by the change $\Delta x = x_1 - a$ in the input thus provides a measure of the "rate of change of f" between the two points a and x_1. Since the quotient $[f(x_1) - f(a)]/(x_1 - a)$ contains no information whatsoever about the function f at any point x between a and x_1, this rate of change again is just an *average* rate of change of f between a and x_1.

    Example. The average rate of change of $f(x) = x^3$ over the interval $[0, 3]$ is

$$\frac{3^3 - 0}{3 - 0} = \frac{27}{3} = 9.$$

Over the interval $[2, 3]$ one obtains

$$\frac{3^3 - 2^3}{3 - 2} = \frac{27 - 8}{1} = 21,$$

which is—no surprise—a different value.

    Let us consider a geometric interpretation of this abstract notion, as follows. The graph of a general (non-linear) function f describes a curve in the coordinate plane. For $x_1 \neq x_2$ the ratio

$$\frac{\Delta f}{\Delta x} = \frac{f(x_2) - f(x_1)}{x_2 - x_1} \tag{IV.2}$$

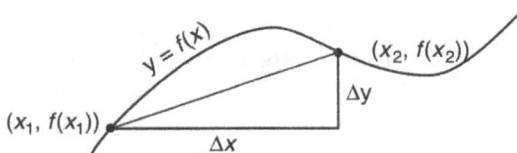

Fig. IV.5 Average rate of change or slope given by $\Delta y/\Delta x$.

can then be interpreted as the slope of the line through the two points $(x_1, f(x_1))$ and $(x_2, f(x_2))$ on the graph (such a line through two distinct points is called a *secant*, to distinguish it from a tangent). (See Fig. IV.5.)

Clearly the graph of f varies quite a bit from that line, that is, the secant only provides crude information about the curve. The quotient (IV.2) is therefore called the *average* slope of the curve between x_1 and x_2.

Our discussion shows that (non-vertical) lines are exactly those curves whose average slopes are constant.

Other Examples of Rates of Change   Average rates of change occur in numerous applications. Here are some additional examples.

Suppose the function T measures the temperature in Celsius degrees at the current location in dependence of the time of day t, e.g., $T(8)$ is the temperature at 8 am. If $T(8) = 16^0$, and $T(11) = 25^0$ the quotient $\Delta T/\Delta t = (25 - 16)/(11 - 8) = 3$ measures the average rate of change in temperature, i.e., the temperature is increasing between 8 am and 11 am at the average rate of 3^0 per hour. Again, the numerical value depends on the chosen temperature scale and units of time.

Volume V, pressure p, and temperature T of an ideal gas are related by the formula $Vp = kT$, where k is a numerical constant. Keeping T fixed, we can view V as a function of pressure that is explicitly given by $V(p) = kT/p$. The ratio

$$\frac{\Delta V}{\Delta p} = \frac{V(p_2) - V(p_1)}{p_2 - p_1}$$

describes the average rate of change of volume between the pressure points p_1 and p_2.

Finally, suppose $P = P(t)$ describes the size of the population in a town in year t. If $P(2010) = 80,000$ and $P(2018) = 92,000$, the ratio $\Delta P/\Delta t = (92,000 - 80,000)/(2018 - 2010) = 1,500$ gives the average rate of growth in population between the years 2010 and 2018, that is, during this period the population grew at an *average* rate of $1,500$ people per year. In

order to compare rates of growth of cities of different sizes, it is more useful
to consider the average *relative* rate of growth, defined by $[\Delta P/\Delta t]/P$,
where usually the value of the population at the beginning of the time period
is chosen. In the case at hand, the relative growth rate between the years
2010 and 2018 is thus given by $1,500/P(2010) = 1,500/80,000 = 0.01875$.
The relative growth rate is most commonly expressed as a percentage, i.e.,
one says that during the relevant period the population grew at an average
rate of $1.875\% \approx 1.9\%$. The use of "percentages" signals that one considers
the *relative* growth rate of the population.

To summarize, we see that "average rates of change" occur in many
different settings. The reader is encouraged to add additional examples
related to her/his own experience and interest.

Returning to the basic factorization

$$f(x) - f(a) = q(x)(x-a),$$

we thus see that the values of the factor $q(x) = [f(x)-f(a)]/(x-a)$, $x \neq a$,
contain important information about the particular process that is modeled
by the function f.

IV.2.2   *Relating Averages over Adjacent Intervals*

Next we want to discuss a basic relationship between average rates of change
over adjacent intervals that will turn out to be significant in later discussions. To illustrate the simple idea, let us consider the average velocities
over two consecutive time intervals $T_1 = [t_0, t_1]$ and $T_2 = [t_1, t_2]$. For example, if the average velocity over T_1 is 40 km/h and over T_2 it is 55 km/h,
then surely the average velocity over the combined time interval $T = [t_0, t_2]$
must be between 40 and 55 km/h. We expect that this relationship holds
in general, i.e., that the average velocity over the *combined* time interval is
at least as large as the smaller of the average velocities over each of the two
time intervals, and at most as large as the larger one of these two average
velocities.

Indeed, this remains correct for average rates of change in general, as
follows. Let us assume that the function f is defined on the two adjacent
intervals $[x_0, x_1]$ and $[x_1, x_2]$, where $x_0 < x_1 < x_2$. For $j = 1, 2$ we denote
the average rate of change of f over the two intervals by $A_j = A_j[x_{j-1}, x_j] = [f(x_j) - f(x_{j-1})]/(x_j - x_{j-1})$.

Lemma 23 *The average rate of change $A = A[x_0, x_2]$ over the combined* *interval $[x_0, x_2]$ satisfies the estimate*

$$\min\{A_1, A_2\} \leq A \leq \max\{A_1, A_2\}.$$

Furthermore, if $A_1 \neq A_2$, then both inequalities are strict.

Proof. The proof is very simple. By writing the definition of average rate of change in product form, one obtains

$$f(x_j) - f(x_{j-1}) = A_j(x_j - x_{j-1}) \text{ for } j = 1, 2.$$

It follows that

$$\min\{A_1, A_2\}(x_j - x_{j-1}) \leq f(x_j) - f(x_{j-1}) \leq \max\{A_1, A_2\}(x_j - x_{j-1}).$$

Now add the two inequalities for $j = 1$ and $j = 2$ and rearrange, using the distributive property. After obvious cancellations, we are left with

$$\min\{A_1, A_2\}(x_2 - x_0) \leq f(x_2) - f(x_0) \leq \max\{A_1, A_2\}(x_2 - x_0).$$

By dividing this last statement by $(x_2 - x_0) > 0$, one obtains the desired result. Furthermore, if $A_1 \neq A_2$, let us assume that $A_1 < A_2$. Then

$$f(x_1) - f(x_0) = A_1(x_1 - x_0) < A_2(x_1 - x_0) = \max\{A_1, A_2\}(x_1 - x_0)$$

and

$$\min\{A_1, A_2\}(x_2 - x_1) = A_1(x_2 - x_1) < A_2(x_2 - x_1) = f(x_2) - f(x_1).$$

To summarize,

$$\min\{A_1, A_2\}(x_1 - x_0) = f(x_1) - f(x_0) < \max\{A_1, A_2\}(x_1 - x_0)$$

and

$$\min\{A_1, A_2\}(x_2 - x_1) < f(x_2) - f(x_1) \leq \max\{A_1, A_2\}(x_j - x_{j-1}).$$

Proceeding as before, carefully keeping track of the strict inequalities, it follows that the final inequalities are now strict. In case $A_2 < A_1$, the proof is analogous, with the obvious modifications. ∎

The result clearly generalizes to any finite collection of adjacent intervals. Furthermore, when the average rates of change over the intervals are strictly increasing, one obtains the following easy consequences.

Corollary 24 *Let $n \geq 2$, and $x_0 < x_1 < ... < x_{n-1} < x_n$. Suppose f is defined on $[x_0, x_n]$, and that the average rates of change A_j over the n successive intervals $[x_{j-1}, x_j]$, $j = 1, 2, ..., n$ satisfy $A_1 < A_2 < ... < A_n$. Then*

(i) $A_1 < A[x_0, x_n] < A_n$ and

(ii) $A[x_0, x_j] < A[x_0, x_{j+1}]$ for $j = 1, 2, ..., n - 1$.

Proof. The proof of (i) is an immediate consequence of the Lemma, generalized to n adjacent intervals, since the hypothesis implies that A_1 and A_n are the minimal, resp. maximal of the rates of change over the n intervals. As for (ii), replacing n with j in (i), one obtains $A[x_0, x_j] < A_j = A[x_{j-1}, x_j]$. If $j < n$, by applying the hypothesis one then obtains $A[x_0, x_j] < A[x_j, x_{j+1}]$. The lower estimate in the Lemma, applied to these last two adjacent intervals, then gives the desired conclusion. ∎

Of course, corresponding results with the inequalities reversed are true as well. See Problem 7 in Exercises 2.4 for details.

IV.2.3   *From Average to Instantaneous Rates of Change*

As we saw in the preceding section, the average slopes of the graph of a function f over different intervals provide only limited information about the function. Many details are simply not captured by such averages. It therefore seems desirable to introduce more refined ways to describe the behavior of the function or of its graph. Recall that in Chapter II we had considered the classical problem of finding *tangents* to curves. In analogy to the (average) rate of change or slope between two distinct points P_1 and P_2 on the graph of a function, the tangents at points P or Q on the graph capture the rates of change or slopes at the single point P, respectively Q, thereby defining the rate of change of a function at single points. (See Fig. IV.6.)

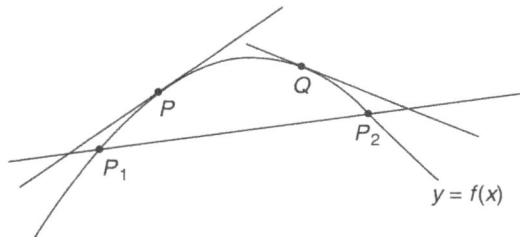

Fig. IV.6   Average slopes and tangent slopes.

The situation is particularly simple in case the function f is a polynomial, or more generally, a rational function. Given the factorization $f(x) - f(a) = q(x)(x - a)$, we saw that the value $q(a)$ gives the slope of the

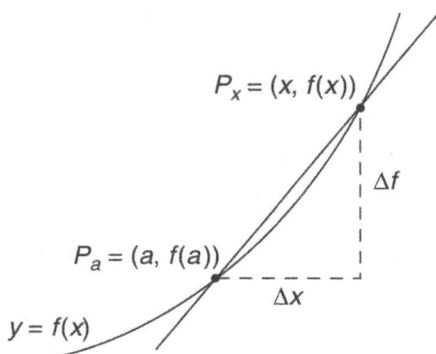

Fig. IV.7 $q(x)$ gives the slope $\Delta f / \Delta x$ of the secant.

tangent line to the graph of f at the point $(a, f(a))$. On the other hand, for $x \neq a$, the value $q(x) = [f(x) - f(a)]/(x - a)$ is the average slope of f between the two distinct points $P_a = (a, f(a))$ and $P_x = (x, f(x))$, as seen in Fig. IV.7.

From this perspective it thus seems natural to interpret $q(a) = D[f](a)$ in exactly the same way as $q(x) = \Delta f / \Delta x$ for $x \neq a$, that is, $q(a)$ is the (average) slope of the graph of f between the two (now identical) points P_a and P_a. The two *distinct* points of intersection of the secant with the graph of f when $x \neq a$ coincide when $x = a$, i.e., we have only *one* point P_a of intersection, which however is *counted twice*, that is, P_a is now a "double point" of intersection between the secant and the curve. According to the discussion in Section II.2.2, this is exactly the defining geometric property of the tangent line to the curve at P_a. When $x = a$, the secant has become the tangent line, and $q(a)$ still measures its slope. Since only **one** point is involved (which however is counted twice), the value $q(a)$, i.e., the slope of the tangent, measures the rate of change of f at the single point $P_a = (a, f(a))$! In more general settings, we say that the derivative $D[f](a) = q(a)$ of f at a measures the "instantaneous" rate of change of f at the point a. In order to highlight this interpretation, in applications one often denotes the derivative by the formal quotient $\frac{df}{dx}$, i.e., $D[f](a) = \frac{df}{dx}(a)$. This notation—which is analogous to $\frac{\Delta f}{\Delta x}$—helps to remind us that the derivative measures a "rate of change".

An analogous interpretation remains valid in all other applications. For example, if $s(t)$ describes the position of an automobile at time t, the derivative $D[s](t_0)$ at time t_0 measures the velocity at that moment, that is, the

instantaneous velocity at time t_0, and we also write $D[s](t_0) = ds/dt(t_0)$.
Similarly, if $V(p) = kT/p$ is the volume of a fixed amount of gas in dependence of the pressure p, the derivative $D[V](p_0) = dV/dp(p_0) = -kT/p_0^2$
describes the instantaneous rate of change of volume with respect to pressure when the pressure has value p_0.[2] Regarding the application to population growth, where $P = P(t)$ measures the size of the population in a
town at time t, it turns out that in all "natural" situations the relevant
function P is not of algebraic type, in particular it is not rational, so the
preceding discussion does not apply directly. As we shall see later, models
of population growth involve exponential functions. Of course, once we
have developed the appropriate concept of derivative, we shall say that the
derivative $D[P](t_0) = dP/dt(t_0)$ measures the instantaneous rate of change
of the population at time t_0. Correspondingly, $D[P](t_0)/P(t_0)$ measures
the (relative) rate of growth at time t_0.

IV.2.4   *Exercises*

1. An airplane departed Albany, NY, at 1:50 p.m., and it landed at Newark
Airport at 2:35 p.m. Determine the average velocity (in miles/hour) of
the airplane on this trip. (You will need to look up a relevant piece of
information that is not given here. Alternatively, use your best estimate.)

2. A motorcycle travels along a highway from 9 am to 10 am with a
constant speed of 70 km/h. Determine the function $s(t)$ that measures the
distance (in km) at time t from the position at 9 am.

3. Determine the average rates of change on the interval $[1, 5]$ of the
functions $F(x) = x^2$ and $E_2(x) = 2^x$.

4. a) Find the average rate of change of the function $f(x) = x^3$ on the
interval $[0, t]$, where $t > 0$. (Note: the answer depends on t.)

b) Show that for any fixed number c the average rate of change of f
(as in a) on the interval $[c, t]$ for $t \neq c$ can be expressed by a polynomial of
degree 2 in t.

c) Find the polynomial given in b). What is its value at $t = c$?

5. The Department of Fisheries estimates that the population of trouts
in a mountain lake grew from about 50,000 to 80,000 from March 1 to
July 31.

a) Determine the average rate of growth per month of the fish population.

---

[2]The derivative $D[V]$ is calculated by applying the reciprocal rule discussed in Section III.3.2.

b) Determine the average relative rate of growth per month of the fish population. Give the answer in percentage form.

6. A train traveled at an average velocity of 70 mph between 10 a.m. and 12 p.m. Thereafter, because of poor condition of the tracks, the train had to slow down and traveled at an average velocity of only 35 mph between 12 and 1 p.m. Determine the average velocity of the train over the whole trip, i.e., between 10 a.m. and 1 p.m.

7. Suppose $x_0 < x_1 < ... < x_n$ and $A_j = A[x_{j-1}, x_j]$ are as in Corollary ( ). Assume $A_1 > A_2 > ... > A_n$. Prove that, in analogy to the Corollary, one has

$$\begin{array}{ll} (i) & A_1 > A[x_0, x_n] > A_n \qquad \text{and} \\ (ii) & A[x_0.x_j] > A[x_0, x_{j+1}] \text{ for } j = 1, 2, ..., n-1. \end{array}$$

8. Let $f(x) = \frac{1}{2}x^4 - 3x$.

a) Determine the difference between the slope of the tangent line to the graph of f at the point $(2, 2)$ and the slope of the line through the points $(2, 2)$ and $(2.1, f(2.1))$.

b) Determine a precise estimate for the difference between the slope of the tangent line to the graph of f at the point $(2, 2)$ and the slope of the line through the points $(2, 2)$ and $(x, f(x))$ that is valid for *all* x between 1 and 3.

9. a) Determine the instantaneous rate of change df/dt of the function f given by $f(t) = 4\sqrt{t}$ at points $t > 0$. (Don't worry about the existence of $\sqrt{t}$. You can always restrict the inputs to $S(\mathbb{Q}^+)$, where $S(t) = t^2$.)

b) What happens to $df/dt(t)$ in part a) as $t \to 0$?

c) Interpret the result in b) geometrically by looking at the graph of f and its tangents.

IV.3 Approximation of Algebraic Derivatives

IV.3.1 *Intuitive Notion of Instantaneous Rate of Change*

The discussion of derivatives for rational functions in Chapter 3, combined with the results of the preceding two sections, provide a detailed analysis of the critical factor q in the basic factorization $f(x) - f(a) = q(x)(x - a)$. In order to avoid any possible misunderstanding, in this section we shall indicate explicitly that the factor q, which is a function of x, depends also on the fixed point a in the domain of f: Changing the point a requires changing the factor q. We shall write q_a for the factor that appears in

$$f(x) - f(a) = q_a(x)(x - a). \tag{IV.3}$$

In the preceding sections we recognized that the values $q_a(x)$ can be interpreted in a uniform and consistent manner as average rates of change of f between the two points a and x. When the two points coincide we have a special situation, but the same conceptual interpretation applies. Instead of the slope of a secant through two *distinct* points, the value $q_a(a)$, i.e., the derivative of f at a, measures the slope of the secant through the *double point* corresponding to a, that is, the slope of the tangent. Tangents are just special cases of secants, and the basic algebraic technique, that is, the factorization (IV.3), treats the two in a unified way. The single formula for $q_a(x)$ in the case of a rational function suggests that there is a strong bond between the derivative $D[f](a) = q_a(a)$, that is, the slope of the tangent, and the average rate of change $q_a(x) = \Delta f/\Delta x$ for $x \neq a$. This is justified also at an intuitive level. For example, regardless of how we actually define and calculate the instantaneous velocity at a single moment in time t_0, we expect that this value is very close to the average velocity over very short time intervals surrounding t_0. Furthermore, the average velocity should get closer and closer to the velocity at time t_0 as the length of the time intervals shrinks to 0. In other words, the instantaneous velocity at t_0 is approximated by average velocities over shorter and shorter time intervals. Similarly, the figure shown in the previous section clearly suggests that as the input x gets closer and closer to a, i.e., when the point P_x moves towards the point P_a on the graph of f, the secant through P_a and P_x turns towards the position of the tangent at P_a. In other words, the slope of the tangent at P_a is approximated by the average slope between P_a and P_x as x approaches a.

Conceptually, we write:

[average slope between P_a and P_x] $\to$ [slope of tangent at P_a] as $P_x \to P_a$,

or even more briefly,

$$q_a(x) \to q_a(a) \ as \ x \to a.$$

IV.3.2   *The Role of Continuity*

What seems quite obvious to the eye is, in fact, easily justified precisely in case of rational functions. Recall that in Section 1.2 we had discovered this approximation process as a simple consequence of the basic factorization, and we referred to it as the continuity of the function q_a at the point $x = a$.

More precisely, since our function f is rational, we know that the factor q_a is rational as well. One therefore can apply Lemma 21. It follows that there exist an interval $I_\delta(a)$ and a constant K, so that

$$|q_a(x) - q_a(a)| \leq K\,|x - a| \text{ for all } x \in I_\delta(a). \qquad \text{(IV.4)}$$

This is the critical estimate that gives precise meaning to the approximation property captured by the statement that $q_a(x) \to q_a(a)$ as $x \to a$. For example, suppose we want to approximate $q_a(a)$ within 10^{-10}. Formula (IV.4) obviously implies that $|q_a(x) - q_a(a)| < 10^{-10}$ for all $x \in I_\delta(a)$ that satisfy $|x - a| < 10^{-10}/K$. Clearly the same argument works if 10^{-10} is replaced by the much smaller number 10^{-100}, or for that matter, by any arbitrarily small number $\varepsilon > 0$. Just choose $|x - a| < \varepsilon/K$ to ensure that $|q_a(x) - q_a(a)| < \varepsilon$. To summarize:

The closer x is to the point a, the closer the average slope $q_a(x)$ will be to the slope of the tangent $q_a(a)$.

More generally, the preceding discussion establishes the following abstract result.

Theorem 25 *Let a be a point in the domain of the rational function f, and denote by $J_{x,a}$ the interval with endpoints a and $x \neq a$. Assume $J_{x,a}$ is in the domain of f. Then there exist an interval I centered at a and a constant K, such that*

$$|average\ rate\ of\ change\ of\ f\ over\ J_{x,a} - D[f](a)|$$
$$\leq K\,|x - a|\ for\ x \in I\ and\ x \neq a.$$

In a less formal way we can say that the average rate of change of f over small intervals (with one of the endpoints at a) approaches the derivative $D[f](a) = f'(a)$ of f at the point a as the lengths of the intervals go to zero. Symbolically, we may write

$$\frac{\Delta f}{\Delta x} \to D[f](a) = \frac{df}{dx}(a) \text{ as } \Delta x \to 0, \text{ or}$$
$$\lim_{\Delta x \to 0} \frac{\Delta f}{\Delta x} = D[f](a) = \frac{df}{dx}(a),$$

where the notation $\lim_{\Delta x \to 0}$ is chosen to refer to the "limiting process" that occurs as $x \to a$, that is, as $\Delta x = x - a \to 0$.

This fundamental approximation process involving rates of change for rational functions is the precursor of the general concept of "limit" that needs to be considered when one studies functions that are not rational or algebraic.

IV.3.3   *Historical Notes on the Approximation Process*

In essence, we have discovered that the algebraic definition of derivative based on the identification of double points via roots of multiplicity 2 or higher, can be replaced by a new, and more powerful *non-algebraic* **approximation process** that views a double point as the "limiting" position of two *distinct* points that move towards each other until they coincide.

This approximation process is the crux of the new ideas developed by Leibniz and Newton. The history of Calculus and Analysis in the 17th century shows that Descartes' algebraic method involving double points was never fully implemented as we have done in the preceding chapter. In particular, we were able to define tangents and establish all the relevant rules for calculating derivatives of any rational function, all by just using simple algebra. Instead, Leibniz and Newton started directly with the deeper and more powerful approximation process. Given that apparently they were not aware of the elementary algebraic approach and of the explicit estimates that provide a direct motivation for the approximation process, the discovery of the approximation process by Leibniz and Newton to solve the tangent problem is particularly remarkable and a lasting testimony to their creativity.

It turns out that in order to formulate and study this approximation process and the related concept of *limit* precisely in case of general functions, requires going beyond the rational numbers, that is, we shall need to introduce and carefully investigate the critical properties of the much larger set of **real** numbers. In the next section we will take an informal look at the exponential function, so that we can clearly recognize the new problems and mysterious numbers that occur in the general approximation process, that cannot be handled by the simple algebraic techniques that were so successful for rational functions. In the next chapter we shall then study real numbers and the exponential function systematically, culminating in the remarkable solution of the tangent problem for exponential functions. Once this case is well understood, it is then just a simple step to handle the general case of *differentiable functions*.

IV.3.4   *Exercises*

1. Let $c_k = 10^{-k}$ for $k = 1, 2, 3, ...$, and denote by A_k the average rate of change of $f(x) = 2x^3 - 3x$ on the interval $[1, 1 + c_k]$.

    a) Use a calculator to evaluate A_k for $k = 1, 2, 3, 4, 5.6$.

    b) Based on the numerical evidence found in a), what to you think is the value of the instantaneous rate of change of f at the point 1?

c) Verify your guess by calculating $D[f](1)$.

2. Do Exercise 1 with the function $g(x) = \frac{1}{x^2+1}$ instead of f.

3. Let $P(x) = x^4$. Find a number $\delta > 0$, so that the difference between the average rate of change of P on the interval $[2, x]$ (or $[x, 2]$ if $x < 2$) and the instantaneous rate of change of P at 2 is less than 10^{-5} for all x with $|x - 2| < \delta$.

IV.4   A Look Beyond Algebraic Functions

IV.4.1   *Introduction*

The discussion in the preceding sections has covered the differential calculus of rational functions. Only elementary algebraic tools were used, beginning with the basic factorization lemma for polynomials and the related concept of *multiplicity* of zeroes. These tools were then generalized in a natural and systematic way to all rational functions, and we discussed the important concept of composition of functions and the Chain Rule, and took a brief glimpse at taking inverses. No new results and concepts needed to be introduced beyond what is learned in typical high school algebra and geometry courses. In particular, we did not require any advanced concepts such as "limits" or "continuity", and no subtle properties of numbers were used beyond the basic arithmetic properties of the rational numbers, that is, the quotients of integers. You may further have noticed that the formulas and other technical aspects really remained quite simple and natural until we got to the product and quotient rules. While the operations of taking products and quotients of functions are of course natural and useful, the complicated algebraic structure of the corresponding differentiation rules is quite surprising indeed, but it is important not to let these "unnatural" rules obscure the simplicity of the fundamental ideas.

In summary, the central ideas appear already at the very beginning, in the setting of the familiar polynomial functions. The crux of the matter is the (algebraic) factorization

$$f(x) - f(a) = q(x)(x - a),$$

where the factor q is just another function of the same type as the original function f, which in principle can be computed explicitly, and that—most importantly—is well defined also at the point a. The value $q(a)$ is then the derivative $D[f](a)$ of f at the point a. Depending on the setting, $q(a) = D[f](a)$ gives the slope of the tangent line at the point $(a, f(a))$, the instantaneous velocity at time a, or, more generally, it can be viewed as an appropriate instantaneous rate of change at the input value a. In particular, we do want to emphasize that many applications to classical topics in the physical sciences, such as velocity and acceleration, as well as to other areas, can be handled by the algebraic methods we have discussed so far, as long as the functions that are used to describe the underlying phenomena are of algebraic type.

In this chapter we discovered a new phenomenon — *continuity* — that goes beyond standard algebra. While no new tools were required beyond the application of a basic estimate for absolute values, the heart of the matter involved absorbing the important implications of this natural estimate, and recognizing the relevant fundamental property that involves an approximation process. As for derivatives, the application of this new idea opens the door to an alternative method, not limited to algebraic functions, for capturing this central concept. As we shall see, when going beyond algebraic functions, the core applications of this new approach (really quite old, as it goes back to the 17th century, just like the algebraic double point method we have discussed so far) will require an extensive expansion of the foundations.

The reality we have to face is that the rational functions and the algebraic techniques we have discussed so far are much too simple and limited in order to describe many of the fundamental phenomena of the real world. In response to this limitation the human mind, in its quest for deeper understanding, has created amazing new functions and abstract concepts that go well beyond the algebraic tools we have considered so far, and that— at its roots—require a sophisticated extension of the concept of number, resulting in the creation of the so-called *real* numbers that generalize the familiar fractions or *rational* numbers. As we shall see, the *real* story of differential calculus—in contrast to the elementary side discussed in previous chapters—begins when we reach beyond the algebraic functions and enter new uncharted territory.

Among the familiar phenomena that transcend algebraic methods are *periodic events*, such as the revolution of planets around the sun, waves in various media (e.g. sound waves or electromagnetic waves), or the fine structure of electrons circling the nucleus of an atom, and problems related to *growth and decay*, as they arise, for example, in the areas of biology (growth of populations), finance (compound interest), or physics (radioactive decay). The relevant simplest mathematical functions that need to be considered—such as trigonometric, exponential, and logarithm functions— have long been known, but they cannot be captured by finite algebraic formulas, concepts, and techniques. To highlight this fact, these functions and their close "relatives" are usually referred to as the elementary *transcendental* functions.

The more complex nature of these *transcendental* functions shows up clearly as soon as one investigates the tangent problem for these functions. Given the central role of the exponential function in applications, as well as in mathematics, we shall focus on this function and take a preliminary look, so as to recognize the new phenomena and difficulties.

IV.4.2   *Preliminary Introduction to Exponential Functions*

We had already mentioned power functions of the form $f(x) = ax^n$, where n is a positive integer. Here the input variable x is in the *base*. This makes the evaluation of such functions quite easy, since only basic arithmetic operations are involved. Furthermore, as we saw in Chapter II.4, finding the derivatives of such functions just involves elementary algebra. The situation is quite different if the base is kept fixed, and the input variable occurs in the *exponent*. Such functions are called *exponential* functions and they arise in numerous important applications.

A typical example involves the calculation of compound interest in the area of finance. Suppose a bank pays interest on a savings account at the rate of 6% per year, compounded annually. This means that at the end of a year the interest earned during the past year is added to the principal. More precisely, if $A(k)$ is the balance on the account at the end of year k, then $A(k + 1) = A(k) + 0.06 \cdot A(k) = A(k)(1 + 0.06)$ (assuming there have been no other deposits or withdrawals). It follows that after t years ($t = 1, 2, ...$) the value of the account is given by

$$A(t) = Q \left(1 + 0.06\right)^t = Q \cdot 1.06^t,$$

where $Q = A(0)$ is the amount deposited at the beginning, i.e., when $t = 0$. Since *annual* compounding is assumed, only *integer* values for t would seem to matter. Still, it is natural to ask how much the initial deposit would have grown after $1/2$ year, or after one month, that is, when $t = 1/12$, and so on. In particular, daily compounding is often used, that is $t = 1/365, 2/365, ...$ The new phenomenon that arises is that evaluation of powers with rational exponents forces us to go beyond the rational numbers. For example, the rules of powers and exponents require that $2^{1/2} \cdot 2^{1/2} = 2^{1/2+1/2} = 2^1 = 2$, so that one has $2^{1/2} = \sqrt{2}$, and we have seen that there is no *rational* number equal to $\sqrt{2}$. So numerical evaluation of exponential functions, even for simple rational inputs, gets quite complicated and forces us to go beyond the rational numbers. Values such as $2^{\sqrt{2}}$ or 2^π are even more complicated. Explicit numerical calculations with exponential functions usually require the use of a scientific calculator.

We shall now review the basic properties of such exponential functions. In particular, we shall emphasize the importance of the relevant *functional equation* that characterizes them, and show how the definitions for different classes of numbers follow from this equation by applying simple general principles.

Exponential Functions for Integer Inputs   Starting at the beginning, the meaning of the power b^n, where b is some number (rational, or real), for positive integers $n = 1, 2, 3, \dots$ is just a short hand notation for repeated multiplication

$$b^n = b \cdot b \cdot \dots \cdot b, \text{ the factor } b \text{ appearing } n \text{ times.}$$

Examples. $3^4 = 3 \cdot 3 \cdot 3 \cdot 3 = 81$, $(1/2)^2 = (1/2) \cdot (1/2) = 1/4$,

$$\pi^3 = \pi \cdot \pi \cdot \pi \approx (3.14)^3 \approx 30.959.$$

Note that $1^n = 1$ and $0^n = 0$ for all $n \in \mathbb{N}$. From now on we shall only consider the case when the base b is different from 0 and 1, as otherwise there would be nothing interesting to say.

If m, n are two positive integers, the basic definition and a simple counting argument show that

$$b^{m+n} = b^m \cdot b^n. \tag{IV.5}$$

Since $mn = n + n + \dots + n$ (m summands n), it also follows that

$$b^{mn} = b^{n+n+\dots n} = b^n \cdot b^n \cdot \dots b^n \text{ (m factors)}$$

$$= (b^n)^m.$$

Another useful formula that can easily be checked states that $(bc)^n = b^n c^n$. However, since two different bases are involved, this formula will not be so relevant for the discussion that follows.

The basic principle that controls the generalization of b^n to exponents u other than just positive integers is the desire to keep matters simple, that is, to stick to the same familiar rules as much as possible. More concretely, if we consider the function $E_b(u) = b^u$, then the rule (IV.5) states that

$$E_b(u + v) = E_b(u)E_b(v) \tag{IV.6}$$

whenever u and v are positive integers. This is the central *functional equation* for exponential functions. It states an internal law of the function under consideration. The basic principle requires that this internal law, which initially was recognized to hold for positive integers as inputs, remains valid for all numbers u and v.

We shall now step by step extend the domain of $E_b(n) = b^n$ from positive integers to other numbers, always staying "within the law", that is, by observing the functional equation (IV.6).

First we want to define $E_b(0)$. Since $b = b^1 = E_b(1) = E_b(0 + 1)$, we apply the law to get $E_b(0 + 1) = E_b(0)E_b(1) = E_b(0)b$. Since $b \neq 0$, the equation $b = E_b(0)b$ implies that we must define $E_b(0) = 1$.

Next we take a positive integer n, and we try to define $E_b(-n)$. Again, according to the law (IV.6),

$$E_b(-n + n) = E_b(-n)E_b(n),$$

and we also know that $E_b(-n + n) = E_b(0) = 1$. The equation $1 = E_b(-n)E_b(n)$ then implies that $E_b(-n)$ must be the multiplicative inverse of $E_b(n) \neq 0$, i.e.,

$$E_b(-n) = [E_b(n)]^{-1}, \text{ also written } \frac{1}{E_b(n)}.$$

In fact, it is this conclusion that justifies the notation b^{-1} for the reciprocal, that is, for the multiplicative inverse $\frac{1}{b}$. So we see that the law requires that $b^{-n} = \frac{1}{b^n}$ for a positive integer n. We have thus extended the definition of $E_b(m)$ to arbitrary integers $m \in \mathbb{Z}$ in the only way that is consistent with (IV.6). It is a routine easy step to now check that the functional equation (IV.6) indeed remains valid for all $u, v \in \mathbb{Z}$.

Rational Exponents   The next extension involves the definition of $E_b(u)$ for a rational number u. Let us start with the simplest case $u = 1/2$. Since $1 = \frac{1}{2} + \frac{1}{2}$, the law requires that

$$b = E_b(1) = E_b\left(\frac{1}{2} + \frac{1}{2}\right) = E_b\left(\frac{1}{2}\right) \cdot E_b\left(\frac{1}{2}\right) = \left[E_b\left(\frac{1}{2}\right)\right]^2.$$

This shows, first of all, that b must be > 0, and we shall assume this from now on. Furthermore, $E_b(1/2)$ must be a number whose square equals b, in other words, $E_b(1/2) = \sqrt{b}$. This explains the notation $b^{1/2} = \sqrt{b}$ that you may have seen before.

Next, we consider $u = 1/n$, where $n \in \mathbb{N}$. In analogy to the case $n = 2$ just considered, the law now requires that

$$b = E_b(1) = E_b\left(n \cdot \frac{1}{n}\right) = E_b\left(\frac{1}{n} + \frac{1}{n} + \ldots + \frac{1}{n}\right) \quad (n \text{ summands})$$

$$= E_b\left(\frac{1}{n}\right) \cdot E_b\left(\frac{1}{n}\right) \cdot \ldots \cdot E_b\left(\frac{1}{n}\right) = \left[E_b\left(\frac{1}{n}\right)\right]^n.$$

We see that $E_b(1/n)$ must be a number that solves the equation $x^n = b$. As we had already seen in case $b = 2$ and $n = 2$, the equation $x^2 = 2$ has no solution in $\mathbb{Q}$. By using a geometric construction, we had seen that the number line (i.e., the real numbers) contains a unique positive solution denoted by $\sqrt{2}$. More generally, we had seen that for any $b > 0$, there is a point on the number line that is a positive solution of $x^2 = b$, which is denoted $\sqrt{b}$. Once we have introduced the key properties of the real numbers $\mathbb{R}$ in the next chapter, we shall prove that given any $b > 0$ and $n \in \mathbb{N}$, there exists a unique positive number $r \in \mathbb{R}$ (that is, a point on the number line) that satisfies the equation $r^n = b$. For the time being we shall simply assume this result. This number is denoted by $r = \sqrt[n]{b}$, and it is called the *nth root of b*. We therefore see that the "law" requires

$$E_b\left(\frac{1}{n}\right) = \sqrt[n]{b},$$

which, in particular, justifies the notation $b^{1/n}$ for $\sqrt[n]{b}$.

Note that when n is even the equation $x^n = b$ has two real solutions, $\sqrt[n]{b}$ and $-\sqrt[n]{b}$; the definition chosen for $E_b(1/n)$ selects the positive solution $\sqrt[n]{b}$.

The case of an arbitrary rational number $u = m/n$ now follows easily, since the law requires that $E_b(m/n) = b^{\frac{1}{n}m} = (b^{\frac{1}{n}})^m = [E_b(\frac{1}{n})]^m$. A slight modification of this argument shows that one also has $E_b(m/n) = \sqrt[n]{E_b(m)}$. To summarize, we have recognized that in order to "follow the law", we *must* define

$$E_b\left(\frac{m}{n}\right) = b^{\frac{m}{n}} = [\sqrt[n]{b}]^m = \sqrt[n]{b^m} \text{ for any } n \in \mathbb{N} \text{ and } m \in \mathbb{Z}.$$

So, following the law (that is, equation (IV.6)), we have now extended the domain of E_b to all rational numbers, subject to assuming the existence of *nth*.roots of positive numbers in $\mathbb{R}$. It would appear that the resulting function still obeys the law. While one could try to make a legal argument for this based on some higher principles (after all, our actions all stayed within the law), mathematicians prefer to check the validity of the functional equation for rational numbers by more precise arguments. In essence, this involves some routine verifications that can safely be skipped, as no surprises appear. We shall therefore assume from now on that if the base b is positive, then the exponential function $E_b(u) = b^u$ is defined for all rational numbers u, although its output is usually not a rational number — something that we just have to accept for the time being. Most importantly, the functional equation

$$E_b(u + v) = E_b(u)E_b(v)$$

and the related equation

$$E_b(u \cdot v) = E_b(u)^v = (b^u)^v$$

hold for all $u, v \in \mathbb{Q}$.

Furthermore if $b, c > 0$, starting from the familiar $(bc)^m = b^m c^m$ for $m \in \mathbb{N}$, one can easily verify that this formula remains correct if m is replaced by an arbitrary $u \in \mathbb{Q}$. In functional notation, we have $E_{bc} = E_b \cdot E_c$ on $\mathbb{Q}$.

The graphs of exponential functions are most easily obtained by means of a graphing calculator. Note that only rational numbers can be processed by a calculator or computer, and that the outputs are again rational numbers, so in general are only approximations for the exact value. Here we show the graph of E_2.

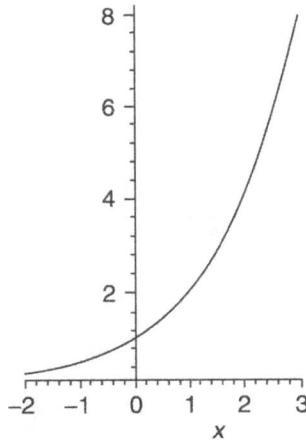

Fig. IV.8   Graph of $y = E_2(x) = 2^x$.

We note that compared to the graphs of polynomials and rational functions, the graph of the exponential function E_2 is very simple — it just looks like a line that has been gently bent in one direction according to a hidden rule. And yet, as we shall see shortly, this graph will reveal amazing new phenomena when one tries to investigate the tangent problem.

IV.4.3   *Tangents for Exponential Functions*

To keep matters simple and concrete, let us consider the exponential function $f = E_2$, that is, $f(x) = 2^x$, that is used to describe a process in which the output doubles whenever the input is increased by one unit. In fact, by the functional equation, f satisfies

$$f(x+1) = 2^{x+1} = 2^x 2^1 = 2f(x) \ \textit{for any } x.$$

It follows that if n is a positive integer, then $f(x+n) = 2^n f(x)$.

We must emphasize that—even though the same operation of "exponentiation" is used—the *exponential* function $f(x) = 2^x$ and the *power* function $y = x^2$, are very different. The latter $y = x^2$ is of *algebraic* type, and its derivative was handled by elementary algebraic methods, while the exponential function $f(x) = 2^x$, as we shall see, forces us to come to grips with amazing new phenomena.

We had already seen the graph of f. Let us now draw a touching line (i.e., a tangent) to the graph at a point P on the graph.

Figure IV.9 certainly suggests that there indeed is a line that fits our intuitive concept of *tangent* line—a line that *touches* the graph but does not *cut* it. Again, as we had seen in Section II.2.2, this geometric feature is made

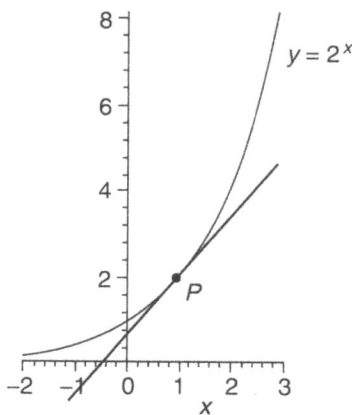

Fig. IV.9   Graph of $y = 2^x$ with a tangent.

precise by observing that small rotations of the tangent reveal that the point of tangency P is indeed a *double* point. In order to investigate the slope of the tangent more in detail, we simplify by choosing $P = (0, 1)$. Proceeding along the familiar path that was so successful in case of polynomials and other algebraic functions, we try to set up the relevant equation and identify the slope for which the equation has a root of multiplicity 2. As usual, consider a line $y = mx + 1$ with slope m through $(0, 1)$. The points of intersection of this line with the graph of f are then the solutions of

$$f(x) = mx + 1, \text{ or } f(x) - f(0) - mx = 0.$$

We therefore look for a factorization

$$f(x) - f(0) = q(x)(x - 0), \text{ i.e., } 2^x - 1 = q(x) \cdot x.$$

This would then lead to the equation $(q(x) - m) \cdot x = 0$. If the slope m equals the value $q(0)$, the equation would then show that the root $x = 0$ has multiplicity greater than 0, although it is not clear at this point whether the multiplicity is actually 2.

Unfortunately, there is no obvious explicitly known factor q defined at $x = 0$ that fits this factorization. In particular, there is no *algebraic* function $q(x)$ that does the job, as that would imply that f itself is algebraic. Furthermore, searching for some explicit expression for q built up from 2^x that would provide an unambiguous natural definition for $q(0)$ turns out to be futile. Of course, as long as $x \neq 0$, the value $q(x)$ is completely determined by the formula

$$q(x) = \frac{2^x - 1}{x} \text{ for } x \neq 0,$$

but this is useless for $x = 0$, since the formula would result in the meaningless expression $0/0$. Hence there is no way to evaluate $q(0)$, which—by analogy to the case of rational functions—would produce the value of the slope of the tangent, i.e., the derivative of $f(x) = 2^x$ at $x = 0$.

We thus see that the (algebraic) factorization method hits a wall, and we are stuck. However, the discussion in the preceding section shows us an alternative method, namely, to try to capture the critical, yet mysterious value $q(0)$ by *approximating* it by $q(x)$ as $x \neq 0$ gets closer and closer to 0. But clearly we are faced with a new problem. In the algebraic case we knew $q(0)$, and hence it was easy to estimate $|q(x) - q(0)|$, justifying the approximation $q(x) \to q(0)$ as $x \to 0$. In the case at hand, q is of course not algebraic, and furthermore, we do not even have any clue for the value $q(0)$. All we have are the values $q(x)$ for $x \neq 0$, and we must determine

if they do approximate some (unknown) number. At least the geometric
version of this idea in the present setting suggests that the missing value
$q(0)$ for the slope of the tangent should be approximated by the slope of
lines through $(0, 1)$ and a second *distinct* nearby point $(x, 2^x)$ on the graph
as $x \neq 0$ approaches 0. (See Fig. IV.10.) In fact, for $x \neq 0$, the slope of
such a line is given precisely by the quotient $q(x)$.

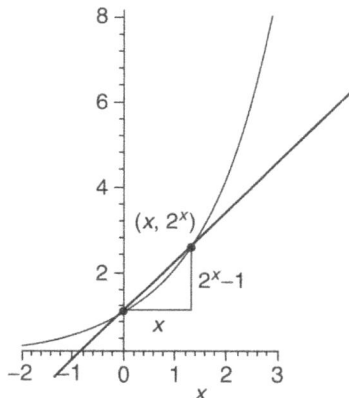

Fig. IV.10   Secant to $y = 2^x$ of slope $q(x) = (2^x - 1)/x$ for $x > 0$.

It certainly looks very plausible that the unknown slope m of the *tangent*
can be approximated by $q(x)$ as the *nonzero* value of x gets closer and
closer to 0. In Fig. IV.10 the line through $(0, 1)$ and $(x, 2^x)$ turns gently in
clockwise direction as $x > 0$ moves closer and closer to 0, so that the point
$(x, 2^x)$ glides towards $(0, 1)$ along the curve that is the graph of $f(x) = 2^x$.
But we do not know the value m for the slope of the tangent, nor do we
even have any obvious guess for it. We are literally shooting in the dark.
Lacking a value for m, there is no way to estimate $|q(x) - m|$ as we were
able to do in case of rational functions. The best we can do is to analyze
the behavior of the average rate of change $q(x)$ as the nonzero value x
approaches 0. Clearly this presents quite a challenge, especially since it is
not easy to calculate the values of $q(x) = (2^x - 1)/x$ for x very small. For
example, what is $2^{1/1000}$, i.e., the 1000th root of 2?

Fortunately, in contrast to the philosophers and mathematicians of the
17th century, we now have powerful computing tools available that make
the *numerical* analysis easy and quick. Of course, this would not solve the
theoretical problem of producing a valid proof, but at least it might give us

some idea of whether the geometrically intuitive argument can be backed up by numerical evidence, and also it might give us an idea of what the final result might look like.

A good programmable scientific calculator would help. A computer that runs one of the powerful computer algebra programs such as *Maple* or *Mathematica* would be even better. In any case, equipped with such tools, we can readily evaluate $q(x)$ for very small non-zero numbers x, and thereby obtain numerical approximations for the elusive slope m. Table IV.1 shows the values $q(x_k)$ for $x_k = 10^{-k}$, $k = 1, 2, ..., 10$, evaluated to ten decimal places.

Table IV.1 Approximation of slope of tangent to 10 digits.

x_k	$q(x_k) = (2^{x_k} - 1)/x_k$
10^{-1}	0.7177346253
10^{-2}	0.6955550056
10^{-3}	0.6933874625
10^{-4}	0.6931712037
10^{-5}	0.6931495828
10^{-6}	0.6931474207
10^{-7}	0.6931472045
10^{-8}	0.6931471829
10^{-9}	0.6931471808
10^{-10}	0.6931471805

It appears that the values $q(x_k)$ approach a number whose decimal expansion begins with 0.69314. Who could have guessed this by just looking at Fig. IV.10? Let us increase the precision by evaluating $q(x_k)$ to 30 digits for $k = 11, ..., 20$. The result is shown in Table IV.2.

Table IV.2 Approximation of slope of tangent to 30 digits.

x_k	$q(x_k) = (2^{x_k} - 1)/x_k$
10^{-11}	0.693147180562347574486828678992
10^{-12}	0.693147180560185535924191277674
10^{-13}	0.693147180559969332067928032084
10^{-14}	0.693147180559947711682301712470
10^{-15}	0.693147180559945549643739080558
10^{-16}	0.693147180559945333439882817368
10^{-17}	0.693147180559945311819497191049
10^{-18}	0.693147180559945309657458628417
10^{-19}	0.693147180559945309441254772154
10^{-20}	0.693147180559945309419634386527

Consistent with the geometric interpretation, the numerical data does provide evidence that the values $q(x)$ approximate some "number", let's call it m_2, as $x \to 0$, that lies between 0.6931471 and 0.6931472, or—more precisely—between 0.69314718055994530 and 0.69314718055994531. However, even though we could narrow the interval that contains m_2 as far as we wish, limited only by the available computing technology, no precise familiar value seems to emerge from this process. For example, no periodicity appears in the decimal expansions displayed above, so it is not clear at all whether m_2 is a rational number.[3] And if m_2 is not rational, what type of "number" is it? Is it some "irrational" number that is the root of a polynomial equation with integer coefficients, analogous to the positive number λ that satisfies $\lambda^2 - 2 = 0$ and which is denoted by $\sqrt{2}$? Or does m_2 even transcend such "algebraic" numbers? We really can not answer these questions at this time.

What is clear, however, is that the tangent problem for the simple natural function $f(x) = 2^x$, with a graph that looks so simple, leads us into new, unknown territory. At the most fundamental level we are not even sure whether our basic concept of number—which includes "irrationals" such as $\sqrt{2}$ beyond the familiar rational numbers—is sufficient to describe the truly complex phenomena that have come to light, and ultimately capture the "correct" value of the slope.

In order to answer some of these questions that are central for an understanding of basic growth phenomena, we need to take a few steps back and first build an appropriate foundation. This foundation should include, in particular, an understanding of the critical properties of the *number system* that we are using, that is, the hidden properties of the number line. Furthermore, we need to develop tools to investigate the *approximation process* that has emerged, first in an elementary and post-facto version in the study of tangents and of instantaneous velocity in the algebraic setting, and now in the far more intriguing form that arises in the study of tangents to the graph of a simple exponential function. In the next chapter we will explore these foundations, and eventually come to the complete solution of the tangent problem for the exponential function, and much more. In contrast to the algebraic methods that we have used so far, things will be quite a bit more complicated, and they will require careful thinking, often involving quite abstract concepts. But we hope that the discussions that

---

[3] Recall that a number is rational if and only if its decimal expansion is finite or periodic. We shall discuss this in detail later on in Section V.2.4.

have led us to this point will help the reader to understand the significance of the problems we try to solve, and furthermore, to realize why new and more sophisticated tools need to be created. Maybe the student will be motivated by realizing that the mathematical abstractions and tools we are going to investigate are among the most amazing creations of the human mind during recent centuries, that have evolved into an indispensable fundamental tool for understanding the world around us.

IV.5 Exercises

1. Let $m = m_2$ denote the elusive number that measures the slope of the tangent to $f(x) = 2^x$ at $(0, 1)$. Show that if the analogous approximation process is worked out at the arbitrary point $(a, 2^a)$ on the graph of f, it leads to the apparent result that the slope of the tangent at this arbitrary point is given by $m_2 2^a$. (Hint: Consider a second point $(a + h, 2^{a+h})$ with $h > 0$ and use $2^{a+h} = 2^a 2^h$ by a basic property of exponentials.)

2. Use a scientific calculator or appropriate computing software to investigate, as in the preceding discussion, numerical approximations to the slope of the tangent to the graph of $g(x) = 10^x$ at the point $(0, 1)$. Try to estimate the first 4 digits of that slope.

Chapter V

The Heart of Real Analysis

V.1 Completeness of the Real Numbers

V.1.1 *Going Beyond the Rational Numbers*

Most of our work so far has used the familiar rational numbers $\mathbb{Q}$, the most elementary extension of the integer numbers $\mathbb{Z}$. While we have occasionally recognized its limitations, for example in the context of finding square roots of positive integers, our focus has been to understand the (algebraic) definition of derivatives and the various rules associated with it, and have deferred dealing with the shortcomings of $\mathbb{Q}$ to a later point. In particular, we simply restricted the domain of the square root function to the subset $\{r^2 : r \in \mathbb{Q}\}$ of $\mathbb{Q}$ in order to avoid having to formally use non-rational numbers. As far as visualizing geometric aspects, this did not cause any problems, since our eyes cannot detect any gaps between the rational numbers on the number line.

However, our somewhat informal look at tangents for the exponential function has revealed that, first of all, the approximation process is going to be central from now on, and secondly, the appearance of mysterious approximating numbers such as 0.69314718055994530094... in what appears to be a very natural problem, force us to look more carefully at the gaps between rational numbers and try to understand the relevant properties of the "complete" number line, that is, how do we make sure that we have covered all the potential gaps?

This is indeed a deep problem, and mathematicians in the 19th century worked very hard in search of satisfactory answers. Several solutions were found, some quite abstract, and others more constructive. For example, the German mathematician Richard Dedekind (1831–1916) defined real numbers in terms of *cuts* (these became known as *Dedekind cuts*), which are an abstract formulation of the most general possible "gap" in the number line. Another definition involved defining numbers in terms of certain sequences of rational numbers (so-called *Cauchy sequences*, after the French mathematician Augustin-Louis Cauchy (1789–1857)), that were known to "converge" abstractly, thus allowing to identify such a sequence with its "limit", and thereby defining a more general type of number abstractly. This particular approach had the added complication that different Cauchy sequences could have the same limit, so one had to allow different "equivalent" sequences to define the same number, sort of analogous to the familiar fact that different fractions can indeed identify the same rational number. Another approach involved using non-terminating decimal expansions. And then there is the formulation involving the existence of the so-called *Least Upper Bound* that we shall use in this book. Fortunately, all these various approaches turned out to be equivalent, in the sense that they all describe the same abstract set (the real numbers $\mathbb{R}$) which is characterized by satisfying certain fundamental rules or axioms. We are already familiar with all but one of these rules from our discussion of rational numbers in Chapter I.1: these are the rules we had labelled Q1 to Q7 in Section I.1.2 that describe the algebraic properties related to addition and multiplication, as well as the rules O1 to O3 governing the order properties discussed in Section I.1.4. An abstract set of elements, in which an addition and multiplication are defined, so that properties Q1–Q7 are satisfied, is known in higher algebra as a *commutative field*. The rational numbers $\mathbb{Q}$ are perhaps the most familiar and natural example of such a field. A somewhat more complicated set is the collection $\mathbb{Q}(x)$ of rational functions that we discussed in Chapter III.1.

If, in addition, there is a notion of order defined among the elements of a field, subject to the rules O1–O3, then we have what is called an *ordered* commutative field. Again, the rational numbers provide the simplest example, while we note that there is no ordering defined in the field of rational functions.

Just as Euclid formulated his famous *axioms* to describe the properties of *lines* and of other basic geometric objects, we now add the requirement that a line—and consequently the "numbers" corresponding to it—satisfies the so-called *completeness axiom*, sometimes also referred to as the *continuity axiom*. A line, as drawn on paper, conveys the intuitive idea of something that can be drawn with a *continuing* stroke of a pen. We idealize by trusting that indeed there are no holes at all in the line. (Note: This really is a major idealization: the concrete physical line and the underlying paper have vast gaps of empty space between the molecules and atoms that make up its matter.) This idealization entails the statement that all holes are completely filled in by points, without any gaps whatsoever. By introducing a ruler, points on the line are related to numbers. The totality of numbers so obtained is the set of all *real numbers*, denoted by $\mathbb{R}$. Every point on the "complete" line corresponds to exactly one *real* number and vice-versa. The rational numbers $\mathbb{Q}$ are just a small proper subset of the set $\mathbb{R}$ of real numbers, though a very important one indeed. On the one hand the rational numbers have a simple representation as fractions built up from integers, on the other hand the rational numbers are densely and evenly distributed along the line. In particular, every real number can be approximated to any desired accuracy by a rational number. More precisely, if $a \in \mathbb{R}$, for each positive integer n one can find a *rational* number s_n, such that the distance between a and s_n is smaller than 10^{-n}, that is, so that $|a - s_n| < 10^{-n}$. We shall discuss this more in detail later on.

V.1.2 *The Completeness Axiom*

We now come to the crux that distinguishes the real numbers $\mathbb{R}$ from the familiar rational numbers $\mathbb{Q}$. We require that our number system satisfies an additional property, generally known as "completeness", that ensures that the number line has no gaps or holes whatsoever. As we shall see, this property has important consequences that will be critical for fully understanding the exponential function and all the more general analysis concepts that go beyond that. As we already mentioned, there are different precise technical formulations of this central *completeness axiom*, depending on the type of abstract "construction" for $\mathbb{R}$ that is emphasized. And fortunately, these formulations are all equivalent, so in the end we deal with just one abstract set $\mathbb{R}$, the so-called *real number field*, that satisfies the rules Q1–Q7, O1–O3, and — most importantly — the completeness axiom. While professional mathematicians must learn about the various constructions of the real numbers that we mentioned above (as well as other constructions), and the various formulations of the completeness axiom, for the beginning student it is best not to complicate matters and to just focus on one particular formulation of this axiom. The version chosen in this book, known as the *Least Upper Bound Property*, is intuitively quite natural, it minimizes technical complications, and it allows to prove all the relevant limit statements without the need to introduce other equivalent formulations.

We shall think of the real numbers $\mathbb{R}$ as *all* the points on the (number) line, that in particular satisfy the rules for order O1–O3. Suppose $S \subset \mathbb{R}$ is a set of numbers. We say that a number M is an *upper bound for S* if $s \leq M$ for every $s \in S$.

The Least Upper Bound (=LUB) Property of $\mathbb{R}$ (A version of the Completeness Axiom). *For every nonempty set $S \subset \mathbb{R}$ that has an upper bound, there exists a number $L_S \in \mathbb{R}$ that has the following properties:*

i) L_S is an upper bound for S.

ii) Any number $c < L_S$ is not an upper bound for S, that is, given $c < L_S$, there exists $s \in S$, such that $c < s$.

Clearly the properties i) and ii) characterize the number L_S as the smallest (or *least*) upper bound for the set S. The situation is visualized on the number line by starting with an upper bound, i.e., a point M to the right of all points in S. We then move the point M to the left as long as it is possible to keep it to the right of S, i.e., we want that the new points $M^\# \leq M$ are still upper bounds for S. Clearly S provides a barrier for this process on the left, and therefore it intuitively looks very plausible that this process must stop at a "smallest" upper bound L_S. As expressed by the *LUB property*, the completeness axiom simply ensures that the real number line indeed matches what our intuition clearly expects. In other words, the completeness property ensures that the process of decreasing the upper bounds as much as possible really comes to an end at a point $L_S \in \mathbb{R}$, rather than at some "hole" in the number line.

We note that the properties i) and ii) imply that there can be **only one** least upper bound L_S for the set S. In fact, if $r \in \mathbb{R}$ is not equal to L_S, then either $r < L_S$, so that r is not upper bound of S according to property ii) for L_S, or $r > L_S$, so that r itself could not be a *least* upper bound for S, since L_S would be an upper bound for S that is smaller than r.

In order to better understand the significance of the LUB property, we shall analyze why the set $\mathbb{Q}$ of rational numbers does **not** have the LUB property. Let us consider the set $A \subset \mathbb{Q}$ defined by

$$A = \{r \in \mathbb{Q} : r > 0 \text{ and } r^2 < 2\}.$$

This set is clearly bounded from above, i.e., it has a an upper bound, for example the number 2 : if $r > 2$, then $r^2 > 4$, so $r \notin A$. We shall now verify that there is NO *rational* number b that satisfies the properties of a "least upper bound" in $\mathbb{Q}$, that is, the rational numbers $\mathbb{Q}$ are not complete in the precise sense described by the completeness axiom. In fact, any positive *rational* number $b \in \mathbb{Q}$ must satisfy $b^2 < 2$ or $b^2 > 2$. (Remember: We can't have $b^2 = 2$ for $b \in \mathbb{Q}$!) One then shows by elementary, though somewhat tedious arguments involving inequalities that the following is true. If $b^2 < 2$, then $(b + r)^2 < 2$ as well for any sufficiently small rational $r > 0$, that is, $b + r$ is a rational number in A that is *larger* than b, so that b is *not an upper bound* for A. And if $b^2 > 2$, then b *is* an upper bound for A, but one also has $(b - r)^2 > 2$ for any sufficiently small rational $r > 0$. Therefore any such number $b - r$ is *smaller* than b and still an upper bound for A, so that b *is not the* **smallest** *upper bound*. We thus have verified that any *rational* b is definitely not a *least upper bound* for A. Hence there is no smallest upper bound for A within the rational numbers. On the other hand, if one considers A as a subset of $\mathbb{R}$, then completeness of $\mathbb{R}$ implies that A has a least upper bound $L_A \in \mathbb{R}$. By essentially the same arguments we just used for a rational b with $b^2 \neq 2$, neither $(L_A)^2 < 2$ nor $(L_A)^2 > 2$ can hold for the least upper bound of A. Therefore, by the order properties, $(L_A)^2 = 2$. In other words, we just verified how the *completeness* of $\mathbb{R}$ implies the "existence" of $\sqrt{2}$ inside the real numbers.

Note that the least upper bound L_S of a set S may or may not be an element of S. For example, the least upper bound L_A of the set A we just considered is not contained in A, which only consists of rational numbers. On the other hand the set $B = \{x \in \mathbb{Q} : x \leq 0\}$ contains its least upper bound 0, which in this case happens to be a rational number!

The LUB property implies that $\mathbb{R}$ also satisfies the analogous *Greatest Lower Bound Property*: If the nonempty set $S \subset \mathbb{R}$ has a *lower* bound l, i.e., if there exists a real number l, so that $l \leq s$ for all $s \in S$, then there exists a (unique) *greatest* lower bound $G_S \in \mathbb{R}$ for S. This means that G_S is a lower bound for S with the property that any number $c > G_S$ is **not** lower bound for S, that is, there exists $s \in S$ so that $s < c$. (See Exercise 14 in Section 7.)

Note that we have not really defined what the real numbers are. Yes, we sort of visualize them as the points on the number line, but that is not a precise definition. As we mentioned earlier, mathematicians have created abstract constructions that allow to precisely define the field of real numbers $\mathbb{R}$ in a variety of ways, but these constructions, and the verification of all the relevant rules, get quite complicated, and not much is really gained for the student who is exposed to all this for the first time. Instead, we take a leap of faith and postulate the existence of a set $\mathbb{R}$ that satisfies all the rules of an ordered commutative field, and, most importantly, also the completeness axiom. Given that much of mathematics deals with abstract formal objects, this non-constructive definition really is not unusual at all.

What ultimately matters are the **rules** that need to be followed. Of course it helps a lot to have a visual image of the abstract set $\mathbb{R}$ before us, namely the number line, and furthermore, to be well familiar with the more concrete structure of the ordered field of rational numbers that satisfies all the rules except the completeness axiom. While adding the completeness axiom has deep consequences, as we shall see, keeping in mind the rational numbers $\mathbb{Q}$, as well as the "complete" number line as a model for $\mathbb{R}$ is most useful.

The fundamental question that needs to be answered is how do we connect the abstract notion of the complete ordered commutative field $\mathbb{R}$ to the number line, or even just to the rational numbers? The answer is obtained by a careful interpretation and application of the basic rules. We start with the additive and multiplicative identities 0 and 1 in $\mathbb{R}$, whose existence is guaranteed by the rules. The order properties imply that $0 < 1$. Adding 1 to both sides gives

$$1 = 0 + 1 < 1 + 1.$$

Repeating the process gives

$$0 + 1 + 1 < 1 + 1 + 1,$$

and so on. Let us define $2 = 1 + 1$, $3 = 2 + 1 = 1 + 1 + 1, ...$, and more generally, for given a natural number n, let us denote by n the sum of n summands 1, that is, $n = 1 + 1 + 1 + ... + 1$ (n summands). The previous statement leads to $n < n + 1$, for each n, so combining all this we get

$$1 < 2 < 3 < ... < n < n + 1 < n + 2 < ...,$$

showing that the process continues without end, thereby identifying a subset $\mathbb{N}^*$ of different elements in $\mathbb{R}$ that behave just like the familiar set $\mathbb{N}$ of natural numbers. Again, what matters are the rules, and given that $\mathbb{R}$, by definition, satisfies the familiar rules, this subset $\mathbb{N}^*$ of $\mathbb{R}$ that we have created behaves just like the familiar natural numbers $\mathbb{N}$ as far as addition, multiplication, and order are concerned. Adding the additive identity 0 of $\mathbb{R}$ to $\mathbb{N}^*$, and then all the additive inverses of the elements in $\mathbb{N}^*$, which exist in $\mathbb{R}$ by the rules of $\mathbb{R}$, one obtains a subset $\mathbb{Z}^*$ of $\mathbb{R}$ that looks and behaves just like the integers $\mathbb{Z}$ (mathematicians say this set $\mathbb{Z}^*$ is *isomorphic* to $\mathbb{Z}$). We thus take the liberty to denote this subset of $\mathbb{R}$ by the same symbol $\mathbb{Z}$, given that as far as the rules are concerned, there is no difference between the familiar $\mathbb{Z}$ and the particular isomorphic subset $\mathbb{Z}^*$ of $\mathbb{R}$ that we just constructed. Since non-zero elements of $\mathbb{R}$ have a multiplicative inverse in $\mathbb{R}$, subject to all the familiar rules, starting from the subset $\mathbb{Z}$, we can construct a subset $\mathbb{Q}^*$ of $\mathbb{R}$ by exactly the same process that we used when we constructed $\mathbb{Q}$ from $\mathbb{Z}$ in Section I.1.2. This set follows exactly all the rules of $\mathbb{Q}$, hence is *isomorphic* to $\mathbb{Q}$, so we simply label it as $\mathbb{Q}$, and call its elements the rational numbers in $\mathbb{R}$. The conclusion is that the rules of $\mathbb{R}$ imply that $\mathbb{R}$ contains subsets that are undistinguishable from the familiar models $\mathbb{N}$, $\mathbb{Z}$, and $\mathbb{Q}$, and hence we denote them by the same symbols. Having reached this point, it is now clear how to establish the connection to the number line. We know how to identify $\mathbb{Q}$ on the number line, and the fact that $\mathbb{Q}$ is dense in $\mathbb{R}$ — a property that we will make precise and prove later on in Section 1.5 by just using the properties of $\mathbb{R}$ — will allow us to conclude that $\mathbb{R}$ can be visualized by the complete number line.

Finally, a remark about notation. Initially, fractions $\frac{a}{b}$ were introduced for $a, b \in \mathbb{Z}$, where $b \neq 0$, as a notation for rational numbers. Recall that $\frac{a}{b}$ is defined by $a \cdot b^{-1}$, where b^{-1} is the multiplicative inverse of the non-zero number b. We later used this same fractional notation, allowing a and b to be rational numbers, or even polynomials, as in the case of the field of rational functions $\mathbb{Q}(x)$. We shall now allow a, b in fractions to be arbitrary *real* numbers, of course preserving the meaning $a/b = a \cdot b^{-1}$. Also, we shall now allow real numbers as coefficients in polynomial and rational expressions and functions, with the analogous notations $\mathbb{R}[x]$ and $\mathbb{R}(x)$. Most importantly, since the results in Chapter II and III only used the algebraic rules Q1–Q7 and the order rules, all these results remain valid if we replace $\mathbb{Q}$ by $\mathbb{R}$.

V.1.3 *Existence of nth Roots*

From now on, whenever we talk about numbers, we shall mean *real* numbers, unless something else is explicitly specified. Of course, the important new property is the fact that set of real numbers $\mathbb{R}$ is **complete**. In particular, every nonempty *bounded* subset S of $\mathbb{R}$ (that is, S has both an upper and a lower bound) has a *least* upper bound, denoted by sup S (supremum of S) and also a *greatest* lower bound, denoted by inf S (infimum of S).

Just as we had seen that completeness implies the existence of a positive real number labeled $b = \sqrt{2}$ that satisfies the equation $b^2 = 2$, one can show the following more general result.

Lemma 26 *For every positive integer n and for any real number $a > 0$, there exists exactly one real number $b > 0$ that satisfies $b^n = a$.*

This number is given by $b = \sup\{x \in \mathbb{R} : x \geq 0 \text{ and } x^n < a\}$. It is called the *nth root* of a, and it is denoted by $b = \sqrt[n]{a}$, or $a^{1/n}$. For $n = 2$ one simply writes $\sqrt[2]{a} = \sqrt{a}$. Note that the symbol $\sqrt{a}$ denotes the *positive* solution of $x^2 = a$; the other (negative) solution of this equation is then the number $-\sqrt{a}$.

The proof of the Lemma follows by essentially the same arguments that we used in the previous section to prove the existence of $\sqrt{a}$ in $\mathbb{R}$. In detail, suppose $x \geq 0$ and $x^n < a$. Then it follows that $x < a + 1$. In fact, if $x \geq (a + 1)$, then $x^n \geq (a + 1)^n > a$. Thus the set $S = \{x \in \mathbb{R} : x \geq 0 \text{ and } x^n < a\}$, which clearly is not empty, has an upper bound, and therefore, by the completeness axiom this set indeed has *Least Upper Bound b*. We want to prove that $b^n = a$. This will follow if we can show that the statement $b^n \neq a$ leads to a contradiction. So let us consider first the case $b^n < a$. Let $r > 0$ and consider $(b + r)^n = b^n + nrb^{n-1} + ...$, by the Binomial Formula, where each of the finitely many remaining terms also has a factor r. Thus $(b + r)^n = b^n + rQ(r)$, where Q is a polynomial. Hence there exists a constant $K > 0$, so that $|Q(r)| \leq K$ for all $0 < r \leq 1$. Let $\varepsilon = a - b^n > 0$. It follows that if $0 < r < \varepsilon/K$, then $(b + r)^n < b^n + (\varepsilon/K)K = b^n + \varepsilon = a$, so that $b + r$ is a number $> b$ that satisfies $(b + r)^n < a$, which would contradict the fact that b is an upper bound for the set S. So we must have $b^n \geq a$. However, if $b^n > a$, an argument similar to the one we just gave shows that $(b - r)^n > a$ for all sufficiently small positive r, so that if $x^n < a$, we would also have $x^n < (b - r)^n$, which implies $x < b - r$, that is, $b - r$ is an upper bound for S. Again, this would contradict that b is the *least* upper bound of the relevant set S. Therefore we must have $b^n = a$. The fact that there is only one such number is trivial, since if b and b^* are two positive numbers that satisfy $b^n = a = (b^*)^n$, we cannot have $b < b^*$, as this would imply $b^n < (b^*)^n$ by the order properties. Similarly, we cannot have $b^* < b$, and therefore we must have $b = b^*$.

Remark. The reader may have noticed that the essence of the argument relates to the continuity of polynomials. In fact, consider the polynomial $P(r) = (b + r)^n$. Suppose $P(0) = b^n \neq a$. Then $\varepsilon = |P(0) - a| > 0$, and by the continuity of P (we can use the basic estimate for $P(r) - P(0)$), one has $|P(r) - P(0)| \leq \varepsilon/2$ for all sufficiently small r, say for all r with $|r|$ less than some positive number δ. For such r it follows that $|P(r) - a| = |(P(0) - a) + (P(r) - P(0))| \geq |(P(0) - a)| - |P(r) - P(0)| > \varepsilon/2$. By the same analysis as above, this ends up contradicting the fact that $b = \sup\{x \in \mathbb{R} : x > 0 \text{ and } x^n < a\}$.

V.1.4   The Archimedian Property

The following result will turn out to be quite useful in many applications.

Lemma 27 *If $c > 0$ is any real positive number, then there exists a natural number n such that $0 < 1/n < c$.*

While this may appear obvious (probably because this is readily checked if c is rational), the *proof* for arbitrary *real* $c > 0$ does in fact involve the completeness axiom. Of course, this result is mainly of interest for very small positive irrational numbers. We shall first verify the following equivalent property, also known as the *Archimedean Property* of the real numbers.

The set $\mathbb{N}$ of natural numbers is NOT bounded in $\mathbb{R}$, that is,

$\mathbb{N}$ *does not have any upper bound.*

Stated differently, the symbol ∞ (= *infinity*), which is commonly used to label "something" that is larger than any natural number, cannot be identified with a real number.

We prove this latter result by contradiction. Assume $\mathbb{N}$ had an upper bound in $\mathbb{R}$. By the LUB property there then exists a real number L that is the *least* upper bound for $\mathbb{N}$. Hence the number $L - 1 < L$ is not an upper bound for $\mathbb{N}$, and therefore there exists $m \in \mathbb{N}$ such that $L - 1 < m$. It then follows by the properties of order that $L < m + 1$. We have thus found a natural number $m + 1 \in \mathbb{N}$ that is *larger* than L, so L could not be an upper bound for $\mathbb{N}$. We end up with a hopeless contradiction. This means that our initial assumption cannot be correct, and therefore $\mathbb{N}$ does not have an upper bound.

Returning to an arbitrary number $c > 0$, the number $1/c$ is also real and positive. As we just saw, there exists a natural number $n > 1/c$. By the order properties, this implies that $0 < 1/n < c$, and we have verified the Lemma. ∎

We note that this Lemma can be used to give a precise formulation for the statement that the "sequence" of numbers $\{1/n : n = 1, 2, 3...\}$ approaches 0, written as $\lim_{n \to \infty} \frac{1}{n} = 0$, namely, given an arbitrarily small number $\varepsilon > 0$, *all but finitely many* of the members of the sequence are $< \varepsilon$. In fact, use the Lemma to find $n_0 \in \mathbb{N}$ with $0 < 1/n_0 < \varepsilon$. Then $n \geq n_0$ implies $1/n \leq 1/n_0$, so that

$$0 < 1/n \leq 1/n_0 < \varepsilon \text{ for \textbf{all} } n \in \mathbb{N} \text{ with } n \geq n_0.$$

There is an apparently slightly more complicated version of this Lemma, as follows.

Given any two positive real numbers ε and L, there exists $n \in \mathbb{N}$, so that $n \cdot \varepsilon > L$.

The proof follows easily from the Lemma. Just note that ε/L is positive as well; by the Lemma, there exists $n \in \mathbb{N}$, so that $1/n < \varepsilon/L$. By applying the multiplicative properties of order, this implies that $1 < \frac{n \cdot \varepsilon}{L}$, and hence $L < n \cdot \varepsilon$.

Again, this is mainly of interest when ε is very small and L very large. This particular version was formulated by Greek geometers and popularized by Archimedes. It was interpreted as stating that any two positive line segments are "comparable", in the sense that sufficiently many copies of a (small) segment add together to exceed any other (long) segment. Since the statement is very easily verified for rational numbers, the fact that Archimedes and other geometers thought to highlight this property suggests that they may have realized something about the complexity of the structure of lines, which much later was formalized by precise definitions of the real numbers.

V.1.5   *Intervals and Related Properties*

We review some standard useful notations that we had already introduced for rational numbers. Given two (real) numbers $a < b$, the *open interval* with *boundary points* a, b equals the set

$$(a, b) = \{\lambda \in \mathbb{R} : a < \lambda < b\}.$$

If one adds the boundary points to (a, b) one obtains the *closed* interval

$$[a, b] = \{\lambda \in \mathbb{R} : a \leq \lambda \leq b\}.$$

In contrast to the earlier discussion in $\mathbb{Q}$, the emphasis here is that intervals now contain **all** *real* numbers between the specified boundary points. Note that both open and closed intervals are clearly bounded from above and below, and that a is the greatest lower bound of both (a, b) and $[a, b]$, and similarly b is the least upper bound of each set. Intervals with *boundary points* a, b are examples of *bounded* sets. More generally, every bounded set of numbers is contained in some bounded interval. Sometimes one considers unbounded intervals such as $\{\lambda \in \mathbb{R} : a < \lambda\}$, which—in analogy to the notation for bounded intervals—we also denote by (a, ∞). As previously noted, ∞ is just a *symbol* and not an element of $\mathbb{R}$; consequently we do *not* call ∞ a *boundary* point of (a, ∞). Correspondingly, the interval $[a, \infty) =$

$\{\lambda \in \mathbb{R} : a \leq \lambda\}$ is a closed interval, as it contains its (only) boundary point a. Similarly, $\mathbb{R}$ itself can be identified with the interval $(-\infty, \infty)$. Note that the interval $(-\infty, \infty)$ is *open* (it does not contain any boundary points), and since there are NO boundary points to include, it is also said to be *closed*. If this sounds strange, think of a door that stands alone, without any frame and wall around it.

The Archimedian Property of the previous section implies the following important relationship between $\mathbb{Q}$ and $\mathbb{R}$ that we had already considered informally earlier on.

Density Theorem. *The rational numbers are everywhere dense in $\mathbb{R}$, meaning that every arbitrarily small non empty open interval (a, b) contains a rational number q.*

By applying this result to the intervals (a, q) and (q, b), and repeating this on and on, one sees that (a, b) in fact contains infinitely many rational points, in other words, the rational numbers are densely spread among the real numbers, so that our (imperfect) eyes may see the real number line completely covered by rational numbers.

To prove this theorem, let us first assume that (a, b) is an interval of length greater than 1, that is, $b - a > 1$. It is then intuitively clear that the interval contains an integer m. But given that our eyes are imperfect, we better verify this fact precisely. Since $\mathbb{N}$ has no upper bound, it readily follows that the integers $\mathbb{Z}$ have no lower bound. Hence we can find $N \in \mathbb{N}$, so that $[a, b] \subset [-N, N]$. Among the finitely many integers from $-N$ to N, let l be the smallest one with $b \leq l$. Then $l - 1 < b$. Since $b - a > 1$ implies that $-1 > -(b-a)$, by adding l to both sides we obtain $l - 1 > l - (b-a) \geq b - (b-a) = a$. Thus $a < l - 1 < b$, i.e., the integer $m = l - 1 \in (a, b)$. For the general case, note that $0 < b - a$. Hence, by the Lemma, there is $n \in \mathbb{N}$, so that $0 < 1/n < b - a$, and by multiplying with the positive number n, one obtains $1 < nb - na$. By the first part of our argument we can find an integer $m \in (na, nb)$, that is $na < m < nb$. Now multiply these inequalities by $1/n = n^{-1} > 0$, resulting in

$$n^{-1}na < n^{-1}m < n^{-1}nb,$$

that is,

$$1 \cdot a < \frac{m}{n} < 1 \cdot b.$$

So the rational number $q = m/n$ is contained in the interval (a, b). ∎

Given a point $a \in \mathbb{R}$, one often needs to identify intervals centered at a that satisfy specific properties. It is convenient to introduce the following notation: given $\delta > 0$, the symbol $I_\delta(a)$ denotes the set $\{x \in \mathbb{R} : |x - a| < \delta\}$, which can also be written in interval notation as $I_\delta(a) = (a - \delta, a + \delta)$. Informally, we shall also say that $U \subset \mathbb{R}$ is a *neighborhood* of a if there exists a positive δ, so that $I_\delta(a) \subset U$. Note that $I_\delta(a)$ is then a neighborhood of any of its points $x \in I_\delta(a)$.

Finally, we discuss another special property of the complete real numbers $\mathbb{R}$ that is often used to prove the existence of specific numbers that are required to satisfy certain properties. For example, suppose we want to find explicit rational approximations for $\sqrt{2}$. We begin by choosing $r_0 = 1$ and $s_0 = 2$, so that $r_0 < \sqrt{2} < s_0$. We then take the midpoint $3/2$ of r_0 and s_0. Since $(3/2)^2 = 9/4 > 2$, we have $\sqrt{2} < 3/2$. Set $r_1 = 1$ and $s_1 = 3/2$, so that $r_1 < \sqrt{2} < s_1$; note that $s_1 - r_1 = 1/2$. By continuing this process we obtain rational numbers r_n and s_n for each $n = 2, 3, ...$, so that $r_n < \sqrt{2} < s_n$ (we cannot have equality since $\sqrt{2}$ is not rational) and $s_n - r_n = 1/2^n$. More in detail, suppose we have found r_{n-1} and s_{n-1} with the desired properties. We then choose the midpoint m_n between r_{n-1} and s_{n-1}. If $m_n < \sqrt{2}$, we set $r_n = m_n$ and $s_n = s_{n-1}$; if $m_n > \sqrt{2}$, we set $r_n = r_{n-1}$ and $s_n = m_n$. In either case we will have $r_n < \sqrt{2} < s_n$, and $s_n - r_n = 1/2(s_{n-1} - r_{n-1}) = 1/2^n$. Each interval $[r_n, s_n]$ is contained in the preceding one. We claim that $\sqrt{2}$ is the only number that is contained in each interval $[r_n, s_n]$. In fact, if $\lambda \neq \sqrt{2}$, we will show that λ is not contained in $[r_n, s_n]$ if n is sufficiently large. Since $c = |\sqrt{2} - \lambda| > 0$, by the Lemma there exists a natural number $n^* > 1$, so that $1/n^* < c$, and therefore one also has $1/2^{n^*} < c$. Since $\sqrt{2} \in [r_{n^*}, s_{n^*}]$, every other number $x \in [r_{n^*}, s_{n^*}]$ satisfies $|\sqrt{2} - x| \leq s_{n^*} - r_{n^*} = 1/2^{n^*} < c = |\sqrt{2} - \lambda|$; it follows that $\lambda \notin [r_{n^*}, s_{n^*}]$, as required.

We generalize this approximation process as follows. Suppose for each $n = 1, 2, 3, ...$ we are given a *closed bounded* interval $[a_n, b_n]$ so that

$$[a_1, b_1] \supseteq [a_2, b_2] \supseteq ... \supseteq [a_n, b_n] \supseteq [a_{n+1}, b_{n+1}] \supseteq ....$$

We call such a sequence a *nested sequence* of closed bounded intervals. The following result states an intuitively obvious property in a precise form. Recall that the symbol $\varnothing$ denotes the "empty set", that is, a set that does not contain any elements at all.

Theorem 28 *If $I_n = [a_n, b_n], n = 1, 2, ...,$ is a nested sequence of closed, bounded intervals in $\mathbb{R}$, then*

$$F = \bigcap_{n=1}^{\infty} [a_n, b_n] \neq \varnothing,$$

i.e., there exists at least one real number $c \in \mathbb{R}$ that is contained in each interval $[a_n, b_n]$.

While this result may appear obvious to you, the situation is not quite so simple. For example, let us take the *open* intervals $I_n = (0, 1/n)$ for $n = 1, 2, 3, ...$, which satisfy $I_n \supset I_{n+1}$ for each n. Since each interval I_n contains only positive numbers, clearly neither 0 nor any *negative* number is contained in $\cap\, I_n$. On the other hand, if c is any *positive* real number, then we know that there is $n^* \in \mathbb{N}$ with $1/n^* < c$, which means that $c \notin (0, 1/n^*)$, and therefore $c \notin \cap\, _{n=1}^{\infty} I_n$. We conclude that

$$\overset{\infty}{\underset{n=1}{\cap}} (0, 1/n) = \varnothing.$$

Similarly, $J_n = [n, \infty)$ for $n = 1, 2, ...$ defines a nested sequence of closed intervals that are NOT bounded. By the Archimedean property it easily follows that $\cap_{n=1}^{\infty} [n, \infty) = \varnothing$. So the hypothesis about the type of intervals in the theorem are indeed essential. More significantly, the theorem is false if we only consider *rational* numbers. For example, recall the rational numbers r_n and s_n introduced in the example before the theorem. Let $I_{n,\mathbb{Q}} = \{x \in \mathbb{Q} : r_n \leq x \leq s_n\}$. Clearly $I_{1,\mathbb{Q}} \supseteq I_{2,\mathbb{Q}} \supseteq I_{3,\mathbb{Q}} \supseteq ...$ is a nested sequence of "non-empty closed bounded intervals of rational numbers". Since $I_{n,\mathbb{Q}} \subset \mathbb{Q}$ by construction, it follows that $\cap_{n=1}^{\infty} I_{n,\mathbb{Q}} \subset \mathbb{Q}$ as well. And since the only number $\lambda \in \mathbb{R}$ that satisfies $r_n \leq \lambda \leq s_n$ for all $n \in \mathbb{N}$ is the number $\sqrt{2}$, and $\sqrt{2}$ is not rational, that is, $\sqrt{2} \notin \mathbb{Q}$, it follows that

$$\overset{\infty}{\underset{n=1}{\cap}} I_{n,\mathbb{Q}} = \varnothing.$$

We see that the validity of the theorem must rest on the completeness of the real number.

To prove the theorem, observe that for any nested sequence of intervals $[a_n, b_n]$, $n = 1, 2, 3, ...$, one must have

$$a_1 \leq a_2 \leq ... \leq a_n \leq a_{n+1} \leq ...b_{n+1} \leq b_n \leq ...b_2 \leq b_1.$$

Therefore the set $A = \{a_1, a_2, ...\}$ of left boundary points is not empty and it is bounded above by any of the right boundary points. It then follows that $\alpha = \sup A \in [a_n, b_n]$ for each n. A more detailed outline of this argument is given in Exercise 16.

To summarize, the *completeness axiom* provides the firm foundation that supports our sometimes faulty and vague geometric intuition, and it ensures that many problems, algebraic or more general, do indeed have solutions within the set of real numbers.

V.1.6   *The Real Numbers are not "Countable"*

There is an additional property of the real numbers $\mathbb{R}$ that is most surprising, and that forced mathematicians to reconsider the meaning of "infinity". While this property is not really relevant for laying the foundations of analysis and calculus, I believe that it ranks among the great discoveries of mankind, just as remarkable as the discovery more than 2000 years ago that $\sqrt{2}$ is not a rational number. And since we have available all the tools necessary to discuss and verify this property, students should learn about it and try to understand it, just as we expect our students to read at least some of the works of the great masters Shakespeare, Goethe, and others, or to absorb and get fascinated by great paintings, such as Leonardo da Vinci's *Mona Lisa* or Pablo Picasso's *Guernica*, to name just a couple of examples.

The essence of this property, discovered by the German mathematician Georg Cantor (1845–1918) in 1874, is the statement that $\mathbb{R}$ exhibits a version of infinity that is much more complex and larger then the familiar notion of infinity that we encounter in the set of natural numbers $\mathbb{N}$.

For most people the collection of natural numbers $\{1, 2, 3, ...\}$ is perhaps their first encounter with an *infinite* set. Note that there is no end to the sequence of natural numbers: once we have reached a particular number, say $n = 10^{1 \; trillion}$, we can always find a larger number, for example $n + 1$. The order properties imply that every time we add 1 to a natural number n, we get a larger number that is different from all the preceding numbers up to n. Thus it is obvious that there is no largest natural number, so the set of natural numbers is not finite. We also saw that even within the larger set of real numbers there is no upper bound for $\mathbb{N}$, which is perhaps not that surprising. Thus we can think of the set $\mathbb{N}$ as the most natural example of an infinite set. It is, in fact, used as the tool to identify more general finite and infinite sets.

More precisely, a set S is said to be *countable* if its elements can be arranged in a suitable order, so that they can be "counted", that is, if one can write $S = \{s_1, s_2, s_3, ...\}$. Clearly every finite set is countable, and the set of natural numbers $\mathbb{N}$ is the prototype of countable sets that are not finite. We can use functions to make this more precise, as follows. First, we say that a set S is *finite* if there exist a natural number N and a function $f : \{1, 2, 3, ..., N\} \to S$ that is one-to-one and onto. It then follows that $S = \{f(1), f(2), ..., f(N)\}$, where all the elements are different from each other, since f is one-to-one. By writing $s_n = f(n)$ for $n = 1, 2, 3, ..., N$, we may then also write $S = \{s_1, s_2, s_3, ..., s_N\}$. Similarly, we say that a set S is *countably infinite* if there exists a function $f : \mathbb{N} \to S$ that is one-to-one and onto. Again, if we write $s_n = f(n)$ for each $n \in \mathbb{N}$, we see that we can write $S = \{s_1, s_2, s_3, ...\}$. Finally, we say that a set S is countable, if S is either finite or countably infinite.

Of course $\mathbb{N}$ itself is countable according to this definition. Just take the identity function $f : \mathbb{N} \to \mathbb{N}$ defined by $f(n) = n$. Let us look at a few other examples. The set $E = \{2, 4, 6, 8, ...\}$ of even numbers is surely not finite, and since it is already arranged in a sequence, it appears to be countable. Using the function version, we can take $f : \mathbb{N} \to E$ defined by $f(n) = 2n$, which is clearly one-to-one and onto. Similar arguments can be used to show that any subset of a countable set is itself countable. Next, we consider the set of integers $\mathbb{Z}$, which is much larger than $\mathbb{N}$. By rearranging $\mathbb{Z}$, we can write $\mathbb{Z} = \{0, 1, -1, 2, -2, 3, -3, ...\}$, which shows that the set of integers is countable as well. The relevant function $f : \mathbb{N} \to \mathbb{Z}$ is defined by $f(2n - 1) = -n + 1$ for odd numbers, and $f(2n) = n$ for even numbers. It is a bit more tricky to rearrange the set $\mathbb{Q}$ of rational numbers so that one recognizes that $\mathbb{Q}$ is also countable, and we leave the details for Exercise 17.

Cantor also proved that the set $\mathbb{A}$ of *algebraic* numbers is countable, and we want to explain the proof. Recall that a real number r is said to be *algebraic* if it is a root (or zero) of a polynomial of degree $n \geq 1$

$$P = a_n x^n + a_{n-1} x^{n-1} + ... + a_1 x + a_0,$$

that is $P(r) = 0$, where P has integer coefficients $a_0, a_1, ..., a_n$, with $a_n \neq 0$. Cantor's key idea was to introduce the *height* $h(P)$ of such a polynomial by

$$h(P) = n + |a_n| + |a_{n-1}| + ... + |a_1| + |a_0|.$$

We note that the height is a natural number. Since P has degree ≥ 1, the lowest height equals 2, for the polynomial $P = 1 \cdot x$ or $-P$. Recall that every rational number m/n is the root of $P(x) = nx - m$, with $h(P) = 1 + |n| + |m|$. Since the coefficients are required to be integers, for a given height $d \geq 2$ there are only finitely many different polynomials with height d. Each such polynomial has degree $n \leq d - 1$, and hence has at most $n \leq d - 1$ roots. Therefore, for each height d there are *finitely* many algebraic numbers. Starting with $d = 2$, we count the corresponding algebraic numbers, then move to $d = 3$ and continue the count by adding the new corresponding algebraic numbers, and so on. While there will be repetitions (a particular algebraic number will usually be the root of several different polynomials, and this has to be taken into consideration when describing the relevant function $f : \mathbb{N} \to \mathbb{A}$, it is now clear that $\mathbb{A}$ is countable. Since $\mathbb{Q}$ is a subset of $\mathbb{A}$, this argument, in particular, provides an alternate proof of the countability of $\mathbb{Q}$.

After having absorbed these examples, one may wonder if perhaps *all* sets are countable. Cantor's amazing discovery shows that this is not true. Here is the precise formulation of Cantor's result.

Theorem 29 *(G. Cantor). The set $\mathbb{R}$ of real numbers is NOT countable.*

Since $\mathbb{R}$ is clearly infinite, its type of infinity must be of a very different nature than the countable infinity of $\mathbb{N}$. More precisely, Cantor's Theorem states that there does not exist a one-to-one and onto function $f : \mathbb{N} \to \mathbb{R}$.

Proof. We are going to prove that any one-to-one function $f : \mathbb{N} \to \mathbb{R}$ cannot be onto, which implies the statement in the Theorem. So suppose we have a function $f : \mathbb{N} \to \mathbb{R}$ that is one-to-one. We want to show that the image $f(\mathbb{N}) \subset \mathbb{R}$ is definitely NOT EQUAL to $\mathbb{R}$. To prove this, we must find at least one real number r that is not in $f(\mathbb{N})$. Let us denote the real number $f(n)$ by c_n for each $n \in \mathbb{N}$. So we can write $f(\mathbb{N}) = \{c_1, c_2, c_3, ...\}$, and since f is one-to-one, $c_n \neq c_m$ for all $n, m \in \mathbb{N}$ with $n \neq m$. We shall now construct a nested sequence of non empty closed bounded intervals I_l, $l = 1, 2, 3, ...$, as follows. Choose any interval $I_1 = [a_1, b_1]$ with $a_1 < b_1$, so that $c_1 \notin I_1$. Next, choose $I_2 = [a_2, b_2] \subset I_1$ with $a_2 < b_2$, so that $c_2 \notin I_2$. This is easily done as follows. If $c_2 \notin I_1$, just choose $I_2 = I_1$. So assume $c_2 \in [a_1, b_1]$; if $c_2 = a_1$, let a_2 be any point with $a_1 < a_2 < b_1$ and take $b_2 = b_1$; and if $a_1 < c_2 \leq b_1$, choose $a_2 = a_1$ and b_2 so that $a_2 = a_1 < b_2 < c_2$. In any case, we have thus found an interval $I_2 = [a_2, b_2] \subset I_1$ with $c_2 \notin I_2$ and $a_2 < b_2$. By the same process we can now find an interval $I_3 = [a_3, b_3] \subset I_2$ with $a_3 < b_3$, so that $c_3 \notin I_3$, and so on. This process can be continued, provided the interval $I_l = [a_l, b_l]$ is always chosen with $a_l < b_l$, which definitely is the case in the construction process that we described. By the Nested Interval Theorem, there exists a real number $r \in \cap_{l=1}^{\infty} [a_l, b_l]$. We claim that such a number r is not an element of $f(\mathbb{N})$, that is, $r \neq c_n$ for any n. In fact, the construction of the nested sequence $\{I_l\}$ was done in such a way that for each n, the number $c_n \notin I_n$, and consequently $c_n \notin \cap_{l=1}^{\infty} [a_l, b_l]$. So $r \neq c_n$, and we have proved that f is not onto.[1] ∎

---

[1] The author learned of this proof, due to Cantor, from the text of Bartle and Sherbert (op. cit.). It differs from another proof given by Cantor, which is more widely known, and

V.1.7 *Exercises*

Solve the inequalities in problems 1 through 5. Write each solution set as an interval.

1. $1 - 6x > 2$
2. $3 + 4x < 1$
3. $-6 < 5 - 2x < 2$
4. $|x + 5| \leq 2$
5. $|5x - 4| < 4$

Simplify the following expressions by eliminating the absolute value sign.

6. $|(-3)(5 - 9)|$
7. $|(-2)^3|$
8. $-|2 - 5|$
9. $|(-1)^{2n}|$, where n is a positive integer.
10. a) Use the fact that $(-1)a = -a$ to verify that the product of two negative numbers is positive.

b) Explain by using a) why there is no *real* solution of the equation $x^2 = a$ for $a < 0$.

11. Suppose $p \in \mathbb{R}$ and $p \geq 0$. Show that if $p \leq \frac{1}{n}$ for each $n \in \mathbb{N}$, then $p = 0$.

12. Modify the argument used to show that $\sqrt{2}$ is not rational to show that $\sqrt{3}$ is not rational either. More generally, show that if an integer $p > 0$ is not a perfect square (i.e., if p is not equal to m^2 for some integer m), then $\sqrt{p}$ is not rational.

13. a) Suppose $r \in \mathbb{Q}$ and set $A(r) = \{a \in \mathbb{Q} : a < r\}$. Prove that $\sup A(r) = r$.

b) Suppose $A \subset \mathbb{Q}$ is bounded above. Does it follow that $\sup A \in \mathbb{Q}$? Give proof or counterexample.

14. In analogy to the LUB property, one can define the Greatest Lower Bound Property of $\mathbb{R}$ as follows: A set S of numbers is bounded from below if there is a number $l \in \mathbb{R}$ so that $l \leq s$ for all $s \in S$. Such l is called a lower bound for S. A number G_S is called a **Greatest Lower Bound** for S if G_S is a lower bound for S, and any number $c > G_S$ is *not* a lower bound. Verify that the real numbers $\mathbb{R}$ satisfy the **Greatest Lower Bound property**, that is, *each nonempty set in $\mathbb{R}$ that is bounded from below has a unique Greatest Lower Bound.* (Hint: If l is a lower bound for S, then $(-l)$ is an upper bound for the set $S^* = \{s : -s \in S\}$.)

---

that was based on what has become known as "Cantor's Diagonal Sequence" argument, which uses decimal expansions of real numbers.

15. Find the least upper bound and greatest lower bound for the set $S = \{1 - \frac{1}{n} : n = 1, 2, 3, ...\}$.

16. Suppose $I_n \subset \mathbb{R}$ is a closed bounded interval $[a_n, b_n]$ for $n = 1, 2, ...$ so that $I_1 \supseteq I_2 \supseteq ... I_n \supseteq I_{n+1} \supseteq ....$ Show that $\bigcap_n I_n \neq \emptyset$ by completing the following steps.

a) Let $A = \{a_n : n = 1, 2, ...\}$. Show that each right end point b_n is an upper bound for A.

b) Explain why $\sup A \leq b_n$ for each $n \in \mathbb{N}$.

c) Show that b) implies that $\sup A \in [a_n, b_n]$ for each $n \in \mathbb{N}$, and consequently $\sup A \in \bigcap_{n=1}^{\infty} I_n$.

d) More generally, prove that the closed interval $[\sup A, \inf B] \subset \bigcap_{n=1}^{\infty} I_n$, where $B = \{b_n : n = 1, 2, ...\}$ is the set of right end points.

17. Consider the set $\mathbb{Q}^+ = \{\frac{m}{n} : m, n \text{ positive integers}\}$. Arrange $\mathbb{Q}^+$ in the following pattern.

line 1: $\quad \dfrac{1}{1}, \dfrac{2}{1}, \dfrac{3}{1}, ..., \dfrac{m}{1}, ...$

line 2: $\quad \dfrac{1}{2}, \dfrac{2}{2}, \dfrac{3}{2}, ..., \dfrac{m}{2}, ...$

line 3: $\quad \dfrac{1}{3}, \dfrac{2}{3}, \dfrac{3}{3}, ..., \dfrac{m}{3}, ...$

....

....

line n: $\quad \dfrac{1}{n}, \dfrac{2}{n}, \dfrac{3}{n}, ..., \dfrac{m}{n}, ...$

....

a) Use this pattern to show that $\mathbb{Q}^+$ is countable. (Hint: Start "counting" in the upper left corner, then take 1/2 and next 2/1, and continue by moving along the diagonals parallel to the first one, skipping any number that has already been covered, and so on.)

b) Show that this implies that $\mathbb{Q}$ is also countable. (Hint: Look at how we saw that $\mathbb{Z}$ is countable.)

V.2   Limits and Continuity

Ever since we established the key estimate $|P(x) - P(a)| \leq K|x - a|$ for polynomials, which led to our first introduction to the concept of continuity of a function, as well as to approximation and limits, we have seen the increasing importance of these new ideas. First of all, they opened a new door to capture derivatives via approximation by average rates of change, a much deeper process than the elementary identification of double roots. Our preliminary investigation of the exponential functions in Section IV.4 revealed that this process is going to be critical in the study of exponential functions, where the algebraic methods were stuck in front of a closed door. And as we shall see in the next section, these ideas are central to extending the domain of exponential functions to all real numbers, allowing us to eventually capture some amazing numbers that appear naturally in the study of tangents for exponential functions. Thus it is time to put these ideas on a solid footing and to establish rules and criteria to prove relevant statements. On the other hand, we shall continue to highlight the intuitive interpretations, and this will be quite sufficient for many readers. Essentially, this is what happened historically. It was only in the 19th century that mathematicians came up with a precise definition of the intuitive concept of limit that had been used informally for well over a century. So we encourage the reader not to dwell too much on the abstract proofs on first reading, instead to continue learning about the amazing story that mathematicians created. However, it is important to make the theoretical back-up available as well, so that in case of any doubts, the reader will have available the tools to resolve any questions, as needed.

V.2.1 *Limits of Sequences*

We shall first consider limits of *sequences* of real numbers. A sequence is a very basic concept. We have already used this, for example, when we considered certain sequences $\{r_n\}$ of rational numbers. Also, recall that a set is countable, if it can be arranged in a sequence. The most basic sequence is just the set of natural numbers $\mathbb{N}$, that often is written as $\mathbb{N} = \{1, 2, 3, 4, ...\}$, which exhibits the numbers as a "sequence" placed in their (natural) order. Starting from this basic sequence, a sequence of real numbers is just a function $f : \mathbb{N} \to \mathbb{R}$. If we introduce $s_n = f(n)$, we can write the sequence, i.e., the function, in the form $\{s_1, s_2, s_3, ...\}$, that is, the function is just represented by a (here infinite) table of its values. Note that we used this same concept when we discussed countable sets in Section 1.7.

Here is the formal definition that makes precise the intuitive statement $s_n \to L$ as n gets larger and larger.

Definition 30 *The sequence $\{s_n : n = 1, 2, 3, ...\}$ of real numbers converges to the limit $L \in \mathbb{R}$, written as $\lim_{n\to\infty} s_n = L$, if for every positive number*

ϵ *there exists a natural number $N(\epsilon)$ that depends on ϵ, so that*

$$|s_n - L| < \epsilon \text{ for all } n \geq N(\epsilon).$$

Alternatively, one may say that for any $\epsilon > 0$ at most finitely many terms of the sequence are at distance $\geq \epsilon$ from L.

The reader should carefully ponder this definition and understand its basic structure: Given a certain allowance in the difference from the limit, as specified by ϵ, one must be able to stay within that allowance for **all** sufficiently large n, in other words, for all terms of the sequence except for finitely many.

Example. It is intuitively obvious that the sequence $\{1/n\}$ converges to 0. Still, we shall prove formally that $\lim_{n \to \infty} \frac{1}{n} = 0$ according to the definition just given. Let $\epsilon > 0$ be given. By the Archimedian Property, there exists a natural number $N = N(\epsilon)$, so that $1/N < \epsilon$. For $n \geq N$ one then has $1/n \leq 1/N < \epsilon$, so that $\left|\frac{1}{n} - 0\right| = \frac{1}{n} < \epsilon$. ∎

The first fact that needs to be verified is the following.

Lemma 31 *If a sequence converges to a limit L, then that limit is determined uniquely.*

Proof. Suppose that $\{s_n\}$ is a sequence that converges to two numbers L_1 and L_2. We shall prove that for any $\epsilon > 0$ one has $|L_1 - L_2| < \epsilon$, which of course implies $|L_1 - L_2| = 0$, that is, $L_1 = L_2$. So, given such an $\epsilon > 0$, by the above definition there exist $N_1(\epsilon/2)$ and $N_2(\epsilon/2)$, so that if $n \geq N_j$, then $|s_n - L_j| < \frac{\epsilon}{2}$ for $j = 1, 2$. Let $N = \max\{N_1, N_2\}$. Pick any number $n \geq N$; one then has

$$|L_1 - L_2| = |(L_1 - s_n) - (L_2 - s_n)| \leq |L_1 - s_n| + |L_2 - s_n| < \frac{\epsilon}{2} + \frac{\epsilon}{2} = \epsilon. ∎$$

Furthermore, we also have the following obvious fact.

Lemma 32 *A convergent sequence is bounded.*

In fact, if $L = \lim_{n \to \infty} s_n$, then all but finitely many terms s_n of the sequence lie in the bounded interval $(L - 1, L + 1)$.

Next, we have the following simple, but very important condition to verify convergence of a sequence.

Theorem 33 *Suppose the sequence $\{s_n\}$ is increasing, that is, $s_n \leq s_{n+1}$ for all n, and bounded from above. Then this sequence converges, and its limit is $\sup\{s_n\}$. An analogous statement holds for decreasing sequences that are bounded from below, in which case the limit is $\inf\{s_n\}$.*

Proof. By the completeness axiom, the set $\{s_n : n \in \mathbb{N}\}$ has a Least Upper Bound L. Suppose $\epsilon > 0$ is given. Then $L - \epsilon$ is not an upper bound for this set, i.e., there exists a term $s_{N(\epsilon)}$ with $L - \epsilon < s_{N(\epsilon)}$. Since the sequence is increasing, and L is, in particular, an upper bound for it, we have

$$L - \epsilon < s_{N(\epsilon)} \le s_n \le L \text{ for all } n \ge N(\epsilon).$$

This implies that

$$|L - s_n| < \epsilon \text{ for all } n \ge N(\epsilon). \ \blacksquare$$

The proof of the result in case of decreasing sequences uses analogous arguments, and is left as an exercise.

Remark. This last result is generally known as the *Monotone Convergence Theorem* for sequences. The term monotone includes both the case of an *increasing* as well as of a *decreasing* sequence.

Finally, here is a basic and most useful class of convergent sequences.

Lemma 34 *Suppose r is a real number with $|r| < 1$. Then $\lim_{n\to\infty} r^n = 0$.*

Proof. Let us first assume that $0 \le r < 1$. We obviously have $r^{n+1} = rr^n \le r^n$, so the sequence $\{r^n\}$ is decreasing. By the Monotone Convergence Theorem this sequence converges, with

$$\lim_{n\to\infty} r^n = q = \inf\{r^n : n = 1, 2, 3, ...\} \ge 0.$$

Since

$$q = \lim r^n = \lim r^{n+1} = \lim rr^n = r \lim r^n = rq,$$

(use ii) in the Lemma below), if $q > 0$, one could divide the equation $q = rq$ by q, resulting in $1 = r$, which contradicts the fact that $r < 1$. Therefore one must have $q = 0$. Finally, in the general case $|r| < 1$, the result we just proved implies that $\lim_{n\to\infty} |r|^n = 0$; simply stating the relevant precise statement for the existence of this limit shows that $\lim_{n\to\infty} r^n = 0$ as well. $\blacksquare$

V.2.2 Natural Properties of Limits

Convergent sequences enjoy several intuitively reasonable properties, which we collect in the following Lemma.

Lemma 35 *Suppose we have two convergent sequences $\{r_n\}$ and $\{s_n\}$. Then*

i) $\lim_{n\to\infty}(r_n \pm s_n) = \lim_{n\to\infty} r_n \pm \lim_{n\to\infty} s_n$;

ii) $\lim_{n\to\infty}(r_n \cdot s_n) = (\lim_{n\to\infty} r_n) \cdot (\lim_{n\to\infty} s_n)$; in particular, if $a \in \mathbb{R}$, then $\lim_{n\to\infty}(a \cdot s_n) = a \cdot (\lim_{n\to\infty} s_n)$;

iii) *if* $\lim_{n\to\infty} s_n \neq 0$, *then also*

$$\lim_{n\to\infty} \frac{r_n}{s_n} = \frac{\lim_{n\to\infty} r_n}{\lim_{n\to\infty} s_n};$$

iv) *if the sequence* $\{c_n\}$ *satisfies* $r_n \leq c_n \leq s_n$ *for all* n, *and* $\lim_{n\to\infty} r_n = \lim_{n\to\infty} s_n = L$, *then* $\{c_n\}$ *also converges, and* $\lim_{n\to\infty} c_n = L$.

For obvious reason the last statement is known as the *Squeeze Theorem*. We also note that by giving an expression for the (presumed) limit of a sequence, it is implied that this sequence converges, and that its limit has the stated value.

Proof. We leave i) and iv) as Exercises, prove ii) in detail, and prove the special case $1/s_n \to 1/L_2$ of iii). The general case then follows by combining this with ii). Suppose $\lim_{n\to\infty} r_n = L_1$ and $\lim_{n\to\infty} s_n = L_2$. Given $\epsilon > 0$, we must show that $|L_1 L_2 - r_n s_n| < \epsilon$ for all sufficiently large n. Since by hypothesis we do control, in particular, $L_1 - r_n$, we introduce a term $-r_n L_2$ to control $L_1 L_2 - r_n L_2$, and then we add it back again, so that the total value does not change. In detail, we have

$$|L_1 L_2 - r_n s_n| = |L_1 L_2 - r_n L_2 + r_n L_2 - r_n s_n|$$
$$= |(L_1 - r_n)L_2 + r_n(L_2 - s_n)| \leq |L_1 - r_n|\,|L_2| + |r_n|\,|L_2 - s_n|.$$

It is now clear how to proceed. For the first term, choose $N_1(\epsilon)$ so that for all $n \geq N_1(\epsilon)$ one has $|L_1 - r_n| < \frac{\epsilon}{2}\frac{1}{1+|L_2|}$. For the second term, since the convergent sequence $\{r_n\}$ is bounded, there is a constant K, so that $|r_n| < K$ for all n. We then choose $N_2(\epsilon) \geq N_1(\varepsilon)$ so that for all $n \geq N_2(\epsilon)$ one has $|L_2 - s_n| < \frac{\epsilon}{2K}$. It then follows that for all $n \geq N_2(\epsilon)$ one has

$$|L_1 L_2 - r_n s_n| < \frac{\epsilon}{2}\frac{|L_2|}{1+|L_2|} + K\frac{\epsilon}{2K} \leq \frac{\epsilon}{2} + \frac{\epsilon}{2} = \epsilon,$$

which proves ii). As for iii), we note that the extra condition $\lim_{n\to\infty} s_n = L_2 \neq 0$ is required, so that the quotient of the limits is defined. As for the quotients $1/s_n$, there could be problems, since some members of the sequence $\{s_n\}$ could be zero. However, this will not happen if n is sufficiently large. More precisely, since $L_2 \neq 0$, we have $|L_2| > 0$; therefore, there exists $N_2(|L_2|)$ so that $|L_2 - s_n| < |L_2|/2$ for all $n \geq N_2(|L_2|)$; for such n we then have

$$|s_n| = |L_2 + (s_n - L_2)| \geq |L_2| - |L_2 - s_n| > |L_2|/2,$$

and hence $s_n \neq 0$. Therefore the sequence $\{1/s_n\}$ is well defined for all such large n, and these are the only members of the sequence that need to be considered. In particular, the above estimate implies that $1/|s_n| < 2/|L_2|$. For such large n we then have

$$\left| \frac{1}{s_n} - \frac{1}{L_2} \right| = \left| \frac{L_2 - s_n}{s_n L_2} \right| < \frac{2}{|L_2|^2} |s_n - L_2|.$$

It is now clear how to proceed, as we are familiar with this sort of estimate. Given any $\varepsilon > 0$, we choose $N'(\varepsilon) \geq N_2(|L_2|)$ so that for all $n \geq N'(\varepsilon)$ one has

$$|s_n - L_2| < (|L_2|^2 / 2) \cdot \varepsilon.$$

The preceding estimate then shows that

$$\left| \frac{1}{s_n} - \frac{1}{L_2} \right| < \varepsilon \text{ for all } n \geq N'(\varepsilon).$$

The proof of $iii)$ then proceeds by essentially algebraic arguments/ estimates, comparable to how we handled $ii)$, although somewhat messier, as we deal with fractions now. Alternatively, use the result we just proved and combine it with $ii)$. ∎

We already know that the rational numbers are dense in $\mathbb{R}$, in the sense that any interval $I_\delta(x)$, no matter how small, contains a rational number, and consequently actually infinitely many such numbers. We shall now give a more precise version, as follows.

Lemma 36 *Let $x \in \mathbb{R}$. Then there exists a sequence of rational numbers $\{r_n : n = 1, 2, 3, ...\}$ with the following properties:*

i) $\{r_n\}$ is strictly increasing and bounded above by x that is, $r_n < r_{n+1} < x$ for all $n \in \mathbb{N}$;

ii) $\lim_{n \to \infty} r_n = x$.

Proof. Given $x \in \mathbb{R}$, by the Density Theorem in Section 1.5, we can find a rational number r_1 in the interval $(x-1, x)$. Let $s_1 = \max\{r_1, x - \frac{1}{2}\}$. Then $x - 1/2 \leq s_1 < x$, and we can find a rational number $r_2 \in (s_1, x)$. Then $s_1 < r_2 < x$, and $0 < x - r_2 < x - s_1 \leq 1/2$. We now repeat this process to obtain rational numbers r_n with $r_{n-1} < r_n < x$ and $x - r_n < 1/n$, for $n = 3, 4, 5, ...$. Just to be sure that this indeed works, let us verify this precisely by induction. We just found r_2 with these properties. Suppose we have found r_n for some $n \geq 2$ satisfying these properties, and let us find r_{n+1} as follows. Let $s_n = \max\{r_n, x - \frac{1}{n+1}\}$. Then $s_n < x$, and we can find a rational number $r_{n+1} \in (s_n, x)$. Since $r_n \leq s_n$, we have $r_n < r_{n+1} < x$, and since $x - \frac{1}{n+1} \leq s_n$, we also have $x - \frac{1}{n+1} < r_{n+1} < x$, which implies $x - r_{n+1} < \frac{1}{n+1}$. So indeed r_{n+1} satisfies the required properties.

We claim that this sequence $\{r_n\}$ does the job. In fact, it surely satisfies *i*). To prove *ii*), let $\epsilon > 0$ be given. By the Archimedian Property there is $N(\epsilon)$ with $1/N(\epsilon) < \epsilon$. Then for $n \geq N(\epsilon)$ one has $1/n \leq 1/N(\epsilon) < \epsilon$ as well, and hence, $x - r_n < 1/n < \epsilon$, as claimed. ■

We note that by appropriate modifications of the arguments one can also find a *decreasing* sequence $\{d_n\} \subset \mathbb{Q}$ with $x < d_{n+1} < d_n$ for all $n \in \mathbb{N}$, so that $\lim_{n \to \infty} d_n = x$.

V.2.3   *The Basics of Infinite Series*

Now that we have a good idea of sequences and a precise definition for what it means for such a sequence to converge to a limit, we can introduce *infinite series*, a special kind of sequence that arises quite often in higher mathematics.

For us, at this point, such infinite series will be used to complete the discussion of the decimal expansion of real numbers that we began in Section I.1.5. Recall the meaning of, for example, 82.32756.

$$82.32756 = 8 \cdot 10^1 + 2 \cdot 10^0 + 3 \cdot 10^{-1} + 2 \cdot 10^{-2} + 7 \cdot 10^{-3} + 5 \cdot 10^{-4} + 6 \cdot 10^{-5}.$$

We had verified that every rational number whose denominator only contains the prime factors 2 and 5 can be written as such a *finite* decimal expansion. On the other hand, most rational numbers do not fall in that category, and their situation is more complicated. In fact, we had already seen new phenomena with the simple number 1/3. Recall that we had verified that for every $n \in \mathbb{N}$ one has

$$\frac{1}{3} = 0.33...3 + \frac{1}{3}10^{-n}$$, where there are n digits 3 to the right of 0.

Since clearly $\lim_{n \to \infty} \frac{1}{3}10^{-n} = 0$, this shows that $\frac{1}{3} = \lim_{n \to \infty} 0.3333...3$ (n digits 3), which we write in the form

$$\frac{1}{3} = \lim_{n \to \infty} \sum_{j=1}^{n} \frac{3}{10^j}.$$

It is tempting to write the right side as 0.333..., suggesting a non-terminating sequence of digits 3. Let us introduce the notation $0.\overline{3}$ as a shorthand notation for such a (non-terminating) decimal expansion where the number 3 is repeated literally forever. So we may write $1/3 = 0.\overline{3}$. It is important to keep in mind that the expression on the right side is just a special notation for the limit of a particular sequence.

More precisely, the preceding discussion shows that the number $1/3$ is the limit of the sequence $\{s_n, n = 1, 2, 3, ...\}$, where

$$s_n = \sum_{j=1}^{n} \frac{3}{10^j}.$$

These numbers s_n are a concrete example of *partial sums* of a so-called *infinite series*. In general, an infinite series (of real numbers) is a (formal) expression

$$\sum_{j=0}^{\infty} a_j, \text{ where } a_j \in \mathbb{R} \text{ for } j = 0, 1, 2, 3, ...$$

Associated with such an infinite series is the sequence of partial sums $\{s_0, s_1, s_2, ..., s_n, ...\}$ defined by

$$s_n = \sum_{j=0}^{n} a_j.$$

Definition 37 *We say that the infinite series $\sum_{j=0}^{\infty} a_j$ converges to the limit $L \in \mathbb{R}$ if the sequence $\{s_n\}$ of partial sums defined above converges to L. In that case we write*

$$\sum_{j=0}^{\infty} a_j = \lim_{n \to \infty} s_n = L.$$

If the sequence of partial sums does NOT have a limit, that is, if it does not converge, we say that the infinite series diverges.

Sometimes an infinite series is also called an infinite *sum*. This is a bad choice of terminology, since a sum only involves finitely many terms. Using the summation symbol $\sum_{j=0}^{\infty}$ is just a formal notation; it reminds us that it represents the limit of certain sequences that are defined in terms of *finite* sums.

Let us mention a very simple condition that MUST be satisfied for any convergent infinite series.

Lemma 38 *If $\sum_{j=0}^{\infty} a_j$ converges, then one must have $\lim_{n \to \infty} a_n = 0$.*

Proof. Note that $a_n = s_n - s_{n-1}$. Since we know that $\lim s_n = L = \lim s_{n-1}$, it follows that

$$\lim_{n \to \infty} a_n = \lim_{n \to \infty} s_n - \lim_{n \to \infty} s_{n-1} = L - L = 0. \quad \blacksquare$$

Therefore the infinite series $\sum_{j=1}^{\infty} a_j$ with $a_j = 1$ for all j diverges. This is of course obvious if one looks at the partial sums $s_n = n$.

It is important to emphasize that this condition is only *necessary*, and that it does NOT imply convergence. A well known example is the so-called *harmonic series*

$$\sum_{j=1}^{\infty} \frac{1}{j} = 1 + \frac{1}{2} + \frac{1}{3} + \frac{1}{4} + \dots.$$

Clearly $\lim_{j\to\infty} \frac{1}{j} = 0$, but we shall now explain why this series does not converge. Consider

$$s_2 = 1 + \frac{1}{2} > 2\frac{1}{2}, \ s_4 = s_{2+2} = s_2 + \left(\frac{1}{3} + \frac{1}{4}\right) > 2\frac{1}{2} + 2\frac{1}{4} = 3\frac{1}{2},$$

$$s_8 = s_{4+4} = s_4 + \left(\frac{1}{5} + \frac{1}{6} + \frac{1}{7} + \frac{1}{8}\right) > 3\frac{1}{2} + 4\frac{1}{8} = 4\frac{1}{2},$$

$$s_{16} = s_8 + \left(\frac{1}{9} + \frac{1}{10} + \dots + \frac{1}{16}\right) > 4\frac{1}{2} + 8\frac{1}{16} = 5\frac{1}{2}.$$

Continuing this process, we see that

$$s_{2^n} > s_{2^{n-1}} + 2^{n-1} \cdot \frac{1}{2^n} > n\frac{1}{2} + \frac{1}{2} = (n+1)\frac{1}{2}.$$

This implies that $\{s_n\}$ is not bounded, and hence the sequence cannot converge. On the other hand, this sequence exhibits some regularity: it is increasing, and since it is not bounded one uses the notation $\lim_{n\to\infty} s_n = \infty$. Since ∞ is not a real number, this notation is consistent with the statement that the sequence $\{s_n\}$ does NOT converge.

Infinite series are an important concept that has many applications in mathematics. One of the most remarkable ones involves the representation of functions such as exponential functions in terms of its Taylor series, an infinite series whose partial sums are certain polynomials. This will be discussed in detail much later on. There is a vast theory of convergent series, but for our purposes it is enough to just consider the following simple, yet most important example of an infinite series, namely the so-called *geometric series*, that is defined by

$$\sum_{j=0}^{\infty} r^j, \text{ where } r \text{ is a fixed real number.}$$

Lemma 39 *The geometric series $\sum_{j=0}^{\infty} r^j$ converges if and only if $|r| < 1$, and its limit in that case is $\frac{1}{1-r}$.*

Proof. Note that when $|r| \geq 1$, then $\lim_{j\to\infty} r^j$ does not exist if $r \neq 1$, and if $r = 1$, then $\lim_{j\to\infty} r^j = 1$, so, in particular, is not 0. Therefore the infinite series does not converge whenever $|r| \geq 1$. Let us now assume that $0 \leq |r| < 1$. By Lemma 28, we know that $\lim_{n\to\infty} r^n = 0$, so that in particular the necessary condition for convergence is satisfied, but we know that this alone is not enough. However, given this particular result, the proof of *convergence* of the geometric series is now a simple consequence of the following formula for the partial sums $s_n = \sum_{j=0}^{n} r^j$, which incidentally holds for all $r \neq 1$. Note that

$$r \cdot s_n = \sum_{j=0}^{n} r \cdot r^j = \sum_{j=0}^{n} r^{j+1} = s_n - 1 + r^{n+1}.$$

A simple rearrangement results in $s_n(1 - r) = 1 - r^{n+1}$ and hence, since $r \neq 1$,

$$s_n = \frac{1 - r^{n+1}}{1 - r}.$$

Since $|r| < 1$, we know that $\lim_{n\to\infty} r^{n+1} = 0$. Therefore, by general properties of limits, it follows that

$$\lim_{n\to\infty} s_n = \frac{1 - \lim_{n\to\infty} r^{n+1}}{1 - r} = \frac{1 - 0}{1 - r}. \blacksquare$$

V.2.4 *Decimal Expansion of Real Numbers*

The discussion in this section is not really relevant for calculus, so the reader eager to move on may safely skip it. On the other hand, we now have the appropriate tools available that make it possible to put this somewhat mysterious topic on a firm foundation.

Infinite Decimals Based on the geometric series, we can now give another proof for the statement that $1/3 = 0.\overline{3}$. In fact, we know that

$$0.\overline{3} = \lim_{n\to\infty} \sum_{j=1}^{n} \frac{3}{10^j} = \lim_{n\to\infty} 3 \left(\sum_{j=0}^{n} \frac{1}{10^j} - 1 \right) = 3 \lim_{n\to\infty} \sum_{j=0}^{n} \frac{1}{10^j} - 3$$

$$= 3 \frac{1}{1 - 1/10} - 3 = 3 \frac{10}{9} - 3 = \frac{10}{3} - 3 = \frac{1}{3}.$$

Very closely related is the infinite decimal expansion $0.\overline{9}$. Note that

$$0.\overline{9} = \lim_{n\to\infty} \sum_{j=1}^{n} \frac{9}{10^j} = \lim_{n\to\infty} 3 \cdot \sum_{j=1}^{n} \frac{3}{10^j}$$

$$= 3 \cdot \lim_{n\to\infty} \sum_{j=1}^{n} \frac{3}{10^j} = 3 \cdot \frac{1}{3} = 1.$$

The equality $1 = 0.\overline{9} = 0.999999...$ is quite baffling for people who don't know much about decimal expansions. However, if you think of $0.\overline{9}$ as the limit of a (convergent) infinite series, there is nothing mysterious about the statement that the value of that limit equals 1.

We shall now consider arbitrary decimal expansions

$$0.a_1a_2a_3..., \text{ where each } a_j \in \{0,1,2,...,8,9\}.$$

This of course means that we consider the infinite series

$$\sum_{j=1}^{\infty} \frac{a_j}{10^j},$$

whose partial sums are $s_n = \sum_{j=1}^{n} \frac{a_j}{10^j}$. Clearly $s_n \leq s_{n+1}$, so the sequence $\{s_n\}$ is increasing. Also, $s_n \leq \sum_{j=1}^{n} \frac{9}{10^j} \leq \lim_{n \to \infty} \sum_{j=1}^{n} \frac{9}{10^j} = 1$, and therefore the sequence is also bounded. Again, by the Monotone Convergence Theorem,

$$\lim_{n \to \infty} \sum_{j=1}^{n} \frac{a_j}{10^j} = \sup\{s_n\} \in \mathbb{R}.$$

We have thus verified that every infinite decimal expansion represents a unique real number.

We now want to show that, conversely, every real number can be represented by an infinite decimal expansion, or, more precisely, as the limit of a particular infinite series. Let $c \in \mathbb{R}$ be an arbitrary real number, and let us assume first that $c \geq 0$. Then there is an integer $m \geq 0$, so that $m \leq c < m+1$. If $c = m$, we are done, since we are familiar with the decimal representation of integers. So assume that $m < c < m+1$. Let us divide the interval $[m, m+1)$ into 10 disjoint intervals $I_j^{(1)} = [m + \frac{j}{10}, m + \frac{j+1}{10})$, $j = 0, 1, 2, 3, ..., 9$. Since $\cup_{j=0}^{9} I_j^{(1)} = [m, m+1)$, c must be contained in exactly one of these intervals, say $c \in I_{j_1}^{(1)}$. If $c = m + \frac{j_1}{10}$, then we are done, and $c = m.j_1$. So let us assume that $m + \frac{j_1}{10} < c < m + \frac{j_2}{10}$. We now repeat the process we just carried out over and over again, as long as c is not the left endpoint of one of the resulting intervals. More precisely, suppose that after n steps we have

$$c \in I_{j_n}^{(n)} = \left[m + \frac{j_1}{10} + \frac{j_2}{10^2} + ... + \frac{j_n}{10^n}, m + \frac{j_1}{10} + \frac{j_2}{10^2} + ... + \frac{j_n + 1}{10^n} \right).$$

Here the numbers $j_l \in \{0, 1, 2, ..., 9\}$, and let us define $s_n = m + \frac{j_1}{10} + \frac{j_2}{10^2} + ... + \frac{j_n}{10^n} = m.j_1j_2...j_n$. If $c = s_n$ we are done, and c has the decimal expansion

$$c = m.j_1j_2...j_n.$$

Otherwise, $s_n < c < s_n + \frac{1}{10^n}$, and we continue by subdividing the interval $[s_n, s_n + \frac{1}{10^n})$ into 10 disjoint intervals $I_j^{(n+1)}$, $j = 0, 1, 2, ..., 9$, of equal length $1/10^{n+1}$. There are two possibilities. Either c turns out to be the left endpoint of one of the intervals, in which case c is represented by a *finite* decimal expansion, or else the process can be continued indefinitely, and we end up with an (increasing) sequence $\{s_n, n = 1, 2, 3, ...\}$ of rational numbers, with

$$s_n < c < s_n + \frac{1}{10^{n+1}} \text{ for each } n \in \mathbb{N}.$$

By construction, s_n is the *nth* partial sum of the infinite series $m + \sum_{l=1}^{\infty} \frac{j_l}{10^l}$; since $|s_n - c| < \frac{1}{10^{n+1}}$ for each n, and since $1/10 < 1$ we know that $\lim_{n \to \infty}(1/10)^n = 0$, it is clear that

$$\lim_{n \to \infty} s_n = c.$$

Therefore the real number $c \geq 0$ is the limit of the infinite series $m + \sum_{l=1}^{\infty} \frac{j_l}{10^l}$, which, as explained above, is represented by the (infinite) decimal expansion $m.j_1 j_2 j_3....$ Finally, if $c < 0$, we apply the previous result to $-c$, and consequently $c = -m.j_1 j_2 j_3....$

Notice that the process we described determines the sequence $\{j_1, j_2, j_3, ...\}$ uniquely, with the sequence being either finite or infinite. However, the equation $1 = 0.\bar{9}$ allows to change the looks of a decimal expansion in two special cases.

Case 1. The sequence is finite, that is the process stops at $j_n > 0$ for some n. Notice that

$$1/10^n = 1/10^n \cdot 0.\bar{9} = 1/10^n \lim_{l \to \infty} \sum_{j=1}^{l} \frac{9}{10^j}$$

$$= \lim_{l \to \infty} \sum_{j=1}^{l} \frac{9}{10^{j+n}}.$$

Therefore, the last (nonzero) term in the decimal representation can be replaced by

$$\frac{j_n}{10^n} = \frac{j_n - 1}{10^n} + \frac{1}{10^n} = \frac{j_n - 1}{10^n} + \lim_{l \to \infty} \sum_{j=1}^{l} \frac{9}{10^{j+n}},$$

in other words we have

$$m.j_1...j_n = m.j_1...j_{n-1}(j_n - 1)999....$$

Case 2. The process goes on indefinitely, but for some n we have $j_n < 9$ and all the digits after j_n are equal to 9. By essentially reversing the process in Case 1, one obtains

$$m.j_1 j_2 \ldots j_n 99999\ldots = m.j_1 j_2 \ldots j_{n-1}(j_n + 1).$$

So if we agree that every finite decimal expansion has to be replaced by the infinite decimal expansion involving only digits 9 after a certain point, as in Case 1 above, then we can say that the (infinite) decimal expansion of a real number is uniquely determined by that number.

Finally, we want to discuss the effect of multiplying an infinite decimal expansion by a number 10^l, where $l \in \mathbb{Z}$ is nonzero. Consider first $l > 0$. If $c = \lim_{n \to \infty} \sum_{j=1}^{n} \frac{a_j}{10^j}$, then the partial sums s_n with $n > l$, can be split into two sums

$$s_n = \sum_{j=1}^{l} \frac{a_j}{10^j} + \sum_{j=l+1}^{n} \frac{a_j}{10^j}.$$

Therefore

$$10^l \cdot \lim_{n \to \infty} \sum_{j=1}^{n} \frac{a_j}{10^j} = \lim_{n \to \infty} \left[10^l \cdot \sum_{j=1}^{n} \frac{a_j}{10^j} \right]$$

$$= \lim_{n \to \infty} \left[10^l \cdot \sum_{j=1}^{l} \frac{a_j}{10^j} + 10^l \cdot \sum_{j=l+1}^{n} \frac{a_j}{10^j} \right].$$

Hence

$$10^l \cdot c = \sum_{j=1}^{l} a_j \cdot 10^{l-j} + \lim_{n \to \infty} \sum_{j=l+1}^{n} \frac{a_j}{10^{j-l}} = a_1 a_2 \ldots a_l + \lim_{n \to \infty} \sum_{j=l+1}^{n} \frac{a_j}{10^{j-l}}.$$

We now replace the counter j by $k = j - l$, so that $j = l + k$, and the above result can be written as

$$10^l \cdot c = a_1 a_2 \ldots a_l + \lim_{n \to \infty} \sum_{k=1}^{n-l} \frac{a_{l+k}}{10^k} = a_1 a_2 \ldots a_l . a_{l+1} a_{l+2} \ldots$$

(decimal point between a_l and a_{l+1}!),

that is, multiplication by 10^l shifts the decimal point l places to the right. If a positive integer m is added to c, the decimal expansion of $10^l m$ simply adds l zeroes to the decimal expansion of m, thereby making room for the digits $a_1 \ldots a_l$, so that the effect on the decimal expansion of $m + c$ is the same. A completely analogous argument shows that multiplication by 10^{-l} shifts the decimal point to the left by l places, for example

$$10^{-3} \cdot 53.2687\ldots = 0.0532687\ldots.$$

Decimal Representation of Rational Numbers Of course, since every rational number is, in particular, a real number, it can be represented by a (possibly infinite) decimal expansion. What we want to show now is that rational numbers are singled out by the fact that their decimal expansion is *repeating*, also referred to as *periodic*. This is a special property, which it may be useful to be aware of, even without going through the detailed verification.

We have already encountered this phenomenon with the number $1/3 = 0.33333....$ Not surprisingly, whenever one deals with rational numbers, things turn out to be a bit messy, so we ask the reader to be patient. On the other hand, this particular property of rational numbers is not really relevant for our investigations of derivatives and exponential functions, etc., so the reader may safely skip this subsection. After all, writing a rational number as a fraction, that is, as a quotient of integers, is usually much simpler than representing it by an infinite repeating decimal expansion, which, in particular requires an understanding of limits and infinite series. One advantage of the decimal expansion is that it makes it easy to compare rational numbers; using fractional notation typically requires extra work to see which number is larger. Furthermore, we should note that all electronic equipment/calculators immediately replace fractions by finite decimals, which necessarily will usually only be approximations, that is particular partial sums of the actual infinite series represented by the complete decimal expansion.

We say that a decimal expansion is repeating (or periodic), if there are a digit j_k and a sequence of l digits, $l \geq 1$, $a_1 a_2 ... a_l$, so that after the digit j_k the rest of the decimal expansion consist of repetitions of the digits $a_1 a_2 ... a_l$, that is, if the decimal expansion looks like

$$c = m.j_1...j_k(a_1 a_2 ...a_l)(a_1 a_2 ...a_l)(a_1 a_2 ...a_l)...,$$

where we have added the brackets just to highlight the repeating group of digits. Note that this includes the case of a *finite* decimal expansion $m.a_1 a_2 ... a_l$, which we can also write as $m.a_1 a_2 ... a_l 000....$ As we had done already in the case $1/3 = 0.\overline{3}$, it is convenient to use the notation $\overline{a_1 a_2 ... a_l}$ to indicate the repeating sequence, that is, the above number c is written as

$$c = m.j_1...j_k\overline{a_1 a_2 ...a_l}.$$

Note that the repeating sequence is not determined uniquely by the given number, for example we have

$$0.21\overline{647} = 0.21647647647... = 0.216\overline{476} = 0.2164\overline{764}.$$

The main result is the following

Theorem 40 *A real number c is rational if and only if its decimal expansion is repeating.*

Proof. We first prove that every repeating decimal expansion represents a rational number. So assume

$$c = m.j_1...j_k\overline{a_1a_2...a_l}.$$

Then

$$10^k c = 10^k m + j_1...j_k.\overline{a_1a_2...a_l} = 10^k m + j_1...j_k + 0.\overline{a_1a_2...a_l}.$$

We shall now prove that $0.\overline{a_1a_2...a_l}$ is a rational number. Solving the above equation for c then shows that c is rational as well. So consider

$$
\begin{aligned}
0.\overline{a_1a_2...a_l} &= \left(\frac{a_1}{10^1} + \frac{a_2}{10^2} + ... + \frac{a_l}{10^l}\right) + \left(\frac{a_1}{10^{l+1}} + \frac{a_2}{10^{l+2}} + ... + \frac{a_l}{10^{l+l}}\right) + ... \\
&= \left(\frac{a_1}{10^1} + \frac{a_2}{10^2} + ... + \frac{a_l}{10^l}\right) + \frac{1}{10^l}\left(\frac{a_1}{10^1} + \frac{a_2}{10^2} + ... + \frac{a_l}{10^l}\right) + ... \\
&= (0.a_1a_2...a_l) \lim_{k\to\infty} \sum_{p=0}^{k} \left(\frac{1}{10^l}\right)^p \\
&= (0.a_1a_2...a_l)\frac{1}{1 - 1/10^l},
\end{aligned}
$$

where in the last step we of course used the limit formula for a geometric series. This shows that $0.\overline{a_1a_2...a_l}$ is a rational number.

The astute reader may raise a question, since when taking the limit of partial sums, we have actually grouped the terms, so that formally the partial sums differ. More precisely, in the case above, rather then considering $\lim_{n\to\infty} s_n$ with $s_n = 0.d_1d_2...d_n$, where the digits d_ν just represent the repeating sequence of the a_j, we consider the (sub)-sequence $s_{lk} = \sum_{p=0}^{k-1}\frac{1}{10^{lp}}\left(\frac{a_1}{10^1} + \frac{a_2}{10^2} + ... + \frac{a_l}{10^l}\right)$ for $k = 1, 2, 3, ...$ The question thus is whether $\lim_{n\to\infty} s_n = \lim_{k\to\infty} s_{lk}$. For arbitrary series, this could indeed be false, as shown by $s_n = \sum_{j=0}^{n}(-1)^j$, which clearly diverges. But note that

$$s_{2k} = 1 + \sum_{j=1}^{k}[(-1) + 1] = 1, \text{ for } k = 1, 2, 3, ...,$$

which satisfies $\lim_{k\to\infty} s_{2k} = 1$, so the result is different. However, it is easy to show that if the sequence $\{s_n\}$ is increasing, as is the case here, then the statement $\lim_{n\to\infty} s_n = \lim_{k\to\infty} s_{lk}$ is indeed correct, even for much more general regroupings. Rather than going through the formal argument (see Exercise 1 in 2.7), just think of an infinitely long stair case of a finite height (thus the heights of the steps must be given by a convergent infinite series). Clearly it doesn't matter if you climb the stair case one step at a time, or by taking two steps at once, or, if you can, three steps at once, or keep changing the number of steps you skip as you move higher and higher.

To complete the proof of the Theorem, we must now prove that if $r \in \mathbb{Q}$, then its decimal expansion is repeating. We shall be guided by the equation $\frac{1}{3} = 0.33...3 + \frac{1}{3}10^{-n}$, with n numbers 3 in the decimal expansion, which we had already considered earlier. To understand the process, let us work on the analogous equation for the number $1/7$. Clearly $1/7 < 1$, so its decimal expansion begins with $0.a_1$. It's pretty obvious that a_1 must be 1, since $2/10 = 1/5 > 1/7$. But looking ahead where we have to find a general procedure to determine the correct digits, let us describe the process. We need to determine how often $1/10$ fits into $1/7$, that is, we need to look at

$$\frac{1}{7} / \frac{1}{10} = \frac{10}{7} = \boxed{1} + \frac{3}{7},$$

that is, we consider the standard division process for natural numbers. The result shows that $\frac{1}{10}$ fits into $\frac{1}{7}$ only one time, with $\frac{3}{7}$ of $\frac{1}{10}$ left over. Thus

$$\frac{1}{7} = \boxed{1} \cdot \frac{1}{10} + \frac{3}{7}\frac{1}{10}.$$

To determine the second digit a_2, that is, how many times does $\frac{1}{10^2}$ fit into the remainder $\frac{3}{7}\frac{1}{10} = \frac{3}{70}$, we proceed the same way, i.e., we consider the quotient

$$\frac{3/70}{1/10^2} = \frac{30}{7} = \boxed{4} + \frac{2}{7},$$

so that

$$\frac{1}{7} = \frac{1}{10} + \frac{\boxed{4}}{10^2} + \frac{2}{7}\frac{1}{10^2}.$$

The next step shows

$$\frac{2/700}{1/10^3} = \frac{20}{7} = \boxed{2} + \frac{6}{7},$$

and therefore

$$\frac{1}{7} = \frac{1}{10} + \frac{4}{10^2} + \frac{\boxed{2}}{10^3} + \frac{6}{7}\frac{1}{10^3}.$$

Continuing in this way we eventually end up with
$$\frac{1}{7} = \frac{1}{10} + \frac{4}{10^2} + \frac{2}{10^3} + \frac{8}{10^4} + \frac{5}{10^5} + \frac{7}{10^6} + \frac{1}{7} \cdot \frac{1}{10^6}$$
$$= 0.142857 + \frac{1}{7}\frac{1}{10^6}.$$

We have come full circle, as we are back at $\frac{1}{7}$, now multiplied with $\frac{1}{10^6}$. So we replace $1/7$ on the right side in the last equation by the result we just obtained, resulting in
$$\frac{1}{7} = 0.142857 + \left[0.142857 + \frac{1}{7}\frac{1}{10^6}\right] \cdot \frac{1}{10^6}$$
$$= 0.142857142857 + \frac{1}{7}\frac{1}{(10^6)^2}.$$

After going through this process n times, one obtains
$$\frac{1}{7} = 0.(142857)(142857)...(142857) + \frac{1}{7}\frac{1}{(10^6)^n},$$

where we have introduced brackets to highlight the block of 6 digits that is repeated n times. We now let $n \to \infty$, and by the general properties of limits one obtains
$$\frac{1}{7} = 0.\overline{142857} + \frac{1}{7} \lim_{n\to\infty} \frac{1}{(10^6)^n} = 0.\overline{142857} + 0.$$

As we see, the process is quite elaborate, even for the simple fraction $1/7$. On the other hand, the technique to find the relevant repeating digits is based on division with remainder. Starting with $1/7$, the standard way to handle this is to consider

$$
\begin{array}{r}
0.142857 \\
7\,|\overline{10} \\
7 \\
\overline{30} \\
28 \\
\overline{20} \\
14 \\
\overline{60} \\
56 \\
\overline{40} \\
35 \\
\overline{50} \\
49 \\
\overline{1}
\end{array}
$$

The process of division resulted in $1/7 = 0.142857$, with remainder $1/10^6$ which also needs to be divided by 7. If you prefer, the division can be done just with natural numbers by first multiplying $1/7$ with 10^6, or any large power of 10, so that one can continue dividing remaining within the natural numbers. The (long) division process then gives

$$10^6 = 142857 \cdot 7 + 1, \text{ or } \frac{10^6}{7} = 142857 + \frac{1}{7}.$$

After dividing by 10^6 one obtains

$$\frac{1}{7} = 0.142857 + \frac{1}{7}\frac{1}{10^6},$$

which is the critical equation that can now be iterated as often as one would like.

Let us explain how the analogous process works for an arbitrary rational number $m/p > 0$. If $0 < p \le m$, then the division process gives $m/p = q \cdot p + \frac{r}{p}$, with $q \in \mathbb{N}$, and where the remainder r is an integer with $0 \le r < p$. Let us now apply the division process as in the detailed example $1/7$ to r/p. At the jth step there will be a certain new remainder r_j with $0 \le r_j < p$. If $r_j = 0$ for some j, then the process stops, and r/p has a finite decimal expansion. So we will assume that the remainders are all $\ne 0$. Since there are only $p-1$ choices possible for a non-zero remainder, at the latest at the pth step the remainder will have to agree with a remainder that occurred before. At that point the result will look like

$$\frac{r}{p} = 0.j_1 j_2 ... j_l a_{l+1} a_{l+2} ... a_{l+k} + \frac{r_l}{p}\frac{1}{10^{l+k}},$$

where $l \ge 0$, $l + k \le p$, and the remainder $r_{l+k} = r_l$ at the $(l+k)$th step agrees with the remainder r_l at the lth step. We thus can rewrite this as

$$\frac{r}{p} = 0.j_1 j_2 ... j_l + \frac{r_l}{p}\frac{1}{10^l} \text{ and } \frac{r_l}{p}\frac{1}{10^l} = 0.a_{l+1}...a_{l+k} \cdot \frac{1}{10^l} + \frac{r_l}{p}\frac{1}{10^{l+k}},$$

so that

$$\frac{r_l}{p} = 0.a_{l+1}...a_{l+k} + \frac{r_l}{p}\frac{1}{10^k}.$$

Now replace $\frac{r_l}{p}$ on the right by the equation we just obtained for $\frac{r_l}{p}$, to obtain

$$\frac{r_l}{p} = 0.a_{l+1}...a_{l+k} + \left[0.a_{l+1}...a_{l+k} + \frac{r_l}{p}\frac{1}{10^k}\right]\frac{1}{10^k}$$

$$= 0.(a_{l+1}...a_{l+k})(a_{l+1}...a_{l+k}) + \frac{r_l}{p}\frac{1}{(10^k)^2},$$

where we used that $0.a_{l+1}...a_{l+k} \cdot \frac{1}{10^k}$ moves the decimal point k places to the left, resulting in $0.00...0a_{l+1}...a_{l+k}$, with k zeroes between the decimal point and a_{l+1}.

After a total of n such steps we have

$$\frac{r_l}{p} = 0.(a_{l+1}...a_{l+k})(a_{l+1}...a_{l+k})...(a_{l+1}...a_{l+k}) + \frac{r_l}{p}\frac{1}{(10^k)^n},$$

where the sequence of digits $(a_{l+1}...a_{l+k})$ occurs n times. Taking the limit as $n \to \infty$ we get

$$\frac{r_l}{p} = 0.\overline{a_{l+1}...a_{l+k}} + \frac{r_l}{p}\lim_{n\to\infty}\frac{1}{(10^k)^n} = 0.\overline{a_{l+1}...a_{l+k}} + 0.$$

While we know that the denominator p gives an upper limit for the possible remainders, so that the repetitions will have to start at the latest at the pth step, for any particular rational number much fewer steps may be required. For example, consider

$$\frac{1}{11} = \frac{100}{11} \cdot \frac{1}{10^2} = \left[9 + \frac{1}{11}\right] \cdot \frac{1}{10^2},$$

where we already repeat the remainder $1/11$ after one division step. We see that

$$\frac{1}{11} = 0.09 + \frac{1}{11}\frac{1}{10^2},$$

which leads to $1/11 = 0.\overline{09}$.

Again, let us emphasize that this result about a special property of the decimal expansion of rational numbers, whose proof is technically quite complicated, can just be viewed as a curious fact that is not at all important for our work. For our purposes it is best to think of rational numbers as fractions, that is, quotients of integers, and not worry about their representation via decimal expansions.

V.2.5 *Limits of Functions*

Now that we have formalized and used the notion of the limit of a sequence, we can easily do the analogous process for limits of functions. Recall that we already introduced the idea of $\lim_{x\to a} P(x) = P(a)$ for a polynomial P, as well as for rational functions. In that context, the whole process was motivated by the critical estimate

$$|P(x) - P(a)| \le K\,|x - a|$$

for all x with $|x - a| \le 1$, where K is a suitable constant. Given any $\epsilon > 0$, the estimate implies that $|P(x) - P(a)| < \epsilon$ for all x with $|x - a| < \epsilon/K$.

It is the same idea as in case of a sequence. One prescribes how close one wants to be to the limit by taking an arbitrarily small ϵ, and then one requires that this closeness is met by all the inputs (the natural number n that identifies the particular element in a sequence, or certain inputs x in the domain of the function) that satisfy a particular condition. The estimate above in case of a polynomial gives a very simple formula to obtain the relevant condition for the input x, given a prescribed tolerance ϵ. In general, there is no such simple formula, instead, the relationship is more of an "existential" type: Given $\epsilon > 0$, there exists..., so that....

In case of polynomials or rational functions, the general version of this relationship is expressed as follows. We say that $\lim_{x \to a} P(x) = P(a)$ if for every $\epsilon > 0$, no matter how small, there exists a number $\delta = \delta(\epsilon) > 0$, (that is, δ depends somehow on the choice of ϵ) so that one has

$$|P(x) - P(a)| < \epsilon \text{ for all } x \neq a \text{ that satisfy } |x - a| < \delta(\epsilon).$$

Note that in this case the limit equals the value of the function at the point a that is considered. Of course, based on the estimate above, the statement $\lim_{x \to a} P(x) = P(a)$ is indeed correct. This particular property is known as the *continuity* of the function P at the point a. Recall that we even had verified earlier that rational functions are continuous at every point in their domain.

Limits get much more mysterious when the limit is not specified as a value of the function under consideration, as we had seen in the case of the discussion of the tangent to the exponential function E_2, where the function

$$q(x) = \frac{2^x - 1}{x}$$

is NOT defined at 0. Recall that numerical data suggested that

$$q(x) \to 0.6931471...\text{as } x \to 0.$$

So how do we capture such strange numbers and verify that they are indeed "limits"?

It took mathematicians in the 19th century quite a long time to distill the following definition of limit. In contrast, our approach here, that was motivated by the critical estimate for polynomials, practically suggests the appropriate definition.

Definition 41 *Suppose f is a function defined for all x in an interval surrounding the point a, but not necessarily at a itself. Then we say that f has a limit L at a, also denoted by $\lim_{x \to a} f(x) = L$, if for every $\epsilon > 0$ there exists a $\delta(\epsilon) > 0$, so that*

$$|f(x) - L| < \epsilon \text{ for all } x \text{ with } 0 < |x - a| < \delta(\epsilon).$$

Note that this is the same relationship as we had introduced for polynomials, except that the function need not be defined at a, and the value L of the limit is not the value $f(a)$, which may not exist at all.

We have the following simple but useful result.

Lemma 42 *Suppose $\lim_{x \to a} f(x) = L$ and that the sequence $\{x_n, n = 1, 2, 3, ...\}$ satisfies $x_n \neq a$ and $\lim_{n \to \infty} x_n = a$. Then $\lim_{n \to \infty} f(x_n) = L$. With the relevant modifications, the result holds also for one-sided limits that are defined below.*

Proof. It is clear that for n sufficiently large x_n is in the interval where f is defined, excluding the point a. We only need to consider such large n. Let $\varepsilon > 0$ be given, and choose $\delta(\varepsilon) > 0$, so that $|f(x) - L| < \varepsilon$ for all x with $0 < |x - a| < \delta(\varepsilon)$. Next, since $\lim_{n \to \infty} x_n = a$, we can choose $N = N(\delta(\varepsilon))$, so that if $n \geq N$ one has $0 < |x_n - a| < \delta(\varepsilon)$. Therefore, for all such n it follows that $|f(x_n) - L| < \varepsilon$. ∎

Given the formal similarities with the limit of a sequence, it is not surprising that one has properties analogous to those we had verified for sequences, as follows.

1) *If f has a limit at the point a, then the value of the limit is determined uniquely.*

2) *If f and g are two functions defined on an interval containing a, but not necessarily at a itself, then*

i) $\lim_{x \to a}(f(x) \pm g(x)) = \lim_{x \to a} f(x) \pm \lim_{x \to a} g(x)$;

ii) $\lim_{x \to a}(f(x) \cdot g(x)) = \lim_{x \to a} f(x) \cdot \lim_{x \to a} g(x)$;

iii) *If $\lim_{x \to a} g(x) \neq 0$, then $\lim_{x \to a} \frac{f(x)}{g(x)} = \frac{\lim_{x \to a} f(x)}{\lim_{x \to a} g(x)}$, where the quotient on the left is defined for all $x \neq a$ sufficiently close to a.*

iv) *If $f(x) \leq h(x) \leq g(x)$ for all $x \neq a$ close to a, and if $\lim_{x \to a} f(x) = \lim_{x \to a} g(x) = L$, then h also has a limit at $x = a$, which equals L.*

We shall prove in detail 2)iii) and encourage the reader to work through the Exercises to prove the remaining statements. Let us just mention that, with appropriate modifications, the structure of the proofs is the same as that of the proofs of the corresponding statements for sequences.

Proof of 2)iii). Suppose $\lim_{x \to a} f(x) = L_f$ and $\lim_{x \to a} g(x) = L_g \neq 0$. By choosing $\varepsilon^* = |L_g|/2 > 0$, there exists a $\gamma > 0$, so that first of all f and g are defined for all x with $0 < |x - a| < \gamma$, and most importantly, that $|g(x) - L_g| < \varepsilon^*$ for all such x. For such x one then has

$$|g(x)| = |L_g + (g(x) - L_g)| \geq |L_g| - |g(x) - L_g| > |L_g| - \varepsilon^* = |L_g|/2 > 0.$$

From now on we shall only consider x with $0 < |x - a| < \gamma$, so that in particular $f(x)/g(x)$ is defined. Now consider

$$\left| \frac{f(x)}{g(x)} - \frac{L_f}{L_g} \right| = \left| \frac{f(x)L_g - L_f g(x)}{g(x)L_g} \right|$$

$$= \left| \frac{f(x)L_g - L_f L_g + L_f L_g - L_f g(x)}{g(x)L_g} \right|$$

$$\leq \frac{|f(x) - L_f|\, |L_g| + |L_f|\, |L_g - g(x)|}{|L_g|^2 / 2}$$

$$\leq K[|f(x) - L_f| + |L_g - g(x)|]$$

for a suitable constant K. This has been the messy algebraic part. The rest is now easy. Given any $\varepsilon > 0$, by the limit property for f and g, we can choose a $\delta(\varepsilon) > 0$, so that $\delta(\varepsilon) \leq \gamma$ and most importantly, so that

$$|f(x) - L_f| < \frac{\varepsilon}{2K} \text{ and } |g(x) - L_g| < \frac{\varepsilon}{2K} \text{ for all } x \text{ with } 0 < |x - a| < \delta(\varepsilon).$$

The previous estimate then implies that

$$\left| \frac{f(x)}{g(x)} - \frac{L_f}{L_g} \right| < \varepsilon \text{ for all } x \text{ with } 0 < |x - a| < \delta(\varepsilon). \blacksquare$$

We shall now consider an example that will suggest that it is sometimes useful to consider *one-sided limits*. Let $g(x) = \frac{|x|}{x}$ for $x \neq 0$. Note that $g(x) = 1$ for $x > 0$, while $g(x) = -1$ for $x < 0$, so clearly g does NOT have a limit at $x = 0$. On the other hand, it makes sense to separate the approach to 0 into the two distinct cases, either approaching from the left ($x < 0$, and we write $x \to 0^-$), or just from the right ($x > 0$, and we write $x \to 0^+$). Clearly we have $\lim_{x \to 0^-} g(x) = -1$ and $\lim_{x \to 0^+} g(x) = +1$.

In general we write $x \to a^-$ if we only consider $x < a$, and correspondingly, $x \to a^+$ if we consider $x > a$ only. The corresponding limits are called one-sided limits (more precisely, left-, resp. right-sided). One then has the following simple fact.

Lemma 43 $\lim_{x \to a} f(x) = L$ *exists if and only if the two one-sided limits exist and are equal to L, that is, if*

$$\lim_{x \to a^-} f(x) = \lim_{x \to a^+} f(x) = L.$$

The proof is very simple, and we give the details to practice using the definition. Since it is pretty obvious that the existence of the limit implies the existence of the one-sided limits, and that their values are equal, we shall focus on the other direction. So assume that the two one-sided limits exist and are equal to L. Let $\epsilon > 0$ be given. By our assumption, there is $\delta^-(\epsilon) > 0$, so that $|L - f(x)| < \epsilon$ for all x with $a - \delta^- < x < a$, and similarly there is $\delta^+(\epsilon) > 0$ so that $|L - f(x)| < \epsilon$ for all x with $a < x < a + \delta^+$. Let $\delta(\epsilon) = \min\{\delta^-, \delta^+\} > 0$. Then we have

$$|L - f(x)| < \epsilon \text{ for all } x \neq a \text{ with } a - \delta(\epsilon) < x < a + \delta(\epsilon),$$

that is, for all x with $0 < |x - a| < \delta(\epsilon)$. ∎

Just as in case of sequences, we have the following important criterion for the existence of limits.

Theorem 44 *Suppose f is defined for all $x < a$ near a, and is increasing, i.e., $f(u) \leq f(v)$ if $u \leq v$. If furthermore f is bounded from above, then*

$$\lim_{x \to a^-} f(x) = \sup\{f(x) : x < a\}.$$

Analogous statements, with appropriate modifications, hold if f is decreasing, and similarly for right-sided limits.

We shall refer to this Theorem as the *Monotone Convergence Theorem* for functions.

Proof. Because f is bounded, the set $\{f(x) : x < a\}$ has a LUB L. Suppose $\epsilon > 0$. Then $L - \epsilon$ is not an upper bound for this set, so there exists $x_0 < a$ with $L - \epsilon < f(x_0)$. Since f is increasing, one has

$$L - \epsilon < f(x_0) \leq f(x) \leq L \text{ for all } x_0 < x < a.$$

So if we choose $\delta^-(\epsilon) = a - x_0 > 0$, then clearly $|L - f(x)| < \epsilon$ for all x with $a - \delta^-(\varepsilon) < x < a$. ∎

V.2.6   *Properties of Continuous Function*

Continuity was introduced in the context of polynomials and, more generally, rational functions. Once we have the precise notion of limit for functions in general, it is natural to say that a function f defined in an interval containing the point a is continuous at a if

$$\lim_{x \to a} f(x) = f(a).$$

Given the general properties of limits mentioned above, it readily follows that one has corresponding properties for continuous functions, as follows.

Theorem 45 *Suppose f and g are functions that are continuous at a. Then*

i) $f \pm g$ is continuous at a;

ii) $f \cdot g$ is continuous at a; in particular, if $c \in \mathbb{R}$, the cf is continuous at a;

iii) if $g(a) \neq 0$, then f/g is continuous at a.

If you just think of the intuitive meaning of continuity, these properties are completely natural. Their proofs follow immediately from the corresponding facts about limits. Let us just note that the hypotheses in iii), together with the continuity of g at a, implies that there is $\delta > 0$, so that $g(x) \neq 0$ for $x \in I_\delta(a)$, and therefore f/g is well defined in a neighborhood of a.

Finally, recall that composition of functions is in fact the most natural operation on functions. Consequently, the following result, too, really is to be expected.

Theorem 46 *Suppose g is continuous at a, with $g(a) = b$, and that f is continuous at b. Then the composition $f \circ g$ is defined in a neighborhood of a and is continuous at a.*

Proof. By hypotheses, f is defined in an interval $I_\gamma(b)$. Since g is continuous at a, there exists a $\delta(\gamma)$, so that $|g(x) - g(a)| < \gamma$ for all x with $|x - a| < \delta(\gamma)$. Hence $g(x) \in I_\gamma(b)$, and therefore $f \circ g$ is defined on $I_{\delta(\gamma)}(a)$. To prove continuity of $f \circ g$, let $\epsilon > 0$ be given. Since f is continuous at b, there exists $\beta(\epsilon) > 0$, which we may assume to be $< \gamma$, so that $|f(y) - f(b)| < \epsilon$ for all $y \in I_{\beta(\epsilon)}(b)$. Since g is continuous at a, given this number $\beta(\epsilon) > 0$, there exists $\delta(\beta)$ so that $|g(x) - g(a)| < \beta(\varepsilon)$, i.e., $g(x) \in I_{\beta(\epsilon)}(b)$, for all x with $|x - a| < \delta(\beta)$. Therefore

$$f(g(x)) - f(g(a)) < \epsilon \text{ for all } x \text{ with } |x - a| < \delta(\beta) = \delta(\beta(\epsilon)).$$

This relationship proves that $\lim_{x \to a} (f \circ g)(x) = (f \circ g)(a)$. ∎

Naturally there also is a corresponding statement involving inverse functions. However, the precise formulation is a bit more complicated, and we shall discuss it in Section 3.4, when we deal with a concrete situation.

Another important property of continuous functions is the following result that makes precise the geometrically intuitive statement that the graph of a continuous function has no holes or gaps.

Theorem 47 *(The Intermediate Value Theorem - IVT) Suppose the function f is continuous on the interval I and that for two points $a < b$ in I one has $f(a) \neq f(b)$. Let λ be any number between $f(a)$ and $f(b)$. Then there exists a point $x_0 \in (a, b)$ with $f(x_0) = \lambda$.*

Proof. Since the graphs of continuous functions have no holes or tears, this result is intuitively clear. Its precise verification requires the completeness of the real numbers, as follows. Let us assume that $f(a) < \lambda < f(b)$, and set $S = \{x \in [a, b] : f(x) \le \lambda\}$. Since $a \in S$, and b is an upper bound for S, S has a *least* upper bound $x_0 = \sup S$. We claim that $f(x_0) = \lambda$. Note that $x_0 > a$. For all sufficiently large integers n, $x_0 - 1/n > a$ as well, and it is not an upper bound for S. Therefore there exists $u_n \in S$ with $x_0 - 1/n < u_n \le x_0$. Since $u_n \to x_0$ as $n \to \infty$, and $f(u_n) \le \lambda$, the continuity of f implies that $f(x_0) = \lim_{n \to \infty} f(u_n) \le \lambda$. In particular this shows that $x_0 < b$. Now suppose $f(x_0) < \lambda$. Again, by the continuity of f, there exists $u \in [a, b]$ with $u > x_0$ and $f(u) < \lambda$. So $u \in S$, and hence x_0 could not be an upper bound for S. This contradiction shows that we must have $f(x_0) = \lambda$. ∎

We already mentioned that if a function g is continuous at the point a, and $g(a) \ne 0$, then there exists an interval $I_\delta(a)$ such that $g(x) \ne 0$ for all $x \in I_\delta(a)$. Similarly, if g is continuous at a, then g is *locally* bounded, that is, there exists a number $K > 0$ and an interval $I_\delta(a)$, so that

$$|g(x)| \le K \text{ for all } x \in I_\delta(a).$$

The proof is very simple. Choose $\delta = \delta(1)$, so that $|g(x) - g(a)| < 1$ for all $x \in I_\delta(a)$. Then

$$|g(x)| = |g(a) + (g(x) - g(a))| \le |g(a)| + |g(x) - g(a)| \le |g(a)| + 1.$$

Note that we had proved this property for rational functions in Section IV.1.2, where it turned out to be a key ingredient in proving that rational functions are continuous at each point of their domain. That result was more elementary than what we just verified, as it did not involve limits, and so on. Rather, it was the relevant estimate for rational functions that led us to introduce the idea of continuity.

Finally, we want to mention the following *global* version of the local boundedness of a continuous function. This result is considerably deeper than the preceding result, as it relies on the completeness of $\mathbb{R}$.

Theorem 48 *Suppose f is continuous on the closed and bounded interval $[a, b]$. Then f is bounded on $[a, b]$.*

Continuity at a boundary point is defined in terms of one-sided limits: f is continuous at a if $\lim_{x \to a^+} f(x) = f(a)$, and similarly at the right endpoint b.

Note that both hypotheses on the interval are critical: $g(x) = 1/x$ is continuous on $(0, 1)$ but not bounded, and similarly, $g(x) = x$ is continuous on $[0, \infty)$ but is not bounded on that interval.

Proof. We shall prove this by contradiction, that is, let us assume that f is NOT bounded on $[a, b]$. Let us divide the interval into two equal closed bounded intervals $I' = [a, \frac{a+b}{2}]$ and $I'' = [\frac{a+b}{2}, b]$. By our assumption, f can not be bounded on both intervals. So pick one of them on which f is NOT bounded and call it I_1. Now repeat the same process starting with I_1 to get a closed bounded interval $I_2 \subset I_1$ of length $\frac{1}{2^2}(b - a)$, so that f is NOT bounded on I_2. Clearly one can repeat this on and on, resulting in a nested sequence of closed bounded intervals I_n, $n = 1, 2, 3, ...$, with $I_{n+1} \subset I_n \subset [a, b]$ and length $(I_n) = \frac{1}{2^n}(b - a)$. Furthermore, f is NOT bounded on each interval I_n. By Theorem 28, $\cap_n I_n$ is not the empty set, so there exists x_0 which is contained in the interval I_n for each n, and clearly $x_0 \in [a, b]$, so that f is continuous at x_0. Therefore f is locally bounded at x_0, i.e., there exist $K > 0$ and $\delta > 0$, so that

$$|f(x)| \le K \text{ for all } x \in I_\delta(x_0) \cap [a, b].$$

Note that if x_0 is one of the boundary points of $[a, b]$, say b, then the above condition is $x \in (b - \delta, b]$. Now choose $N(\delta) \in \mathbb{N}$, so that $1/2^{N(\delta)} < \delta/(b-a)$. By construction, f is NOT bounded on $I_{N(\delta)}$, but since $I_{N(\delta)} \subset I_\delta(x_0) \cap [a, b]$ (check it out!), and we just saw that $|f(x)| \le K$ for all $x \in I_\delta(x_0) \cap [a, b]$, and therefore, in particular, for all $x \in I_{N(\delta)}$, f IS bounded on $I_{N(\delta)}$. This hopeless contradiction implies that our assumption that f is NOT bounded on $[a, b]$ cannot be correct. ∎

V.2.7   *Exercises*

1. a) Let $\{s_n : n = 1, 2, ...\}$ be a sequence. A subsequence of $\{s_n\}$ is a sequence $\{s_{n_k} : k = 1, 2, ...\}$, where $\{n_k \in \mathbb{N} : k = 1, 2, ...\}$ is a sequence.

b) Prove that if $\lim_{n \to \infty} s_n = L$, then every subsequence $\{s_{n_k}\}$ also converges with $\lim_{k \to \infty} s_{n_k} = L$.

2. a) Does the sequence defined by $s_n = (-1)^n$ converge? Does it have any convergent subsequences?

b) Does the sequence defined by $c_n = (-1)^n \cdot \frac{1}{n}$ converge? If yes, find its limit.

3. Prove the Monotone Convergence Theorem in case of a decreasing sequence that is bounded from below.

4. Find $\lim_{n \to \infty} d_n$, where $d_n = \sum_{j=0}^n (-\frac{2}{3})^j$.

5. Find the fraction that represents the number $r = 2.427\overline{153}$.

6. Find the decimal expansion of 2/13.

7. Prove that $q = 0.\overline{a_1 a_2 ... a_l}$ is rational by considering $10^l q$.

8. Prove the Squeeze Theorem for sequences. (Part iv in Lemma 31.)

9. Prove the limit properties 2) i) and ii) in Section 5 for functions.

10. Determine the limits

$$a) \quad \lim_{x \to 25} \frac{x - 25}{\sqrt{x} - 5}, \qquad b) \quad \lim_{r \to 2} \frac{4 - r^2}{r^3 - 8}.$$

11. Define $g(x) = \begin{cases} \frac{x-1}{x+3} & \text{for } x \neq -3 \\ -4 & \text{for } x = -3 \end{cases}$. At which points $a \in \mathbb{R}$ is g continuous? Explain!

12. We had seen that $\sum_{n=1}^{\infty} \frac{1}{n}$ does not converge. Prove that $\sum_{n=1}^{\infty} (-)^{n+1} \frac{1}{n}$ converges. Hint: Let $\{s_n\}$ be the sequence of partial sums. Show that $0 < s_{2n+2} < s_{sn}$ for each n and look at the subsequence $\{s_{2n}\}$.

13. Determine $\lim_{x \to 3} \frac{x^3 - 27}{x - 3}$. (Hint: Think of a particular derivative!)

14. Give an alternate proof of the Intermediate Value Theorem by completing the following outline. Suppose $f(a) < \lambda < f(b)$. Let c_1 be the midpoint of $[a, b]$. If $f(c_1) = \lambda$ we are done. Otherwise, either $f(a) < \lambda < f(c_1)$ or $f(c_1) < \lambda < f(b)$; this identifies an interval $[a_1, b_1]$ of length $\frac{1}{2}(b - a)$ with $f(a_1) < \lambda < f(b_1)$. Continuing this process, construct a nested sequence of intervals $[a_n, b_n]$, $n = 1, 2, 3, ...$, with $f(a_n) \leq \lambda \leq f(b_n)$. The rest is up to you.

15. a) Show that $x^3 - 7 = 0$ has a solution between 1.9 and 2.

b) Use the intermediate value theorem to show that every polynomial of odd degree has at least one zero in $\mathbb{R}$.

16. Suppose $g : [0, 1] \to [0, 1]$ is continuous at every point in $[0, 1]$. Show that there exists at least one point $x \in [0, 1]$ with $g(x) = x$. (Hint: Apply the Intermediate Value Theorem to $f(x) = g(x) - x$.)

17. Let $I = [a, b]$ be a closed bounded interval and suppose $g : I \to \mathbb{R}$ is continuous and $g(x) > 0$ for all $x \in I$. Show that there is $\delta > 0$, so that $g(x) \geq \delta$ for all $x \in I$.

V.3   Exponential Functions for Real Numbers

In Chapter IV.4.2 we introduced the basics of the exponential function $E_b(x) = b^x$. We noticed at that time that even for rational inputs the definition required going beyond $\mathbb{Q}$, as the central functional equation

$$E_b(u + v) = E_b(u)E_b(v)$$

implies that $b^{1/n}$ has to be a number that satisfies $(b^{1/n})^n = b$, that is, $b^{1/n}$ is the nth root of b for any $n \in \mathbb{N}$. In Section 1.3 we proved that such

nth roots always exist within the real numbers $\mathbb{R}$, thereby completing the definition of E_b for rational inputs. The next important step is to extend the definition of E_b to arbitrary real numbers as inputs.

V.3.1 *Properties of Exponential Functions*

We first list a few important properties of exponential functions that easily follow from the functional equation and the definitions discussed earlier. While we only consider rational inputs at this time, all results will remain valid for arbitrary real numbers as inputs, as will be considered in Section 3.3.

We fix the base $b > 0$. First of all, $E_b(u) \neq 0$ for all u. This follows from $E_b(-u) \cdot E_b(u) = E_b(-u + u) = E_b(0) = 1$. Furthermore, the definition of $E_b(m/n)$ implies that $E_b(u) > 0$. Now assume $b > 1$. Then $\lambda = E_b(1/n) > 1$ for each $n \in \mathbb{N}$, because $\lambda \leq 1$ would imply that $b = \lambda^n \leq 1$, which contradicts the assumption $b > 1$. Consequently, if $m > 0$, then $E_b(m/n) = \lambda^m > 1$, that is, $E_b(u) > 1$ for all (rational) $u > 0$, and therefore $E_b(-u) = 1/E_b(u) < 1$. If $0 < b < 1$, analogous properties follow by using $E_b(-u) = 1/E_b(u) = E_{1/b}(u)$, where now $1/b > 1$. Let us summarize these results.

If $b > 1$, then $E_b(u) = b^u > 1$ for all $u > 0$, and $E_b(u) = b^u < 1$ for $u < 0$.

If $0 < b < 1$, then $0 < E_b(u) = b^u < 1$ for all $u > 0$, and $E_b(u) = b^u > 1$ for $u < 0$.

The graphs of exponential functions are most easily obtained by means of a graphing calculator. Note that only rational numbers, or rational approximations to irrational numbers can be processed by a calculator or computer. Figure V.1 shows the graph of E_2.

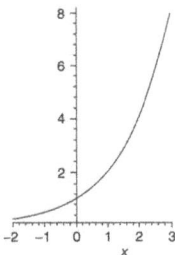

Fig. V.1 Graph of $y = E_2(x) = 2^x$.

The graphs of E_b for $b > 1$ look very similar to the graph of E_2. Here are three more such graphs.

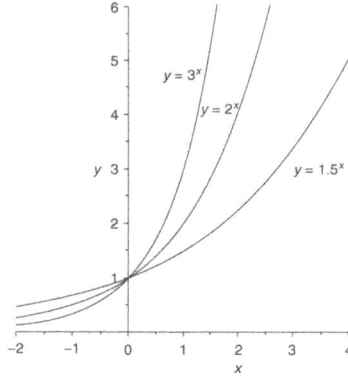

Fig. V.2   Exponential functions with base $b > 1$.

The graphs suggest the following fact.
If $b > 1$, the exponential function is *strictly increasing*:

$$\text{If } u < v, \text{then } E_b(u) < E_b(v).$$

In fact, write $v = u + \gamma$, where $\gamma > 0$. Then $1 < E_b(\gamma)$, and multiplication on both sides of this inequality with the positive number $E_b(u)$ gives $E_b(u) < E_b(u)E_b(\gamma) = E_b(u + \gamma) = E_b(v)$.

If $0 < b < 1$, the situation is reversed: it follows that E_b is strictly *decreasing*, i.e., if $u < v$, then $E_b(u) > E_b(v)$.

Here are some graphs of exponential functions with base $b < 1$.

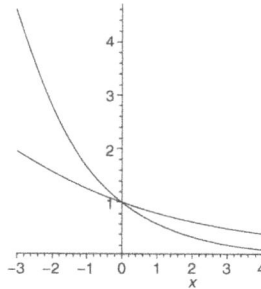

Fig. V.3   Exponential functions with base $b < 1$.

Since $E_b(-u) = b^{-u} = (1/b)^u = E_{1/b}(u)$, the graph of E_b $(b < 1)$ is obtained by reflecting the graph of $E_{1/b}$ $(1/b > 1)$ on the y-axis, that is, by replacing u with $-u$.

Finally we note the following property that is intuitively obvious from the graphs of the exponential functions.

$$E_b(u) \to 1 \ as \ u \to 0.$$

Given our discussion of limits in Section 2.5, we write this property in the form

$$\lim_{u \to 0} E_b(u) = 1. \tag{V.1}$$

Let us verify this precisely. Again, while we only use rational inputs at this point, the proof will work exactly the same way for arbitrary real numbers as inputs. We want to show that the distance between $E_b(u)$ and 1 can be made arbitrarily small by choosing u sufficiently close to 0. Assume $b > 1$. Then the number 1 is a lower bound for the set $\{E_b(1/n) : n \in \mathbb{N}\}$, and therefore the *greatest* lower bound λ for this set, i.e., the infimum $\lambda = \inf\{E_b(1/n) : n \in \mathbb{N}\}$ must be ≥ 1. Since $\lambda \leq b^{1/n}$ implies $\lambda^n \leq b$ for all n, the set $\{\lambda^n : n \in \mathbb{N}\}$ is bounded from above. Below we show that if $\lambda > 1$, then this set is NOT bounded, and consequently we must have $\lambda = 1$. Hence, given any $\varepsilon > 0$, the number $1 + \varepsilon$ is *not* a lower bound for $\{E_b(1/n)\}$, so that there exists a natural number n_0 such that $E_b(1/n_0) < 1 + \varepsilon$. Since for $u > 0$ the values $E_b(u)$ decrease as $u \to 0$, we have $1 < E_b(u) \leq E_b(1/n_0) < 1 + \varepsilon$ for all $0 < u \leq 1/n_0$ as well. By taking reciprocals, it then follows that $1 > E_b(-u) > 1/(1 + \varepsilon)$ for all such u. Since $1/(1 + \varepsilon) = 1 - \varepsilon/(1 + \varepsilon) > 1 - \varepsilon$, one obtains altogether that

$$1 - \varepsilon < E_b(u) < 1 + \varepsilon \ \text{for all } u \text{ with } |u| < 1/n_0, \tag{V.2}$$

that is,

$$|E_b(u) - 1| < \varepsilon \ \text{for all } u \text{ with } |u| < 1/n_0. \ \blacksquare$$

This shows that no matter how small ε is chosen, the distance between $E_b(u)$ and 1 will be smaller than ε provided u is sufficiently close to 0. The reader should review the preceding arguments carefully and understand the reasoning well, as this is the first verification of a non-elementary limit statement.

Note that since $1 = E_b(0)$, we have $\lim_{u \to 0} E_b(u) = E_b(0)$, that is, the function E_b is *continuous* at 0.

If $b < 1$, note that $b^u = 1/E_{1/b}(u)$, where now $1/b > 1$; the result then follows by applying the estimate (V.2) to $E_{1/b}(u)$ and taking reciprocals.

We also identify some properties of exponential functions for large input values. If $b > 1$, the graph suggests that b^x grows larger than any fixed large number M if x is chosen sufficiently large. In fact, if $\delta = b - 1 > 0$, then $b = 1 + \delta$. It readily follows from the binomial theorem that for $n \in \mathbb{N}$ one has $b^n = (1 + \delta)^n > 1 + n\delta$. Given M, it follows that if $n > M/\delta$, then $b^n > M$. Since E_b is strictly increasing ($b > 1$), one has $b^u > M$ for all $u \geq n > M/\delta$. One writes

$$\lim_{x \to \infty} b^x = \infty \ \text{when} \ b > 1.$$

We emphasize once again that ∞ is *not* a number. The preceding statement is just a convenient shorthand notation to refer to a quantity (b^x in the case at hand) that grows without any bound as x gets larger. In particular, we just verified that if $b > 1$, the set $\{b^n : n \in \mathbb{N}\}$ is not bounded, a fact that we used in the proof of the previous statement.

Since $b^{-x} = 1/b^x$, it then follows that $b^{-x} \to 0$ as $x \to \infty$. We write this symbolically as $\lim_{x \to \infty} b^{-x} = 0$, a statement that is also written as

$$\lim_{x \to -\infty} b^x = 0 \ \text{when} \ b > 1.$$

By considering numerical examples, one recognizes that exponential functions b^x with base $b > 1$ in fact grow much faster than any particular power function x^k, no matter how large the exponent k is chosen. More precisely, if $b > 1$ and k is any fixed positive integer, then

$$\frac{b^x}{x^k} \to \infty \ \text{as} \ x \to \infty \ , \ \text{or} \ \lim_{x \to \infty} \frac{b^x}{x^k} = \infty.$$

This latter result can also be established for $x \in \mathbb{N}$ by using the binomial theorem. (See Exercise 7 in Section 3.5.)

V.3.2   *Definition of $E_b(x) = b^x$ for Real Numbers x*

Recall that the collection of all real numbers is visualized by the (continuous) number line, which has no gaps whatsoever. The graph of an exponential function E_b (for rational inputs) is given by a curve that is obtained by gently bending the "line" consisting of the rational numbers according to a particular rule. Extending the definition of E_b to all real numbers should result in a graph that to the eye looks exactly like the graph corresponding to rational inputs, where now the curve is assumed to be completely filled in without any gaps whatsoever, i.e., the curve should satisfy the same "continuity property" as the real line $\mathbb{R}$. In other words, we expect that the exponential function E_b defined on $\mathbb{R}$ is continuous at every point.

Let us consider the case with base $b > 1$, so that E_b is strictly increasing. If $x \in \mathbb{R}$, the point $(x, E_b(x))$ on the graph should arise as the "least upper bound" of the set of points $\{(r, E_b(r)) : r \in \mathbb{Q} \text{ and } r < x\}$, where the order relation on the curve is the natural one coming from the order relation on the number line. This would ensure that $\lim_{r \to x^-} E_b(r) = E_b(x)$. More precisely, given $x \in \mathbb{R}$, choose a natural number n with $x < n$. For any $r \in \mathbb{Q}$ with $r < x$ one then has $r < n$ as well, and hence $E_b(r) < E_b(n)$. Therefore the set $S(x) = \{E_b(r) : r \in \mathbb{Q} \text{ with } r < x\}$ has the upper bound $E_b(n)$, and consequently, by the completeness axiom, it has a *least* upper bound $\sup S(x)$ in $\mathbb{R}$.

Definition 49 *Assume $b > 1$. For $x \in \mathbb{R}$ one defines*

$$E_b(x) = \sup\{E_b(r) : r \in \mathbb{Q} \text{ with } r < x\}.$$

An analogous definition involving the greatest lower bound is made in case $0 < b < 1$.

Remark. In case $q \in \mathbb{Q}$ to begin with, b^q is already defined, and since E_b is increasing, b^q clearly is an upper bound for $S(q) = \{E_b(r) : r \in \mathbb{Q}$ with $r < q\}$. Indeed, b^q is the *least* upper bound of this set; this follows from (V.1), since $E_b(q - 1/n) = E_b(q)E_b(-1/n) \to E_b(q)$ as $n \to \infty$. In other words, if q is rational, one has $\sup\{E_b(r) : r \in \mathbb{Q} \text{ with } r < q\} = b^q$, so that the above definition is consistent with the original definition of the exponential function for rational inputs.

To be precise, when we first introduced the exponential function E_b, it was assumed that the base b was a (positive) rational number. Now that we have enlarged $\mathbb{Q}$ to $\mathbb{R}$, we of course want to allow any positive *real* number as base. In that more general case, the definition of E_b first for natural numbers, then integers, and ultimately rational numbers, which all rely on the functional equation, works just the same as when the base is rational. So from now on we allow the base to be any positive real number.

Next, we want to discuss a somewhat more concrete interpretation of the definition of $E_b(x) = b^x$ for arbitrary *real* x involving approximation by $E_b(r_n)$ for a suitable sequence of rational numbers $\{r_n : n = 1, 2, 3, ...\}$ with $r_n \to x$.

Lemma 50 *Suppose $b > 1$ and $x \in \mathbb{R}$. Let $\{r_n\}$ be the increasing sequence given by Lemma 32. One then has*

$$E_b(x) = \lim_{n \to \infty} E_b(r_n).$$

The point of this lemma is that it is perhaps easier to visualize $E_b(x)$ via the approximating sequence then in terms of the more abstract LUB Property.

Proof. Let q be a rational number $> x$. Since the function E_b is strictly increasing (on rational numbers), the sequence $\{E_b(r_n) : n = 1, 2, 3, ...\}$ is increasing, and furthermore it is bounded above by $E_b(q)$. By Theorem 29, one has $\lim_{n \to \infty} E_b(r_n) = \sup\{E_b(r_n)\}$, and it is easily seen that this last number agrees with $\sup\{E_b(r) : r < x\} = E_b(x)$. ∎

With the obvious modifications in the preceding proof one sees that b^x is also the limit of $\{E_b(s_n)\}$, where $\{s_n\}$ is a decreasing sequence of rational numbers with limit x.

In case $0 < b < 1$, we know that E_b is strictly decreasing for rational numbers. In this case, $E_b(x) = \inf\{E_b(r) : r \in \mathbb{Q} \text{ and } r < x\}$, and we see by arguments analogous to those we just gave that $E_b(x)$ is the limit of the (now decreasing) sequence $E_b(r_n)$.

To summarize the essence: at the intuitive level, the function $E_b(u) = b^u$ has been defined in such a way that its graph is a curve that is obtained by simply bending the number line so that it fits through all the points $\{(r, b^r) : r \in \mathbb{Q}\}$. The continuity *axiom* (i.e., completeness) of $\mathbb{R}$ and the definition of E_b for arbitrary real numbers ensure that there are no "gaps" in the graph, just as there are no gaps in the number line. And furthermore, the extension of E_b to arbitrary real numbers as input that we just described will ensure that the function remains continuous on $\mathbb{R}$. We shall verify this precisely in the next section.

V.3.3   *Basic Properties of E_b on $\mathbb{R}$*

The most important result for the function E_b is that the fundamental functional equation remains valid for all real numbers.

Theorem 51 *The functional equation*

$$E_b(u + v) = E_b(u) \cdot E_b(v)$$

holds for all $u, v \in \mathbb{R}$.

Proof. By Lemma 31 we can choose increasing sequences of rational numbers $\{u_n\}$ and $\{v_n\}$ with $\lim_{n \to \infty} u_n = u$ and $\lim_{n \to \infty} v_n = v$. Then $\{u_n + v_n\}$ is an increasing sequence with $\lim_{n \to \infty}(u_n + v_n) = u + v$. By Theorem 28, we have $E_b(u+v) = \lim_{n \to \infty} E_b(u_n + v_n)$. Since we know that the functional equation holds for rational inputs, one has $E_b(u_n + v_n) = E_b(u_n)E_b(v_n)$. Finally, by property ii) in Lemma 30,

$$\lim_{n \to \infty} [E_b(u_n)E_b(v_n)] = \lim_{n \to \infty} E_b(u_n) \cdot \lim_{n \to \infty} E_b(v_n) = E_b(u) \cdot E_b(v). \blacksquare$$

We had established that if $b > 1$, then $E_b(r) > 1$ for rational numbers $r > 0$. We can now easily show that this holds for arbitrary real numbers $u > 0$ as well. In fact, choose an increasing sequence $\{r_n\}$ of rational numbers with $\lim_{n \to \infty} r_n = u$. Since $u > 0$, there exists $N(u)$, so that for $n \geq N(u)$ one has $r_n > 0$ as well. For such n we know that $E_b(r_n) > 1$, and since $E_b(r_n) < E_b(u)$, we must have $E_b(u) > 1$ as well.

It then follows that $E_b(-u) < 1$. If $b < 1$, the situation is reversed: $E_b(u) < 1$ for $u > 0$, which then implies $E_b(-u) > 1$.

Next we show that for $b > 1$ the function E_b is strictly increasing on $\mathbb{R}$. In fact, if $u < v$, then $v = u + \delta$, where $\delta > 0$. Hence

$$E_b(v) = E_b(u + \delta) = E_b(u) \cdot E_b(\delta) > E_b(u) \cdot 1.$$

Similarly, if $b < 1$, then E_b is strictly decreasing on $\mathbb{R}$.

By arguments that are identical to the case when the input was limited to rational numbers, one can now show that $\lim_{u \to 0} E_b(u) = 1 = E_b(0)$, where $u \in \mathbb{R}$, for any $b > 0$, that is, E_b is continuous at 0.

The functional equation then easily implies that E_b is continuous at every point $a \in \mathbb{R}$, as follows.

Note that $E_b(x) = E_b((x - a) + a) = E_b(x - a) \cdot E_b(a)$. Set $u = x - a$. Then $x \to a$ means $u \to 0$, so that

$$\lim_{x \to a} E_b(x) = \lim_{u \to 0} [E_b(u)E_b(a)] = \lim_{u \to 0} [E_b(u)] \cdot E_b(a) = 1 \cdot E_b(a) = E_b(a). \blacksquare$$

Finally, we extend two other equations that are known for rational values to arbitrary real inputs.

Lemma 52 *For any $u, v \in \mathbb{R}$ one has $E_b(u)^v = E_b(uv)$.*

Proof. Note that $E_b(u) = c > 0$, so c qualifies as a base for the exponential function. Thus $E_b(u)^v = E_c(v)$. For the proof, we of course use the fact that the Lemma is correct for $u, v \in \mathbb{Q}$. Let $\{r_n\}$ and $\{s_n\}$ be increasing sequences of rational numbers with $\lim_{n \to \infty} r_n = u$ and $\lim_{n \to \infty} s_n = v$. Then $E_c(v) = \lim_{n \to \infty} E_c(s_n)$. Let us fix s_n. Since $E_c(s_n) = c^{s_n}$ and s_n is rational, the function $f(x) = x^{s_n}$ is a familiar power function with rational exponent, which is continuous on $\{x > 0\}$, and therefore at the point $c > 0$. Since E_b is continuous, we have $\lim_{k \to \infty} E_b(r_k) = E_b(u) = c$, and therefore

$$\lim_{k \to \infty} E_b(r_k)^{s_n} = E_b(u)^{s_n}.$$

Since the Lemma holds for rational inputs, we have $E_b(r_k)^{s_n} = E_b(r_k s_n)$, and since $\lim_{k \to \infty}(r_k s_n) = u s_n$, it follows that

$$E_c(s_n) = E_b(u)^{s_n} = \lim_{k \to \infty} E_b(r_k)^{s_n} = \lim_{k \to \infty} E_b(r_k s_n) = E_b(u s_n).$$

Since the exponential functions E_c and E_b are continuous, we now take the limit as $n \to \infty$ to obtain

$$E_b(u)^v = E_c(v) = \lim_{n \to \infty} E_c(s_n) = \lim_{n \to \infty} E_b(u s_n) = E_b(uv). \blacksquare$$

Lemma 53 *If $a, b > 0$, then $E_a \cdot E_b = E_{ab}$, that is, $a^x b^x = (ab)^x$ for $x \in \mathbb{R}$.*

Proof. Fix $x \in \mathbb{R}$ and choose an increasing sequence $\{r_n\}$ of rational numbers with $\lim_{n \to \infty} r_n = x$. By Section IV.4.2 we know that $a^{r_n} b^{r_n} = (ab)^{r_n}$ for each n. Now take the limit as $n \to \infty$ and use the continuity of the exponential functions and properties of limits to obtain the desired conclusion. $\blacksquare$

Given the importance of the exponential function, both in applications as well as for the development of key concepts in mathematics, let us summarize the basic properties of these functions. The reader is urged to review these carefully and to make sure that she/he fully understands them.

Properties of Exponential Functions of Real Numbers.

(1) For any $b > 0$ the exponential function $E_b(x) = b^x$ is defined for all real numbers, with $E_b(0) = 1$, and $E_b(1) = b$.

(2) E_b satisfies the functional equations

$$E_b(u + v) = E_b(u) E_b(v) \text{ and}$$
$$E_b(u)^v = E_b(uv) \text{ for all } u, v \in \mathbb{R}$$

In particular, the first equation implies that for any $n \in \mathbb{N}$, $E_b(\frac{1}{n})$ is the nth root of b, that is $[E_b(\frac{1}{n})]^n = b$.

(3) $E_b(x) > 0$ for all $x \in \mathbb{R}$, so that, in particular, $E_b(x)$ is never zero. $E_b(-u)$ is the multiplicative inverse of $E_b(u)$, which is also denoted by $[E_b(u)]^{-1}$ or $\frac{1}{E_b(u)}$.

(4) E_b is continuous at every point $a \in \mathbb{R}$.

(5) If $1 < b$, then E_b is *strictly increasing* (i.e., $b^c < b^{c'}$ whenever $c < c'$), and if $0 < b < 1$, then E_b is *strictly decreasing* (i.e., $b^c > b^{c'}$ whenever $c < c'$). Note that this implies that if $b \neq 1$, then E_b is *one-to-one*, that is, if $c \neq c'$, then $b^c \neq b^{c'}$.

(6) For any $a, b > 0$ one has $E_a \cdot E_b = E_{ab}$.

V.3.4   *Inverse Functions and Logarithms*

Inverse Functions   In Chapter III.2.4 we introduced the inverse function f^{-1} of a function $f : \Omega \to f(\Omega)$ that is one-to-one. Recall that $f^{-1} : f(\Omega) \to \Omega$ is defined as follows. If $b \in f(\Omega)$, by the assumption on f there exists a *unique* $a \in \Omega$ with $f(a) = b$, and one defines $f^{-1}(b) = a$. A familiar example involves the squaring function $S : \mathbb{R}^{\geq 0} \to f(\mathbb{R}^{\geq 0})$, defined by $S(x) = x^2$, whose inverse $S^{-1}(x) = \sqrt{x}$ is the square root function. Based on the completeness of $\mathbb{R}$, we now know that $S(\mathbb{R}^{\geq 0}) = \mathbb{R}^{\geq 0}$, that is, for every real number $b \geq 0$ there exists a number $a \geq 0$ with $a^2 = b$, which is denoted by $a = \sqrt{b}$. Similarly, for any $n \geq 3$, the function $P_n(x) = x^n$ is one-to-one on $\mathbb{R}^{\geq 0}$, and by Lemma 26 we know that $P_n(\mathbb{R}^{\geq 0}) = \mathbb{R}^{\geq 0}$. Its inverse is the nth root function $\sqrt[n]{x} = x^{1/n}$. Our goal is to study the inverse function of the exponential function E_b, which is one-to-one for any base $b \neq 1$.

But before we get into that, we want to establish a few general properties of inverse functions. Geometrically, if $\Omega \subset \mathbb{R}$, and f is real-valued, then f is one-to-one if its graph satisfies the "horizontal line test": any horizontal line meets the graph of f at most in one point. See the figures below.

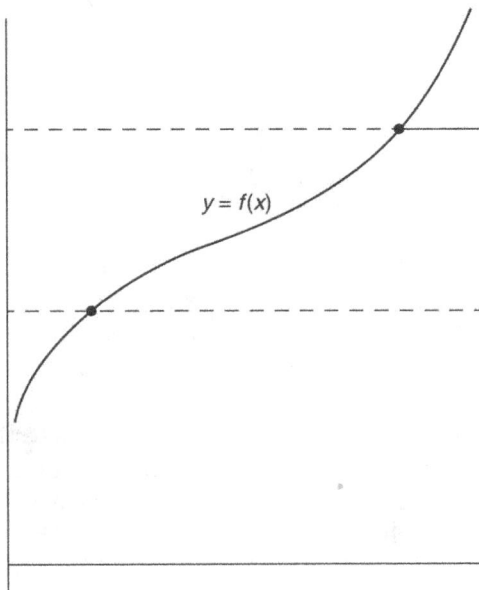

Fig. V.4 f is one-to-one.

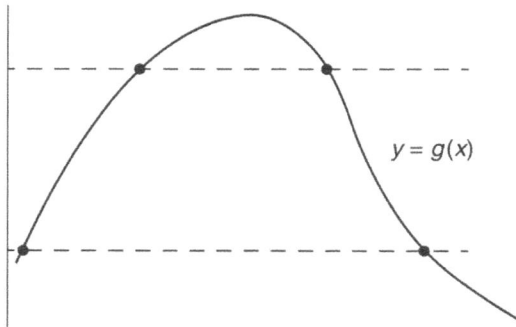

Fig. V.5 g is NOT one-to-one.

Next, we want to examine the relationship between the graphs of a function f and of its inverse g in general.

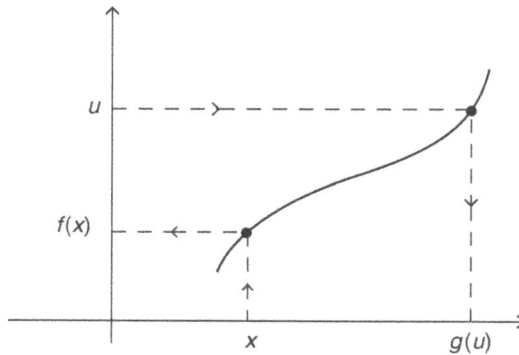

Fig. V.6   Graph of f and the reverse relation.

According to Fig. V.6, the graph of the function $y = f(x)$ can also be used to describe the inverse function $g = f^{-1}$ of f. Just start with a point u on the vertical axis that lies in $f(\Omega)$. The horizontal line $y = u$ intersects the graph of f in precisely one point (x, u) (*horizontal* line test!); then $u = f(x)$ and $g(u) = x$. In order to be consistent with the convention to mark the input value of a function on the horizontal coordinate axis, usually labeled x-axis, we need to interchange the coordinate axis and switch notation, i.e., we interchange x with u and write $g(x) = u$. As seen below, geometrically, interchanging the coordinates of a point (a, b) leads to the point (b, a) that is visualized by reflecting (a, b) on the line $y = x$.

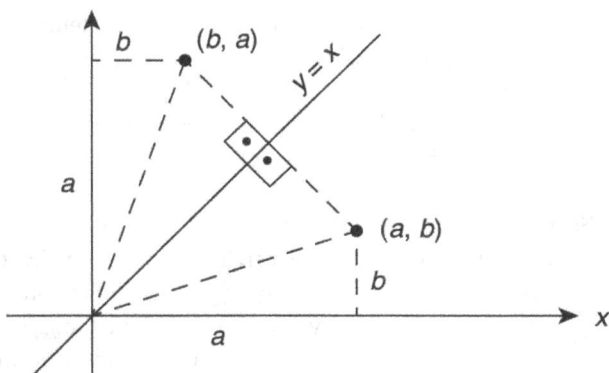

Fig. V.7   Reflection of (a, b) on the line $y = x$.

Applying this reflection to each point $(x, f(x))$ on the graph of f gives the set of points $\{(f(x), x), x \in \Omega\} = \{(u, g(u)), u \in f(\Omega)\}$. This set is evidently the graph of g displayed in the usual form, that is, the input variable is displayed on the horizontal axis and labeled u instead of the commonly used x. Lastly one may replace u with x to make the notation consistent.

To summarize:

The graph of the inverse function g of a one-to-one function f is obtained by reflecting the graph of f on the line $y = x$.

Next, we want to verify an important property of the image of a continuous one-to-one function.

Theorem 54 *Suppose f is strictly increasing (and hence one-to-one) on the open interval $I \in \mathbb{R}$ and that it is continuous on I, that is, at each point $a \in I$. Then $J = f(I)$ is an open interval in $\mathbb{R}$, and the inverse $g : J \to I$ of f is continuous on J. The same conclusion holds if f is strictly decreasing on I.*

The rigorous proof of this result is somewhat delicate, as it does require explicit use of the completeness of the real numbers. However, at the intuitive geometric level the result is clear, as follows. The hypotheses imply that the graph of f is a "continuous curve" without any cuts or holes that satisfies both the vertical and horizontal line tests. We just saw that the graph of the inverse g is obtained by reflecting the graph of f on the line $y = x$, a process that does not make any changes to the graph but just places it in a different way in the coordinate system. Hence the graph of the inverse g is also a "continuous curve" without any cuts or holes, i.e., g is continuous as well.

Proof of Theorem 54. To prove that $J = f(I)$ is an open interval, let $b = f(a) \in J$. Since I is open, there is $\epsilon > 0$ so that $I_{2\epsilon}(a) \subset I$. Hence $f(a-\varepsilon)$ and $f(a+\varepsilon)$ are in J, and by the Intermediate Value Theorem (IVT) discussed in Section 2.6, the whole interval $(f(a-\varepsilon), f(a+\varepsilon))$ is contained in J. This shows that J is open, and furthermore, again by the IVT, it also follows that J is an interval. Finally, we must show that the inverse $g : J \to I$ of f is continuous at every point $b \in J$, i.e., $\lim_{y \to b} g(y) = g(b)$. Note that $g(b) = a \in I$. Since I is assumed open, a is not a boundary point of I. We prescribe the (arbitrary) closeness of $g(y)$ to the expected limit $g(b) = a \in I$ by fixing an arbitrarily small closed interval $[a - \varepsilon, a + \varepsilon] \subset I$ with $\varepsilon > 0$. Let us show that for all y that are sufficiently close to b one has $g(y) \in (a - \varepsilon, a + \varepsilon)$. Since f is increasing, $f(I_\varepsilon(a)) = (f(a - \varepsilon), f(a + \varepsilon))$, and this latter interval is open and contains the point b, since $f(a - \varepsilon) < f(a) = b < f(a + \varepsilon)$. Now choose $\delta > 0$ sufficiently small, so that the interval $J_\delta(b)$ is contained in $(f(a - \varepsilon), f(a + \varepsilon))$. Then $g(J_\delta(b)) \subset I_\varepsilon(a)$, i.e.,

$$|g(y) - g(b)| < \varepsilon \text{ for all } y \text{ with } |y - b| < \delta,$$

as required. ∎

Remark. In case f is strictly decreasing, the proof is essentially the same, subject to the obvious necessary modifications.

Remark. One can show that a function f that is continuous and one-to-one on an interval I is in fact either strictly increasing on I or strictly decreasing on I. However, the proof (see Exercise 9) is somewhat intricate, and since we shall apply the Theorem mainly to functions that are known to be either increasing or decreasing, the Theorem we stated is quite sufficient for our applications.

Logarithms   We shall now apply the general properties regarding inverse functions to exponential functions. We had seen that all exponential functions

$$E_b(x) = b^x \text{ with } 0 < b \neq 1$$

are one-to-one and continuous on the whole real line $\mathbb{R}$. Also, since $E_b(x) > 0$ for all x, the image $E_b(\mathbb{R}) \subset \{y \in \mathbb{R} : y > 0\}$. Let us denote this latter set by $\mathbb{R}^+$. Suppose $b > 1$. Then E_b is strictly increasing on $\mathbb{R}$,

$\lim_{x\to\infty} E_b(x) = \infty$, and $\lim_{x\to-\infty} E_b(x) = 0$. Therefore, for any small $\varepsilon > 0$ and large N, there are A and $B \in E_b(\mathbb{R})$ with $0 < A < \varepsilon$ and $N < B$. By the Intermediate Value Theorem, the whole interval (A, B) is contained in the image of E_b. This shows that $E_b(\mathbb{R}) = \mathbb{R}^+$. The same conclusion holds if $0 < b < 1$, where now E_b is strictly decreasing.

Therefore, by the general results in the previous section, E_b has an inverse $(E_b)^{-1} : \mathbb{R}^+ \to \mathbb{R}$ that is continuous at every point.

This inverse function of E_b is called the **logarithm function to the base** b, and it is denoted by $\log_b$. Thus, if $u = E_b(x) = b^x$, then $x = \log_b u$, and vice-versa. So one has the equations

$$u = b^{\log_b u} \text{ and } \log_b(b^x) = x.$$

Usually, when considering exponential and logarithm functions one assumes that the base b is greater than 1, so that both functions are (strictly) increasing. By applying the reflection principle to the graph of $E_2(x) = 2^x$, one obtains the graph of the inverse function $y = \log_2 x$ by reflection, as shown in Fig. V.8.

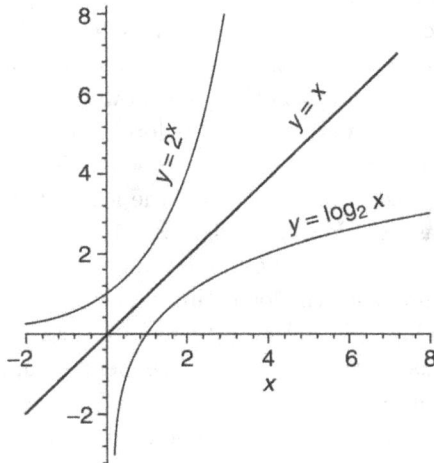

Fig. V.8   Graph of $y = \log_2 x$.

We summarize the basic properties of logarithm functions $\log_b$, where we assume that the base $b > 1$.

i) $\log_b$ *is defined on* $\mathbb{R}^+ = (0, \infty)$ *and is strictly increasing and one-to-one;*

ii) $\log_b(1) = 0$;

iii) $\log_b(uv) = \log_b(u) + \log_b(v)$;

iv) $\log_b(u^a) = a \log_b(u)$.

The properties iii) and iv) are the functional equations of the logarithm that follow from the corresponding equations for E_b. (See Basic Properties **2**.) For example, in order to verify iii), set $x_1 = \log_b(u)$ and $x_2 = \log_b(v)$. Then $u = b^{x_1}$ and $v = b^{x_2}$, and therefore

$$uv = b^{x_1}b^{x_2} = b^{x_1+x_2}$$

by the functional equation for the exponential function E_b. This latter equation implies that

$$\log_b(uv) = x_1 + x_2 = \log_b(u) + \log_b(v).$$

Property iv) is left as an exercise.

Properties iii) and iv) were used before the era of computers to facilitate large scale computations. Extensive tables of logarithms had been compiled, so that large numbers could be readily replaced by their logarithms, and vice versa. By using logarithms, *multiplication* of two large numbers could be replaced by the simpler operation of *adding* the corresponding logarithms. A computing device known as a *slide rule*, which was widely used by scientists and engineers before handheld electronic calculators became available in the 1980s, is also based on logarithms. Because our number system is based on powers of 10, the logarithms most widely used were the ones to base 10. For example, $\log_{10} 100 = 2$, $\log_{10} 1000 = 3$, and so on. Given the importance of the binary number system in today's digital world, one would think that the logarithm function to base 2 should be the more useful one today. However, it turns out that the so-called *natural* logarithm, whose base is the special transcendental number $e = 2.71828...$, is the one that is most widely used. We shall discuss the origin of this mysterious number e in detail in Section 4.4, when we examine the tangent problem for exponential functions.

V.3.5 *Exercises*

1. Evaluate the expressions 5^{-3}, $27^{1/3}$, $4^{-2} \cdot 2^4$, $32^{3/5}$, $6^6 \cdot 6^{-4}$, $(2^{\sqrt{3}})^{\sqrt{3}}$, $(25^{1/6})^3$ by applying appropriate functional equations. (Do NOT use a calculator!)

2. Use a calculator to find approximate values for $10^{0.6}$, $3^{\sqrt{3}}$, π^π.

3. Simplify as much as possible:

$$i) \quad \frac{b^{-3}b^5b^{1/2}}{b^{3/2}b^{-2}} \qquad ii) \quad \frac{\sqrt[5]{c}\sqrt[2]{c^4}}{c^{-2/5}\sqrt[3]{c^2}} \qquad (b, c > 0).$$

4. Use a graphing calculator to plot the functions $f(x) = x^{10}$ and $g(x) = 2^x$ in one window for $-1 \leq x \leq 5$.

a) Which function grows faster?

b) What are the solutions of the equation $f(x) = g(x)$? (Use graphing techniques.)

c) Are you sure to have found *all* solutions in b)? How does the graph relate to the statement that exponential functions grow faster than power functions?

d) Investigate the behavior of f and g by changing the viewing window.

5. Use a scientific calculator to estimate the value of $x^{20} \cdot 2^{-x}$ as $x \to \infty$, i.e., as x gets larger and larger.

6. We had verified that E_b is continuous at 0. Prove that E_b is continuous at each point $a \in \mathbb{R}$. (Hint: Use the functional equation.)

7. a) Suppose $b > 1$, i.e., $b = 1 + \delta$, where $\delta > 0$. Fix a positive integer k. Show that if $n \geq k + 1$, then $b^n > 1 + \binom{n}{k+1}\delta^{k+1}$, where $\binom{n}{k+1} = \frac{n!}{(k+1)!(n-k-1)!}$ is the binomial coefficient. (Hint: Expand $(1 + \delta)^n$ by the binomial theorem.)

b) Show that there exists a constant $c_k > 0$, such that $\frac{n!}{(k+1)!(n-k-1)!} \geq n^{k+1} \cdot c_k$ for all $n \geq 2k$.

c) Use a) and b) to show that $\frac{b^n}{n^k} \to \infty$ as $n \to \infty$.

8. A roast is taken out of an oven at 350^0 F at time $t = 0$ and set on the counter to cool off. Its temperature (in degrees F) after t hours is given by $T(t) = 350 - Q \cdot (1 - 2^{-t})$ for some constant Q.

a) Make a rough sketch of the graph of T. (Do NOT use a graphing calculator!)

b) Give an interpretation of the number Q. (Hint: See what happens after a long time....)

9. Show that if $f : I \to \mathbb{R}$ is continuous and one-to-one on the open interval I, the either f is strictly increasing on I, or f is strictly decreasing on I.

V.4   Derivatives of Exponential Functions

As we already recognized in Section IV.4.3, the study of tangent lines and
of the related notion of instantaneous rate of change for the exponential
function leads to new phenomena that transcend the elementary algebraic
methods that we had used up to that point. In particular, we saw that
there is no analogue of the basic elementary factorization that was central
to identify tangents in the case of rational functions. The algebraic factorization led us to discover that the derivative can also be captured by
an approximation process. A first attempt to utilize this for the exponential function made us realize that new important concepts needed to be
introduced to capture the mysterious numbers that are the outcome of the
approximation. We built the relevant foundations in the previous three
sections, and we are now ready to investigate the existence and properties
of the derivatives of exponential functions in great detail. The central role
of the exponential function is evidenced by the fact that the insights gained
along the way will naturally lead us to formulate the general concept of
differentiable function.

V.4.1   *Tangents for $E_2(x) = 2^x$*

Let us begin with the concrete exponential function $E_2(x) = 2^x$ that we
already considered in Section IV.4.3. Inspection of the graph of this function suggests that it has a (non vertical) tangent line at every point. (See
Fig. V.9.) Again, the point P of intersection is a double point (just rotate
the tangent slightly to reveal the two points), but in this case there is no
(algebraic) technique to identify the slope of the tangent, that is, of that
line which intersects the graph of E_2 in a double point.

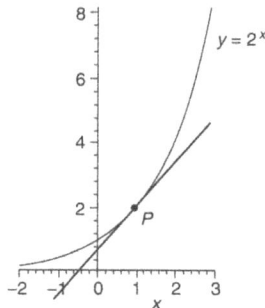

Fig. V.9   Graph of $y = 2^x$ with a tangent.

The basic problem thus is to find the slope of such tangents. Given a fixed number a, we consider the factorization

$$E_2(x) - E_2(a) = q_a(x)(x - a)$$

where $q_a(x)$ is uniquely defined for all $x \neq a$ by the average rate of change

$$q_a(x) = \frac{E_2(x) - E_2(a)}{x - a}.$$

In contrast to the algebraic case, there is no explicit formula for q_a that is defined for $x = a$, so there is no obvious way to capture the derivative of E_2 at $x = a$. However, motivated by the results in the algebraic case we discussed in Section IV.2, we consider the double point as the limiting position of two distinct points that approach each other. For $x \neq a$, the average slope between the two corresponding points on the graph of E_2 is given by $q_a(x)$. We are thus led to investigate whether there is a value L that arises as the "limit" of $q_a(x)$ as $x \to a$. In analogy to the algebraic case, that limit L would give us the desired slope of the tangent to the graph of E_2 at the (double) point $(a, E_2(a))$. Note that this problem is quite a bit more complicated than the approximation in the algebraic case. Since we do not know any value of L a priori—note that we can not evaluate q_a at $x = a$—we can not attempt to estimate $|q_a(x) - L|$ directly. Instead, we need to determine the existence of the "limit" by carefully studying the behavior of $q_a(x)$ when $x \neq a$ gets closer and closer to a.

Before proceeding with the analysis, let us simplify the problem as follows. Set $h = x - a$, so that $x = a + h$ and $x \to a$ corresponds to $h \to 0$. Then notice that by the functional equation of the exponential function one has

$$\begin{aligned}
E_2(a + h) - E_2(a) &= 2^{a+h} - 2^a = 2^a 2^h - 2^a \\
&= 2^a[2^h - 1] = E_2(a)[E_2(h) - E_2(0)] \\
&= E_2(a)[q_0(h) \cdot h],
\end{aligned}$$

where in the last step we have used the factorization $E_2(h) - E_2(0) = q_0(h)(h - 0)$ at the point $a = 0$ for $h \neq 0$. Since $E_2(a + h) - E_2(a) = q_a(a + h) \cdot h$, it follows that

$$q_a(a + h) = E_2(a)q_0(h) \text{ for all } h \neq 0. \tag{V.3}$$

This shows that finding the derivative of E_2 at the arbitrary point a is reduced to finding the derivative for the case $a = 0$. More precisely, assume for the moment that the limiting process for $q_0(h)$ as $h \to 0$ indeed leads to a meaningful result that we denote by

$$\lim_{h \to 0} q_0(h).$$

Equation (V.3) then shows that the corresponding limit process for q_a at the arbitrary point a is also meaningful, and that it results in

$$\lim_{h \to 0} q_a(a + h) = E_2(a) \lim_{h \to 0} q_0(h).$$

Geometrically, this means that the slope of the tangent to the graph of E_2 at the point $x = a$ equals $E_2(a) = 2^a$ multiplied with the slope of the tangent at $x = 0$. As in the algebraic case, let us denote the slope of the tangent line at $(a, E_2(a))$, that is, the derivative of the function E_2 at the point $x = a$, by $D[E_2](a)$, or also by $E_2'(a)$. Thus

$$D[E_2](a) = \lim_{h \to 0} q_a(a + h).$$

This is consistent with the corresponding results in the algebraic case discussed in Section 4.2, where $D[f](a) = q_a(a) = \lim_{h \to 0} q_a(a + h)$ for a rational function f. The preceding arguments show that one has the formula

$$D[E_2](a) = E_2(a) D[E_2](0), \text{ or}$$
$$(2^x)' = 2^x \cdot derivative \text{ } at \text{ } 0.$$

So, in order to determine the derivative of E_2 at arbitrary points, it is enough to study in detail the derivative at $a = 0$.

V.4.2   The Tangent to E_2 at $x = 0$

The relevant factorization for E_2 at the point 0 is given by $2^x - 1 = q_0(x)x$. Figure V.10 shows that for $x \neq 0$ the slopes $q_0(x)$ of the secant lines through the points $(0, 1)$ and $(x, 2^x)$ decrease as x decreases from a *positive* value to 0, and they continue to decrease as $x < 0$ continues to decrease.

Since the point $a = 0$ is fixed in this section, we shall simplify notation by using q instead of q_0. Let us recall the explicit numerical data for the values of $q\left(10^{-k}\right)$ that we had already considered in Section IV.4.3.

Table V.1 confirms that the numbers $q(10^{-k})$ decrease as k gets larger and larger, i.e., as $10^{-k} \to 0$, and that they seem to approximate a number whose decimal expansion begins with 0.69314.... So both the geometric visualization and the numerical data provide evidence that the slope is decreasing as $x \to 0^+$, that is, as $x > 0$ decreases to 0 from the right side.

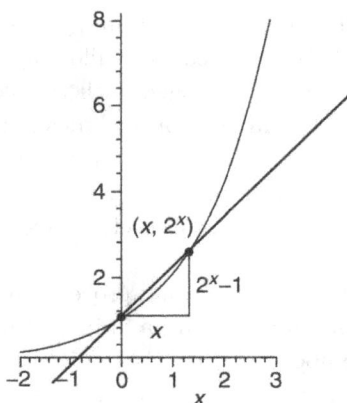

Fig. V.10 Secant to $y = 2^x$ of slope $q_0(x) = (2^x - 1)/x$ for $x > 0$.

Table V.1

x_k	$q(x_k) = (2^{x_k} - 1)/x_k$
10^{-1}	0.7177346253
10^{-2}	0.6955550056
10^{-3}	0.6933874625
10^{-4}	0.6931712037
10^{-5}	0.6931495828
10^{-6}	0.6931474207
10^{-7}	0.6931472045
10^{-8}	0.6931471829
10^{-9}	0.6931471808
10^{-10}	0.6931471805

However, this evidence does not provide a rigorous proof. Since this fact is so critical in the proof of the existence of the relevant limit, and since the exponential function is the first non-algebraic function that we study in detail, it is important to provide a complete proof of this fact, which we state as follows

Lemma 55 *The average rate of change $q(x)$ of E_2 on the interval $[0, x]$ is strictly increasing for $x > 0$. Similarly, for $x < 0$ the average rate of change $q(x)$ on the interval $[x, 0]$ is increasing as well.*

Note that this Lemma states the critical property that we just mentioned. Don't get confused by the fact that we had stated that $q(x)$ *decreases* as $x > 0$ *decreases* to 0. In the Lemma, we reverse direction: here

$x > 0$ moves to the right, so we expect that $q(x)$ increases. (As you climb up a mountain, your altitude increases, while when you climb down, the altitude decreases, but nothing changed with the slope of the mountain.) The precise proof of this Lemma is a bit technical, and it is quite alright to skip it on first reading, so as not to get distracted from the main arguments involved in the proof of the existence of the relevant limit. We therefore postpone the proof until the end of this section, and continue the discussion assuming the Lemma.

We shall now verify that the limiting process suggested by the geometric visualization and the numerical data is indeed meaningful, that is, we shall identify a real number c_2, which is the limit of $q(x)$ as $x \to 0$ from the positive side, i.e., $c_2 = \lim_{x \to 0+} q(x)$, in the precise sense discussed in Section 2.5. (The subscript 2 is in reference to the base 2 of the exponential function that is considered here.) In fact, by the Lemma, $q(x)$ is strictly increasing for $x > 0$, i.e., it decreases as x decreases to zero from the right. Clearly

$$q(h) = \frac{2^h - 1}{h} > 0 \text{ for } h > 0,$$

so that the set $S^+ = \{q(h) : h > 0\}$ is bounded from below, and therefore, by the Monotone Convergence Theorem in Section 2.4, $\lim_{h \to 0+} q(h)$ exists and equals $\inf S^+$, which we take to be our number c_2. Let us quickly recall the essence of the proof. Since c_2 is the *greatest* lower bound, for any positive ϵ the number $c_2 + \epsilon$ is NOT a lower bound for S^+. So there exists $h(\epsilon) > 0$ such that $c_2 \le q(h(\epsilon)) < c_2 + \epsilon$. Since $q(h)$ is getting smaller as h decreases, it then follows that one even has

$$c_2 \le q(h) \le q(h(\epsilon)) < c_2 + \epsilon \text{ for all } h \text{ with } 0 < h < h(\epsilon),$$

and this proves that

$$\lim_{h \to 0+} q(h) = c_2, \tag{V.4}$$

The numerical data in Table V.1 suggests that $c_2 = 0.69314....$

In order to complete the discussion and obtain a more precise numerical estimate for c_2 we need to also consider the values $q(h)$ for *negative* h that approach 0. Here is the relevant numerical data.

Note that as $x_k < 0$ increases towards zero, $q(x_k)$ increases as well. Again the data confirms what is visible geometrically. Below we shall verify exactly that $q(h) \le c_2$ for $h < 0$. Assuming this fact, the data in Tables V.1 and V.2 provide the estimate

$$q(-10^{-9}) = 0.6931471803... \le c_2 \le 0.6931471808... = q(10^{-9}).$$

Table V.2

x_k	$q(x_k) = (2^{x_k} - 1)/x_k$
-10^{-1}	0.6696700846
-10^{-2}	0.6907504562
-10^{-3}	0.6929070095
-10^{-4}	0.6931231584
-10^{-5}	0.6931447783
-10^{-6}	0.6931469403
-10^{-7}	0.6931471565
-10^{-8}	0.6931471781
-10^{-9}	0.6931471803
-10^{-10}	0.6931471805

This estimate determines the first 9 digits of c_2. More digits of c_2 can be captured by increasing the computing technology, that is, by increasing the number of significant digits in the calculations, and evaluating $q(-x_k)$ and $q(x_k)$ for numbers x_k closer and closer to zero.

The numerical evidence in Table V.2 clearly suggests that

$$\lim_{h \to 0^-} q(h) = c_2, \tag{V.5}$$

where the notation $h \to 0^-$ encodes that h approaches 0 from the left, i.e., through *negative* numbers. We will now show that this last equation is indeed correct. Note that for $h > 0$ one has

$$q(-h) = \frac{2^{-h} - 1}{-h} = 2^{-h} \frac{1 - 2^h}{-h} \tag{V.6}$$

$$= 2^{-h} \frac{2^h - 1}{h} = 2^{-h} q(h).$$

Since the exponential function is continuous, we have $\lim_{h \to 0} 2^h = 1$, and consequently also $\lim_{h \to 0^+} 2^{-h} = 1$. By general properties of limits, the limit of a product is the product of the limits and therefore, applying this in (V.6), one obtains

$$\lim_{h \to 0^+} q(-h) = \lim_{h \to 0^+} 2^{-h} \cdot \lim_{h \to 0^+} q(h) = 1 \cdot c_2. \tag{V.7}$$

Since $h \to 0^-$ is the same as $-h \to 0^+$, the statement above proves equation (V.5), and since the two one-sided limits coincide, we can combine the equations (V.4) and (V.5) into the single statement

$$\lim_{h \to 0} q(h) = \lim_{h \to 0} \frac{2^h - 1}{h} = c_2 = 0.6931471805.... \tag{V.8}$$

As already mentioned earlier, by careful numerical work with the help of a computer one can evaluate the decimal expansion of c_2 to any desired degree of accuracy. While this result confirms the geometric visualization and the numerical data we obtained, we want to emphasize that the proof we just gave, which in particular defined $c_2 = \inf\{q(h) : h > 0\}$, is theoretical, and does not depend on any numerical data. Computing technology is only used in order to obtain the decimal expansion of c_2 to any desired level of accuracy, subject to the limitation of the technology.

Note that while the *decimal expansion* of c_2 looks quite mysterious, the *geometric meaning* of c_2 is very simple and not at all mysterious:

c_2 *is the slope of the tangent to the graph of* $E_2(x) = 2^x$ *at the point* $(0,1)$*!*

The number c_2 can thus be readily visualized by the length of the short vertical line segment shown in Fig. V.11.

This limit c_2 identified in equation (V.8) represents the "missing value $q(0)$" for q at $h = 0$. Note that to define the value $q(0) = c_2 = \lim_{h\to 0} q(h)$ is the only choice that makes the function q continuous at the point 0! Geometrically, it is that particular number that is approximated by slopes of secants, just as in the case of algebraic functions discussed in Section IV.2.2. In analogy to the familiar case of algebraic functions we thus say that the function $y = E_2(x)$ is *differentiable at* $x = 0$ (i.e., there exists a well-defined tangent at the point $(0,1)$), and that the value $c_2 = \lim_{h\to 0} q(h)$ is the *derivative* $(E_2)'(0)$ of E_2 at $x = 0$. As we already observed in the preceding section, the functional equation for E_2 then implies that E_2 is differentiable at each point $a \in \mathbb{R}$, and that

$$(E_2)'(a) = \lim_{h\to 0} q_a(a + h) = E_2(a) \cdot c_2,$$

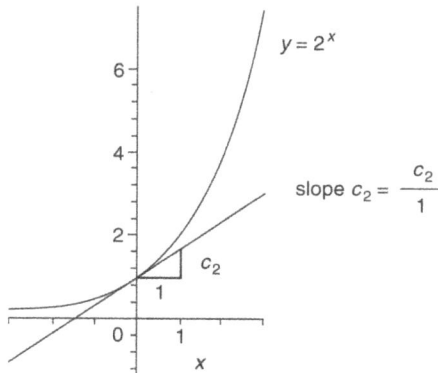

Fig. V.11   Visualization of c_2, the slope of the tangent at $(0,1)$.

where c_2 is defined by (V.8). We thus have solved the tangent problem at every point of the graph of $E_2(x) = 2^x$.

To complete the rigorous proof of the preceding statements, we must now prove Lemma 55. Maybe the reader will feel motivated to go through the details now, after we have recognized the importance of the fact that the function $q(x)$ is strictly increasing for $x \neq 0$.

Proof of Lemma 55. Let us consider the case $x > 0$ first. We must prove that $q(x_1) < q(x_2)$ whenever $0 < x_1 < x_2$. Recall that $q(x) = \frac{E_2(x) - E_2(0)}{x - 0}$ is the average rate of change $A([0, x])$ of E_2 over the interval $[0, x]$. We shall use the Corollary 19 to Lemma 18, proved in Chapter IV.2.2. For a fixed $\delta > 0$ and any $x \in \mathbb{R}$ we consider the average rate of change of E_2 over the interval $[x, x + \delta]$:

$$A([x, x + \delta]) = \frac{E_2(x + \delta) - E_2(x)}{\delta} = E_2(x) \frac{E_2(\delta) - 1}{\delta},$$

where we have used the functional equation for E_2. Since $E_2(x)$ is strictly increasing, it follows that $A([x, x + \delta])$ is strictly increasing as well as a function of x. We shall first prove the desired inequality for two rational numbers $0 < r_1 < r_2$. We can assume that r_1 and r_2 are fractions with the same denominator n, so that $r_l = m_l/n$, $l = 1, 2$, where $0 < m_1 < m_2$ are integers. We shall consider the intervals $[\frac{j}{n}, \frac{j}{n} + \frac{1}{n}]$ for $j \in \mathbb{Z}$. By what we just observed, with $\delta = 1/n$, $A([\frac{j}{n}, \frac{j}{n} + \frac{1}{n}])$ is strictly increasing in j, so that the hypothesis of the Corollary are satisfied with $x_j = \frac{j}{n}$ and for any finite collection of successive intervals $[\frac{j}{n}, \frac{j}{n} + \frac{1}{n}] = [\frac{j}{n}, \frac{j+1}{n}]$. Note that $x_0 = 0$. Part $ii)$ of that Corollary then implies that

$$A([0, x_j]) < A([0, x_{j+1}]) \leq A([0, x_{m_2}])$$

for $j = 1, 2, 3, ..., m_2 - 1$. In particular, for $j = m_1$, one obtains

$$q(r_1) = A([0, x_{m_1}]) < A([0, x_{m_2}]) = q(r_2).$$

Finally, given two real numbers with $0 < x_1 < x_2$, choose a strictly *decreasing* sequence $\{r_n\}$ of rational numbers with $\lim r_n = x_1$ and a strictly *increasing* sequence $\{s_n\}$ of rational numbers with $\lim s_n = x_2$. Given $x_2 - x_1 = \epsilon > 0$, choose $N(\epsilon) \in \mathbb{N}$ so large that one has

$$|r_n - x_1| < \epsilon/2 \text{ and } |s_n - x_2| < \epsilon/2 \text{ for all } n \geq N(\epsilon).$$

For such n one then has

$$s_n - r_n = (x_2 - x_1) + (s_n - x_2) - (r_n - x_1)$$
$$\geq \epsilon - |s_n - x_2| - |r_n - x_1| > \epsilon - \varepsilon/2 - \varepsilon/2 = 0.$$

Therefore,

$$x_1 < r_{n+1} < r_n < s_n < s_{n+1} < x_2 \text{ for } n \geq N(\epsilon),$$

which implies, by what we just verified for rational numbers, that

$$q(r_{n+1}) < q(r_n) \leq q(r_{N(\epsilon)}) < q(s_{N(\epsilon)}) \leq q(s_n) < q(s_{n+1}).$$

Since $q(x)$ is continuous on $x > 0$, we can take the limit $n \to \infty$, resulting in

$$q(x_1) = \lim_{n \to \infty} q(r_n) < \lim_{n \to \infty} q(s_n) = q(x_2)$$

as claimed. The proof for $x < 0$, i.e., for $x_1 < x_2 < 0$, works the same way, subject to appropriate modifications as needed. ■

V.4.3 *Other Exponential Functions*

The discussion in the preceding section readily generalizes to exponential functions $E_b(x) = b^x$ with an arbitrary base $b > 0$. By introducing appropriate modifications as needed, the detailed proof we just gave in case of the base $b = 2$ carries over to the general case, and we shall skip the details. In particular, for each such $b > 0$ the limit

$$\lim_{h \to 0} \frac{E_b(h) - E_b(0)}{h} = \lim_{h \to 0} \frac{b^h - 1}{h}$$

exists in $\mathbb{R}$, and we denote it by c_b. This includes the trivial case $b = 1$, with $c_1 = 0$. The geometric interpretation of this limit c_b is that it is the limit of the slopes of secants through the two distinct points $(0, 1)$ and (h, b^h), $h \neq 0$, on the graph of E_b, and that it defines the slope of the tangent to the graph of E_b at the point $(0, 1)$, just as we had seen in the case of rational functions. We call c_b the derivative $E_b'(0) = D[E_b](0)$ of the exponential function E_b at 0. Furthermore, for any other point $a \in \mathbb{R}$ one has

$$E_b'(a) = \lim_{h \to 0} \frac{E_b(a + h) - E_b(a)}{h} = E_b(a) \cdot c_b.$$

Note that for $b = 1$ one trivially has $c_1 = 0$, but otherwise the numerical values of the numbers c_b are quite surprising and intriguing. For example, with the help of a computer one readily obtains the following numerical approximations:

$$c_3 = 1.098612...$$

$$c_4 = 1.386294...$$

$$c_5 = 1.609437...$$

$$c_{10} = 2.302585....$$

These numbers c_b with mysterious decimal representations thus appear very naturally as soon as one considers the tangent problem for exponential functions. We repeat that their geometric meaning is clear:

c_b *equals the slope of the tangent to the graph of*
$E_b(x) = b^x$ *at the point* $(0, 1)$.

We shall soon find another interpretation of these numbers that gives further insight than just the (approximate) decimal expansion.

V.4.4 The "Natural" Exponential Function

Surely the strange values c_b that we identified prompt the question whether there exists a base b for which the slope of the tangent to $y = b^x$ at $x = 0$ is some simple natural number, say the number 1. Since $c_2 = 0.693... < 1$ and $c_3 = 1.098... > 1$, it appears reasonable that there is a base $b^{\#}$ somewhat smaller than 3 for which $c_{b^{\#}} = 1$ exactly. A simple "rescaling" argument will allow us to verify this precisely and to determine this particular base $b^{\#}$ in terms of the number c_2, that is, in terms of the derivative of E_2 at 0.

Starting with $c_2 = \lim_{h \to 0}(2^h - 1)/h$, we can use logarithms to express the values of c_b for arbitrary bases $b > 0$ in terms of c_2, as follows. Since $\log_2$ is the inverse function of $E_2(x) = 2^x$, one has $b = E_2(\log_2 b) = 2^{\log_2 b}$, and hence, by the properties of exponential functions,

$$b^h = (2^{\log_2 b})^h = 2^{h \log_2 b} \quad \text{for any } h.$$

Therefore,

$$c_b = \lim_{h \to 0} \frac{b^h - 1}{h} = \lim_{h \to 0} \frac{2^{h \log_2 b} - 1}{h}.$$

We now rescale: instead of considering $h \to 0$ we consider $t = h \log_2 b \to 0$. This is analogous to a change in units in measurements of physical quantities, for example, changing from meters to centimeters (or feet) to measure a distance that goes to 0. Let us assume that $b \neq 1$, so that $\log_2 b \neq 0$. Substituting $h \log_2 b = t$, so that $h = t/\log_2 b$, one obtains

$$\frac{2^{h \log_2 b} - 1}{h} = \frac{2^t - 1}{t/\log_2 b} = \log_2 b \, \frac{2^t - 1}{t}.$$

Since $t \to 0$ precisely when $h \to 0$, it now follows that

$$c_b = \lim_{h \to 0} \frac{2^{h \log_2 b} - 1}{h} = (\log_2 b) \lim_{t \to 0} \frac{2^t - 1}{t} = \log_2 b \cdot c_2.$$

Since $c_1 = 0$ and $\log_2 1 = 0$, this last equation also holds for $b = 1$. We thus see that

$$c_b = c_2 \cdot \log_2 b \quad \text{for any } b > 0.$$

Rather than starting with a familiar base b for an exponential function, such as $b = 2$, thereby ending up with the mysterious and awkward value $c_2 = 0.69314718...$, we can now turn matters around and prescribe a convenient value for the slope c_b. The preceding equation then allows us to determine the corresponding—and possibly quite strange—base b. Of particular interest is to find the base $b^\#$ that satisfies $c_{b^\#} = 1$—clearly as simple as it gets—so that the exponential function with the corresponding base $b^\#$ satisfies the differentiation formula

$$(E_{b^\#})' = c_{b^\#} \cdot E_{b^\#} = 1 \cdot E_{b^\#} = E_{b^\#}.$$

This base $b^\#$ is uniquely determined by

$$1 = c_{b^\#} = c_2 \log_2 b^\#, \text{ i.e., } \log_2 b^\# = \frac{1}{c_2},$$

and hence

$$b^\# = E_2(\log_2 b^\#) = 2^{1/c_2} = 2^{1/0.6931478...} = 2.7182818...,$$

where the last number is of course obtained with the aid of a calculator. This number $b^\#$ turns out to be ubiquitous in mathematics, at a par with the number π. It is most commonly denoted by the letter "e". The corresponding exponential function $y = e^x$ is called the *natural exponential function*, and we shall also denote it by the letter "E" (no subscript!), that is, E is the function that is defined by $E(x) = e^x$ for all real numbers x. It is important to clearly understand that this number

$$e = 2.7182818...$$

has been identified as that unique base for which the tangent at the point $(0, 1)$ to the graph of the corresponding exponential function $E(x) = e^x$ has slope 1. In particular, the number $e = 2^{1/c_2}$ satisfies

$$\lim_{h \to 0} \frac{e^h - 1}{h} = 1.$$

Later on we shall discover other formulas that directly express the number e as a limit. Let us mention in passing that e—just as π—is **not** an *algebraic* number, and hence, in particular, e is not rational.

V.4.5   *The Natural Logarithm*

The natural exponential function $E(x) = e^x$ with base e was singled out among all possible exponential functions so as to satisfy the property

$$E'(x) = E(x) \text{ for all } x \in \mathbb{R}.$$

Just as any other exponential function with base $b \neq 1$, $E(x)$ is one-to-one on its whole domain $\mathbb{R}$, and hence is invertible. Its inverse function is the logarithm $\log_e$ to the base e. This particular function is called the *natural* logarithm, and it is denoted by $\ln x$, or $\log x$ (no indication of base). Accordingly, the number e is also referred to as the *base of the natural logarithm*. Its domain is the set of all positive real numbers. The inverse relationship between $E(x) = e^x$ and $\ln x$ is captured by the formulas

$$e^{\ln y} = y \quad for \ y > 0, \ and$$
$$\ln(e^x) = x \quad for \ x \in \mathbb{R}.$$

We now revisit the expression for the number c_b in terms of $\log_2 b$ that we had obtained earlier by using the *natural* logarithm instead. Just as before, by replacing $b = e^{\ln b}$ and then $h \ln b = t$, it follows that

$$c_b = \lim_{h \to 0} \frac{b^h - 1}{h} = \lim_{h \to 0} \frac{e^{h \ln b} - 1}{h}$$
$$= \lim_{t \to 0} \frac{e^t - 1}{t / \ln b} = \ln b \cdot \lim_{t \to 0} \frac{e^t - 1}{t}$$
$$= \ln b \cdot 1.$$

We have therefore identified the number c_b as the natural logarithm $\ln b$ of b, i.e., c_b is that unique number that satisfies $b = e^{c_b}$. In particular,

$$c_2 = 0.693147... \ = \ln 2, \text{ and } e^{0.693147...} = 2.$$

More generally, the above formula shows how to calculate approximations of $\ln b$ for any $b > 0$ by considering the limit

$$\ln b = \lim_{h \to 0} \frac{b^h - 1}{h}.$$

The formula for the derivatives of exponential functions now takes the form

$$D[E_b](x) = (b^x)' = \ln b \cdot b^x,$$

or

$$E_b'(x) = \ln b \ E_b(x).$$

The natural exponential function $E(x) = e^x$ and its inverse $y = \ln x$ occur so often in mathematics and in many applications that most *scientific* calculators have special function keys for them.[2] The reader should get familiar with evaluating these functions by practicing with a suitable calculator. Most numerical work involving these functions is now handled with the aid of such scientific calculators, which have replaced the use of tables or slide rules from decades ago.

V.4.6   *Derivatives of Logarithms*

In Section 5.3, after we have introduced the general notion of a differentiable function, we will discuss the general rule to obtain the derivative of the inverse of a differentiable function that is one-to-one. In fact, we had already considered a preliminary version of this rule in Section III.2 in case of rational functions. The general rule will, in particular, apply to the inverse of exponential functions, that is, to logarithm functions. However, given the importance of exponential and logarithm functions, we shall now discuss an intuitive geometric argument to find the derivative of logarithm functions. This will give us the opportunity to review the basics of the geometry of inverse functions that we discussed in Section 3.4 and apply them in an important case.

Recall from Section 3.4 that the graph of the inverse of the function E is obtained by reflection of the graph of E on the line $y = x$. Since the tangent to the graph of E at $(0,1)$ has slope 1 and hence is parallel to the line $y = x$, as seen in Fig. V.12, reflection of that tangent on the line $y = x$ gives the line through $(1,0)$ with that same slope 1. Clearly this line is the tangent to the graph of $y = \ln x$ at that point. Again, we say that $y = \ln x$ is differentiable at $x = 1$, i.e., the graph has a tangent line at that point, and we call the slope of that tangent the derivative of $\ln$ at $x = 1$. This geometric argument thus suggests that $D[\ln](1) = (\ln)'(1) = 1$.

A similar simple relationship holds true at all other points (a, b) on the graph of E, where $b = e^a$. As seen below, the tangent at that point with slope $e^a = \Delta_2/\Delta_1$ is reflected to the tangent to the graph of the natural logarithm at the point $(b, a) = (b, \ln b)$.

A look at the triangles shown in Fig. V.13 reveals that the slope of the reflected tangent at $(b, \ln b)$ is $\Delta_1/\Delta_2 = 1/e^a = 1/b$.

---

[2]Sometimes one of the functions is accessed by entering the inverse key before the other function key, for example e^x is obtained by entering a number for x followed by [inv] + [ln].

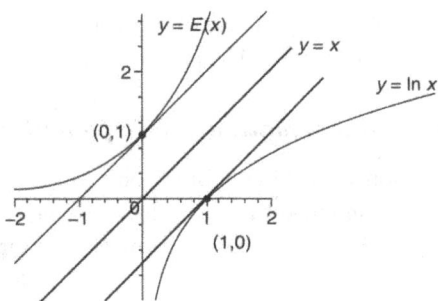

Fig. V.12   Reflection of the tangent of $y = e^x$ at $(0, 1)$.

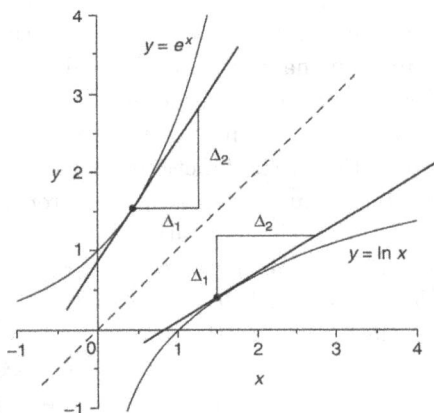

Fig. V.13   Slope of tangent to $y = \ln x$ at $b = e^a$.

We have thus verified by an intuitive geometric argument that the function given by $y = \ln x$ has a tangent, that is, it is differentiable at all points $x > 0$, and that

$$D[\ln](x) = (\ln x)' = \frac{1}{x} \text{ for all } x > 0.$$

Finally we observe that the geometric arguments that we used for the natural logarithm can be adapted to a logarithm function $\log_b$ with an arbitrary base $b \neq 1$. The conclusion is that each such function is differentiable at each point $x \in \mathbb{R}^+$ (i.e., it has a tangent at each point on the graph), and that

$$D[\log_b](x) = \frac{1}{\ln b \cdot x} \text{ for } x > 0.$$

We repeat that in Section 5.3 we shall verify these statements as an application of the general rule for derivatives of inverse functions.

V.4.7   *The Differential Equation for Exponential Functions*

We conclude this section by taking a brief look at how the solution of the tangent problem for exponential functions leads to significant applications. Each such function $y = E_b(x)$ satisfies the simple "differential equation"

$$y' = k\,y$$

for some constant k. Note that the trivial case $b = 1$, with $E_1(x) = 1^x = 1$, is covered as well, with the corresponding constant $k = 0$.

More generally, a relationship expressed by some equation that involves an (unknown) function and its derivative is called a (first order) differential equation. Many phenomena in nature are modeled by such differential equations, which often involve "higher order derivatives" and/or more than one variable. It is one of the principal tasks of mathematical analysis to determine specific properties of the functions that satisfy a given differential equation. In the case at hand, we shall prove later in Section VI.3 that *any* function f that satisfies such an equation, i.e., $f'(x) = k\,f(x)$, is necessarily equal to some constant multiple $C\,E_b(x)$ of a particular exponential function. The base b must satisfy $\ln b = k$, so that $b = e^k$.

To illustrate how differential equations and properties of their solutions can be used to estimate and make predictions about processes whose behavior can be modelled by such equations, let us consider a population of bacteria whose size at time t is given by the function $P = P(t)$, which we shall assume to have a tangent, that is, an instantaneous rate of change at every time t, whose slope defines the derivative $D[P](t) = P'(t)$. Under suitable external conditions, the instantaneous rate of change $dP/dt(t) = P'(t)$ of the population at time t is proportional to the size of the population $P(t)$ at that time. This property is consistent with common sense and can be verified experimentally in numerous settings. This means that there exists a constant k so that

$$P'(t) = kP(t).$$

It then follows from the preceding discussion that $P(t)$ must be described by an exponential function. This same process applies to many other "populations" that are growing according to an "exponential model". This information can be used to answer questions about the particular population under consideration, and to make predictions about the future.

Example. *The population of a city is projected to grow at a rate of 2% annually over the next five years. Assuming the population is 450,000 in 2020, estimate the size of the population in 2025.*

We measure time t in years and denote the population at time t by $P(t)$. The statement about the growth rate means that the instantaneous rate of change $P'(t)$ at time t equals 2% of the population $P(t)$. This translates into the equation

$$P'(t) = 0.02P(t).$$

As we mentioned earlier, solutions of this equation are functions of the form $P(t) = CE_b(t)$ for suitable constants C and b. In the present case we must have $\ln b = 0.02$, or $b = e^{0.02}$. If we count the years so that $t = 0$ corresponds to the year 2020, then $C = P(0) = 450,000$. Therefore, according to this model, the population after t years will be

$$P(t) = 450,000 \, (e^{0.02})^t = 450,000 \, e^{0.02t}.$$

Hence the population in 2025 is estimated by $P(5) = 450,000 \, e^{0.02 \cdot 5} = 450,000 \, e^{0.1} = 450,000 \cdot 1.1052 \approx 497,000$.

Other questions can readily be answered based on this formula for the population. For example, suppose the city continues to grow at the same rate. After how many years will the population reach about 1 million? The number of years t asked for must satisfy $1,000,000 = 450,000 \, e^{0.02t}$, or $e^{0.02t} = 1/.45 = 2.2222$. Therefore $0.02t = \ln 2.2222 \approx 0.7985$, and consequently $t = 0.7985/0.02 \approx 40$. It follows that assuming the same growth rate in future years, the population will reach the one million mark in approximately 40 years, that is, around the year 2060.

We shall study further applications of this sort in Chapter VI.

V.4.8   *Exercises*

1. a) Find the derivative of $y = 3^x$.

   b) Determine the equation of the tangent lines to the graph of $y = 3^x$ at the points where $x = 0$ and where $x = 1$.

2. Find the equation of the tangent line to the graph of $y = 2^x$ that goes through the point $(2, 0)$. (Note that the given point is NOT on the graph. Make a sketch including the (unknown) point of tangency.)

3. At what point does the tangent to the graph of $y = 2^x$ have slope 1?

4. Use a scientific calculator to evaluate e^2 and $\sqrt{e}$. (If your calculator does not have a separate key for the exponential function, try [inv] + [ln], or else check the manual.)

5. Simplify the following expressions *without using any calculators!*

a) $\ln(e^3)$,   b) $\ln\left(\dfrac{1}{e}\right)$,   c) $5\ln 4 + 7\ln\dfrac{1}{2}$,   d) $e^{1/\ln(e^2)}$.

6. a) Evaluate $\ln 10$, $\log_{10} e$, and the product $\ln 10 \cdot \log_{10} e$ by using a calculator.

b) Show by using the properties of logarithms that for any $b > 0$ with $b \neq 1$ one has

$$\ln b \cdot \log_b e = 1.$$

c) More generally, verify the equation $\log_a b \cdot \log_b a = 1$ for any $a, b > 0$ and $\neq 1$.

7. Find the equation of the tangent to the graph of $y = \ln x$ at the point where $x = 2$.

8. a) Use numerical approximations with a calculator to find an approximate value for c_6, the derivative of $y = 6^x$ at $x = 0$.

b) We know that $c_6 = \ln 6$. Use a scientific calculator to evaluate $\ln 6$. Check whether your answer agrees (approximately) with the answer in a).

9. Suppose f is an exponential function, and let $k > 0$ be a constant. Introduce the rescaled function f_k by $f_k(x) = f(kx)$. Show that $f_k'(a) = kf'(ka)$ at the point $x = a$. (Hint: Replace $kh = t$, and hence $1/h = k/t$, in the average rates of change that approximate $f_k'(a)$, as in the proof of $c_b = \log_2 b \cdot c_2$.)

10. Suppose a population of bacteria of size $P(t)$ at time t (in hours) grows according to the differential equation $\frac{dP}{dt} = kP$. Determine k if the population doubles in 12 hours.

V.5   Differentiable Functions

V.5.1   *From Exponential to Differentiable Functions*

Now that we have a firm understanding of tangents and derivatives for exponential functions, it is just a simple step to generalize what we have learned to arbitrary functions. Recall from Section 2.5 that a function f defined in an interval containing the point a is said to be continuous at a if f has a limit $\lim_{x \to a} f(x)$ at the point a, and if that limit equals $f(a)$.

Note that a function that has a limit at $x = a$ and that is also defined at a is not necessarily continuous at a. For example, let

$$k(x) = \begin{cases} (x^2 - 4)/(x + 2) & \text{for } x \neq -2 \\ 0 & \text{for } x = -2 \end{cases}.$$

For $x \neq -2$, k is a rational function which is not defined at $x = -2$; the above definition extends this function by simply adding -2 to the domain by specifying an output for the input $x = -2$. Recall that when calculating a limit for $x \to a$, only values of the function for $x \neq a$ enter into the picture, so it does not matter if the function happens to also be defined at a. In the case at hand, after factoring the numerator $x^2 - 4 = (x - 2)(x + 2)$, one can cancel the common factor $(x + 2)$ in numerator and denominator as long as $x \neq -2$. One therefore sees that $\lim_{x \to -2} k(x) = \lim_{x \to -2}(x - 2) = -4$, where the last limit statement is trivial since $x - 2$ is a polynomial and hence the limit equals the value at $x = -2$. Since $\lim_{x \to -2} k(x) = -4 \neq 0 = k(-2)$, clearly k is not continuous at $x = -2$. On the other hand, k is continuous at all other points $x \neq -2$, since k agrees with a rational function for all $x \neq -2$. If you think that this example is somewhat artificial you are right. Since all rational functions are continuous at each point in their domain, a function that is discontinuous at some point needs to be built up in some "artificial" way, that is, it can not just be given by a single basic algebraic formula. Note that our example fails to be continuous because we made a poor choice for $k(-2)$. Clearly the discontinuity disappears if one changes that value, i.e., if one *defines* $k(-2) = \lim_{x \to -2} k(x) = -4$.

The same situation applies with the critical function that determines the slope of the tangent of the exponential function E_2, or more generally, E_b for arbitrary base $b > 0$, at the point $(0, 1)$ on the graph. This function is

$$q(x) = \frac{b^x - 1}{x}, \text{ defined for } x \neq 0.$$

A highly non-trivial theoretical argument verified that this function does have a limit at $x = 0$ which is a real number that we have denoted by c_b.[3] Eventually we discovered that $c_b = \log b$, where log is the natural logarithm function. We can add 0 to the domain of this function q, thereby obtaining a new function $\widehat{q}$, by defining $\widehat{q}(0) = \log b$. Thus $\widehat{q}$ is defined by

$$\widehat{q}(x) = \begin{cases} q(x) = \frac{b^x - 1}{x} & \text{for } x \neq 0 \\ \log b & \text{for } x = 0 \end{cases}.$$

Since

$$\lim_{x \to 0} \widehat{q}(x) = \lim_{x \to 0} q(x) = \log b = \widehat{q}(0),$$

we see that the function $\widehat{q}$ is continuous at 0. Of course, $\widehat{q}$ is continuous at all points $x \neq 0$ by general rules about limits and continuous functions,

---

[3]We worked out the details explicitly in case $b = 2$ in Section 4. As mentioned there, the case of general b uses the same arguments, with the appropriate modifications.

while the continuity at 0 required the highly non-trivial verification of the existence of the relevant limit.

We see from these examples that the existence of the relevant limit is the essential property for continuity. Once it is established that there is a limit as $x \to a$, one simply defines, or redefines the value of the function at the point a by the value of the limit to make the function continuous at $x = a$.

We can therefore rephrase the key result obtained for the exponential function E_b as follows. For each $a \in \mathbb{R}$ there is a factorization

$$E_b(x) - E_b(a) = \widehat{q_a}(x)(x - a),$$

$$\text{where } \widehat{q_a}(x) = E_b(a) \cdot \widehat{q_0}(x - a) \text{ is continuous at } a.$$

Note that this property is correct for any rational function at each point a in its domain. In this case, the corresponding factor q was obtained by algebra, and is in fact a rational function defined at a as well, and hence it is (automatically) continuous at a.

We thus have identified the critical property that is satisfied by all rational functions, and, most importantly, also by all exponential functions, and that relates to the slope of tangents, that is, what we have referred to as the derivative of the function under consideration. Earlier we had already used the term *differentiable* to refer to the property that a function has a tangent at a particular point. We shall now use the information we just obtained to make a precise definition of this property, as follows.

Definition 56 *Suppose f is defined on an interval centered at the point a. We say that f is differentiable at a if there exists a factorization*

$$f(x) - f(a) = q_a(x)(x - a),$$

where the function q_a is continuous at $x = a$. The value $q_a(a)$ is then called the derivative of f at a and it is denoted by $D[f](a)$ or $f'(a)$.

Note that since the values $q_a(x)$ are completely determined by f for $x \neq a$, the requirement of continuity at $x = a$ and the uniqueness of limits imply that the value $q_a(a)$ is determined uniquely by the differentiable function f.

We know that every rational function satisfies this property at each point a in its domain. Recall that we had verified this property by simple algebra, plus an elementary estimate that made explicit the intuitive idea of continuity. Consequently, we can now say that *a rational function is differentiable* according to the above definition at each point of its domain.

The same is true for exponential functions, although the proof of this property was considerably deeper than the rational function case. In fact, it is this proof that required the introduction of deep new ideas involving the completeness of the real numbers and the general notion of limit. Given the central role of this result, we state it as a separate Theorem.

Theorem 57 *For each $b > 0$ the exponential function $y = E_b(x) = b^x$ is differentiable at all points $x \in \mathbb{R}$, with derivative*

$$D[E_b](x) = \log b \cdot E_b(x).$$

Note that if f is differentiable at a, then its derivative $D[f](a) = q_a(a) = \lim_{x \to a} q_a(x)$; since for $x \neq a$ the value $q_a(x)$ is the average rate of change of f between a and x, the derivative is approximated by average rates of change, and hence it measures the instantaneous rate of change of f at a, just as in the algebraic case. In other words, a function that is differentiable at a according to the precise Definition 56 does indeed have a tangent to its graph at the point $(a, f(a))$. Geometrically, the derivative $D[f](a)$ equals the slope of this tangent line. In fact, we can turn matters around and define the tangent to the graph of a differentiable function at the point $(a, f(a))$ as that line through this point that has slope $D[f](a)$.

Warning. Do NOT jump to the conclusion that the derivative $D[f]$ of a differentiable function is a continuous function at a, just because the factor q_a, whose value $q_a(a)$ is the derivative $D[f](a)$, is continuous at a. While derivatives of rational (or exponential) functions are again rational (resp. exponential) functions, and hence are continuous at every point of their domain, the general case is more complicated. In order to clear up possible confusion, consider a function f that is differentiable at every point a in an interval I. To find its derivative at the point $z \in I$, we consider the relevant factorization $f(x) - f(z) = q_z(x)(x - z)$. The definition requires that q_z, where z is fixed, is a function with input x that is continuous in x at the point z, but it does not tell us anything about how the function q_z depends on z. Since the derivative $D[f](z) = q_z(z)$ does depend on the function q_z that changes with z, we really have no information at all in this general case about how $D[f](z)$ depends on z. It is a bit tricky to give an explicit example of a function that is differentiable on an interval, but whose derivative fails to be continuous at some point. We will give such an example in Section 6.

On the other hand, differentiability implies that the function f itself is continuous, as follows.

Corollary 58 *If f is differentiable at $x = a$, then f is continuous at $x = a$.*

Proof. Note that $f(x) = f(a) + q_a(x)(x-a)$. Since the constant function $f(a)$ and the factor $(x-a)$ are clearly continuous at a, and q_a is continuous at a by the hypothesis, standard properties of continuous functions (see Section 2.6) imply that f is continuous at a as well. ∎

On the other hand, there are continuous functions that are NOT differentiable. For example, consider the function $f(x) = |x|$ that is continuous at $x = 0$. Its graph is shown in Fig. V.14. The factorization $f(x) = q(x)x$ implies that $q(x) = f(x)/x = |x|/x$ for $x \neq 0$; we had seen earlier in Section 2.6 that the function $A(x) = |x|/x$ does not have a limit as $x \to 0$, and hence can definitely not be extended to be continuous at 0, so that f is not differentiable at 0.

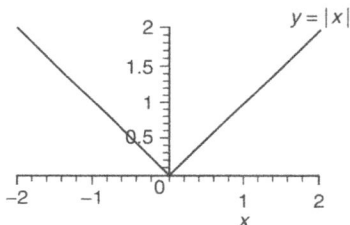

Fig. V.14 $f(x) = |x|$ is not differentiable at $x = 0$.

Finally, we state the following characterization of differentiability that is the one that has been used for centuries in most calculus texts.

Theorem 59 *f is differentiable at the point a if and only if*

$$\lim_{x \to a} \frac{f(x) - f(a)}{x - a}$$

exists. That limit is then equal to the derivative $D[f](a) = f'(a)$.

Proof. Just notice that $f(x) - f(a) = q_a(x)(x - a)$ is equivalent to $q_a(x) = [f(x) - f(a)]/(x - a)$ for $x \neq a$. Then q_a extends to a continuous function at $x = a$ if and only if $\lim_{x \to a} q_a(x)$ exists.

Remark. The traditional introduction to derivatives begins with the above formulation of differentiability and derivative of a function. Of course, in order to make sense of this, one needs to first discuss the deep notion of limit. Once this is done, one can then begin examining simple examples such as $y = x^2$, or, more generally, polynomials. In contrast, the approach chosen here begins with simple algebra, handles polynomials and rational functions without any need to use limits, and then leads up to more general functions, where it becomes evident that new deep ideas are needed.

V.5.2 *Local Linear Approximation*

Suppose the function $y = f(x)$ is differentiable at $x = a$. In particular this implies that the graph of f has a well defined tangent at $(a, f(a))$ whose slope is given by $f'(a)$. The equation of the tangent line is given by the linear function

$$L_{f,a}(x) = f(a) + f'(a)(x - a).$$

In the algebraic case, the tangent line is distinguished among all possible lines through the point $(a, f(a))$ by the fact that it intersects the graph of f at the point $(a, f(a))$ with multiplicity at least two. In particular, the error $\mathcal{E}_a(x) = f(x) - L_{f,a}(x)$ between the function and its tangent satisfies $\mathcal{E}_a(x) = k(x)(x - a)^2$, and therefore it goes to 0 *much* faster than $(x - a)$. This last property remains true in the general *differentiable* case, that is, the rate of decrease of the error $\mathcal{E}_a(x)$ is *faster* than $(x - a)$, although it may, in general, be slower than $(x - a)^2$. To understand this better, note that

$$\mathcal{E}_a(x) = f(x) - L_{f,a}(x) = f(x) - f(a) - f'(a)(x - a).$$

Since f is differentiable, one has $f(x) - f(a) = q_a(x)(x - a)$, with q_a continuous at $x = a$ and $q_a(a) = f'(a)$. Therefore

$$\mathcal{E}_a(x) = q_a(x)(x - a) - f'(a)(x - a)$$
$$= [q_a(x) - q_a(a)](x - a).$$

Since $[q_a(x) - q_a(a)] \to 0$ as $x \to a$, when x is sufficiently close to a the error $\mathcal{E}_a(x)$ is considerably smaller than $x - a$, in the sense that the *relative* error $\mathcal{E}_a(x)/(x - a)$ goes to zero as well, i.e.,

$$\lim_{x \to a} \frac{\mathcal{E}_a(x)}{x - a} = \lim_{x \to a} [q_a(x) - q_a(a)] = 0.$$

Figure V.15 shows the relationship geometrically.

So the graph of f and its tangent are very close indeed near the point $(a, f(a))$, and the approximation gets better as $x \to a$. The intuitive geometric interpretation of this approximation suggests that one can say that a function is differentiable at $x = a$ precisely when its graph looks essentially like a line (that is, the tangent) near that point, as seen in Fig. V.16.

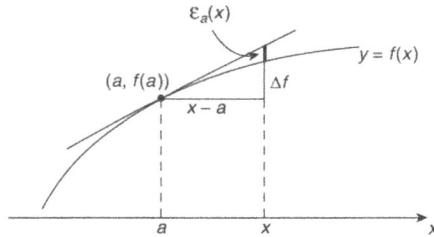

Fig. V.15   Error $\mathcal{E}_a(x)$ between the function and its tangent.

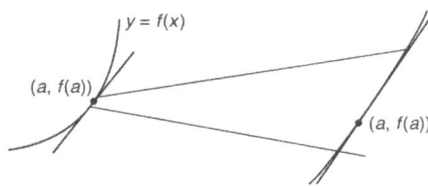

Fig. V.16   Graph of function and tangent near $(a, f(a))$.

In fact, if one looks with a magnifying glass into a very small neighborhood of the point $(a, f(a))$, the graph of f and the line become de facto indistinguishable.

This discussion verifies one part of the following characterization of differentiability.

Theorem 60 *The function f is differentiable at a if and only if f is well approximated near a by a linear function $L_{f,a}(x) = f(a) + m(x - a)$, in the sense that the "error" $\mathcal{E}_a(x) = f(x) - L_{f,a}(x)$ is of the form*

$$\mathcal{E}_a(x) = g(x)(x - a), \text{ where } g \text{ is continuous at } x = a \text{ with } g(a) = 0.$$

If this condition is satisfied, the slope m of the approximating line is uniquely determined and agrees with the derivative $D[f](a)$.

As noted earlier, the crux of the condition on g is the requirement that
$$\lim_{x \to a} g(x) = 0.$$

To complete the proof of the theorem, note that if the approximation of f by $L_{f,a}$ holds for some m, with the error $\mathcal{E}_a(x)$ satisfying the condition stated in the theorem, then

$$f(x) - f(a) = m(x - a) + \mathcal{E}_a(x)$$
$$= [m + g(x)](x - a).$$

The factor $q_a(x) = m + g(x)$ is then continuous at $x = a$ with $\lim_{x \to a} q_a(x) = m + g(a) = m$, so that f is indeed differentiable at a, with $f'(a) = q_a(a) = m$ as required.

In the days before hand-held calculators became widely available, the linear approximation of a differentiable function was often used to approximate (unknown) values of a function near a particular point at which the value is known, or in order to estimate errors in experimental work. (See Exercises 3 and 4 in Section 8 below.) For example, for the exponential function $E_2(x) = 2^x$ with $E_2'(x) = \ln 2 \, 2^x$, one knows that $E_2(2) = 4$, and one can then estimate

$$2^{2.1} \approx E_2(2) + E_2'(2)(2.1 - 2)$$
$$= 4 + 4 \ln 2 \cdot (0.1)$$
$$= 4.2773,$$

or, more generally,

$$2^{2+h} \approx 4 + 4(\ln 2) \cdot h \quad \text{for small } h.$$

Since the derivative of $y = e^x$ at $x = 0$ is equal to 1, the linear approximation for this function at $x = 0$ is particularly simple and gives the practical estimate

$$e^x \approx 1 + x \text{ for small } x.$$

Aside from such practical estimations, the most important application of the linear approximation is the conceptual understanding that is summarized in the following statement.

Differentiability is equivalent to good local linear approximation.

Stated differently, one can say:

Locally the graph of a differentiable function looks like a (non-vertical) line.

This property is the foundation for the important principle that whatever is correct for *linear* functions should remain correct—locally—for differentiable functions in general. As we shall see later, this principle turns out to be useful for understanding key properties of differentiable functions. Furthermore, it is this idea of good local linear approximation that is critical for understanding differentiability in more general contexts, for example in case of functions of *more than one* variable.

Based on the property stated above, it is easy to recognize from the graph of a continuous function where that function fails to be differentiable. For example, if the graph has a corner at a point P, as in case of $f(x) = |x|$ at $(0,0)$, the graph, shown in an earlier figure, surely does not look like a line near P, and consequently the function represented by the graph is NOT differentiable at that point. Of course, we had explicitly verified earlier that this particular function is NOT differentiable at 0.

Another problem occurs if the graph of a function does have a tangent that is *vertical*, so that no slope is defined. Since derivatives evaluate the slope of tangents, a function cannot be differentiable at such points.

Example. Consider the function $g(x) = x^{1/3}$. Since $|g(x)| \leq |x|^{1/3}$, one surely has $\lim_{x \to 0} g(x) = 0$, so that g is continuous at $x = 0$. Also, g is continuous at all points $x \neq 0$, since g is the inverse function of $f(u) = u^3$, which is continuous and strictly increasing on $\mathbb{R}$. The tangent to the graph of g at $(0,0)$ is given by the y-axis (see Fig. V.17), i.e., it is vertical.

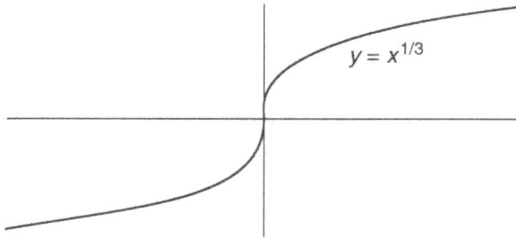

Fig. V.17   Graph of $y = x^{1/3}$ with vertical tangent at $(0,0)$.

By factoring
$$x^{1/3} = q(x)x \text{ for } x \neq 0, \text{ with } q(x) = x^{-2/3},$$
one sees that $q(x)$ has no limit as $x \to 0$. Consequently, $q(x)$ cannot be extended as a continuous function to $x = 0$, that is, $g(x) = x^{1/3}$ is **not** differentiable at $x = 0$. On the other hand, relating this example to the rule about inverse functions we had briefly introduced in Section III.2.4 (and that we will discuss in general in the next section), since g is the inverse of $f(u) = u^3$, one knows that g is differentiable at all points $b = u^3$ with $f'(u) = 3u^2 \neq 0$, that is, at all points $b = u^3 \neq 0$, and that $g'(b) = 1/f'(u) = 1/(3u^2) = 1/(3b^{2/3}) = \frac{1}{3}b^{-2/3}$. However, the rule cannot be applied at the point 0, since $f'(0) = 0$. The function f has a horizontal tangent at $(0,0)$, and consequently, by reflection, its inverse g has a *vertical* tangent at that point.

To summarize, if near a point $P = (a, f(a))$ the graph of a function f looks very much like a **non**-vertical line (which is then part of the tangent line), one may safely conclude that the function f is differentiable at $x = a$.

Finally, we use the function g just considered to define the function $F = g^4$. By the chain rule, which we had verified in Section III.2.3 for rational functions, and which carries over to differentiable functions (see next section), or simply because $F(x) = x^{4/3}$ is algebraic, F is differentiable at all points different from 0. In contrast to g, however, it turns out that F is differentiable at 0 as well. Just note that $F(x) - F(0) = x^{4/3} = x^{1/3} \cdot x$, where the factor $g(x) = x^{1/3}$ is continuous at 0. So F is indeed differentiable at 0, with $F'(0) = g(0) = 0$. However, F is not *algebraically* differentiable at 0, since the error term $\mathcal{E}_0(x) = F(x) - 0 = g(x)x$ does not have a zero of multiplicity 2 at 0.

V.5.3   *Chain Rule and Inverse Function Rule*

We had emphasized early on that *composition* of functions, as well as the concept of the *inverse* of a function that is one-to-one, are the most natural operations associated to functions in general settings. We had established the differentiation rule for compositions of polynomials, the so-called *Chain Rule*, in Section II.5.2, and later extended it to rational functions in Section III.2.3. We also took a brief look at the corresponding rule for inverse functions, such as $g(x) = \sqrt{x}$ for $x > 0$, in Section III.2.4. We shall now extend these rules to general differentiable functions.

Chain Rule for Differentiable Functions   Recall that we had realized in the last chapter that differentiable functions are, locally, "essentially" like linear functions. Since we know the chain rule for linear functions (a special case of polynomials), at an intuitive level the extension to general differentiable functions is then clear. Consider the *linear* approximations L_f and L_g of f and g at the relevant points; one has $(L_f \circ L_g)' = (L_f)' \cdot (L_g)'$, and since $L_f \circ L_g$ is again linear, it seems reasonable that it is a good linear approximation to $f \circ g$, and therefore we expect $(f \circ g)' = f' \cdot g'$ to hold as well. In essence, since differentiability is equivalent to good local linear approximation, what is correct for linear functions remains correct (locally) for arbitrary differentiable functions.

The precise verification of the preceding intuitive argument is just as easy. Suppose that g is differentiable at a and f is differentiable at $b = g(a)$.

We then have the factorizations

$$f(y) - f(b) = q_f(y)(y - b) \text{ and } g(x) - g(a) = q_g(x)(x - a),$$

with q_f continuous at b and q_g is continuous at a. As in the case of rational functions, by direct substitution of $y = g(x)$ and $b = g(a)$, it then follows that

$$(f \circ g)(x) - (f \circ g)(a) = f(g(x)) - f(g(a)) = q_f(g(x)) \cdot (g(x) - g(a))$$
$$= q_f(g(x)) \cdot q_g(x)(x - a).$$

This shows that there exists a factorization $(f \circ g)(x) - (f \circ g)(a) = q(x)(x - a)$ for the composition $f \circ g$, where the critical factor q is given by

$$q(x) = q_f(g(x)) \cdot q_g(x).$$

Since g and q_g are continuous at a, and q_f is continuous at $b = g(a)$, the rules for continuous functions now imply that q is continuous at a as well. Therefore $f \circ g$ is differentiable at a, and

$$D[f \circ g](a) = q(a) = q_f(g(a)) \cdot q_g(a)$$
$$= D[f](g(a)) \cdot D[g](a)$$
$$= D[f](b) \cdot D[g](a).$$

We have thus verified the **Chain Rule**, that is,

$$D[f \circ g] = [D[f] \circ g] \cdot D[g], \text{ or even shorter,}$$
$$(f \circ g)' = f' \cdot g',$$

for arbitrary differentiable functions. In the second formulation one of course needs to be careful to choose the input values correctly. In particular, the formula for $(f \circ g)'(a)$ cannot involve $f'(a)$, since f and hence f' need not be defined at all near the point a. Instead, for the composition to be defined, f must be defined near the point $b = g(a)$. Correspondingly, the appropriate input for the derivative f' is this point b as well. In fact, the formula $(f \circ g)'(a) = f'(b) \cdot g'(a)$ that we just obtained exhibits the only meaningful way to choose the input values in each factor.

Let us consider some examples.

i) Recall that $E_b(x) = b^x = e^{(\ln b)x} = E((\ln b)x)$. By the Chain Rule $E_b'(x) = E'((\ln b)x) \cdot ((\ln b)x)' = E((\ln b)x) \cdot \ln b = \ln b \cdot b^x$. Of course this agrees with the rule we had established in Section 4.3. But since the differentiation rule $E' = E$ for the natural exponential function is so easy to remember, it may be helpful to see that the general case follows from this "natural" case by a simple application of the chain rule.

ii) $(e^{x^2})' = e^{x^2}(x^2)' = e^{x^2}(2x)$.

iii) $F(x) = 3 \cdot (5^x)^4 - 2(5^x)^3$. We note that this function is the composition of 5^x with the polynomial $f(u) = 3u^4 - 2u^3$. Hence

$$D[F](x) = D[f](5^x) \cdot D[5^x] = (12u^3 - 6u^2) \circ (5^x) \cdot (\ln 5) \cdot 5^x$$
$$= [12(5^x)^3 - 6(5^x)^2] \cdot (\ln 5) \cdot 5^x.$$

iv) $H(x) = \ln(x^2 + 1)$. Recall that $D[\ln](u) = 1/u$ for $u > 0$. Hence,

$$H'(x) = \frac{1}{x^2 + 1}(2x).$$

NOTE. Unless explicitly asked for, no further (algebraic or other) simplification should be attempted! The answers obtained above clearly reflect the structure of the differentiation rules that have been applied; any further transformations would make it difficult to recognize the steps that have been taken.

Inverse Functions Next we shall discuss the simple formula for the derivative of the inverse of a differentiable function. In particular, we will prove the formula for the derivative of the natural logarithm $\ln x$—which is the inverse of the exponential function $E(x) = e^x$—that we just used in an example in the preceding section, and that we had established in Section 4.6 by a geometric argument.

Let us first discuss the intuitive argument that is at the heart of the proof. Since differentiable functions are locally essentially linear, we consider first the case of a linear function $L(x) = mx + c$. Such a function is one-to-one precisely if $m \neq 0$, i.e., if $L' \neq 0$. In this case the inverse $g(y) = \frac{1}{m}y - c/m$ is again linear, and $g' = 1/L'$. Turning to an arbitrary invertible function f (i.e., f must be one-to-one) that is differentiable at a, we consider its linear approximation $L_{f,a}(x)$, and we assume that $f'(a) = L'_{f,a} \neq 0$. Let $b = f(a)$. By reflecting the graphs of f and $L_{f,a}$ (note that the latter is the tangent line of f at the point $(a, f(a))$), we obtain the graph of the inverse g of f together with its tangent line at the point $(f(a), a) = (b, g(b))$, as shown in Fig. V.18.

As seen in Fig. V.19, if the line L_1 has slope $m_1 = d/c$, the reflected line L_2 has slope $m_2 = c/d = 1/m_1$.

Since the reflection clearly preserves the geometric properties of the linear approximation, i.e., the relationship between the graph of the function and its tangent line, one obtains the following **Inverse Function Rule:**

If $f'(a) \neq 0$, then the inverse g of f is differentiable at $b = f(a)$, with

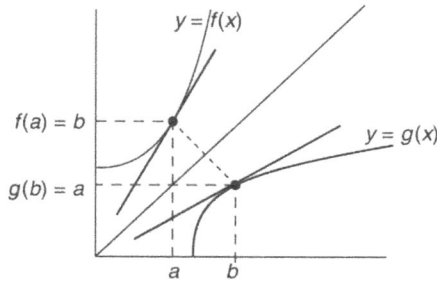

Fig. V.18   Graph of f and its inverse, with tangents.

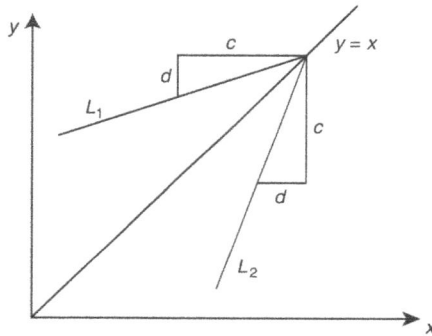

Fig. V.19   The slopes of the line L_1 and of its reflection L_2.

$$D[g](b) = \frac{1}{D[f](g(b))} = \frac{1}{D[f](a)}.$$

While the intuitive geometric argument is quite convincing, this result is easily verified precisely by considering the factorization

$$f(x) - f(a) = q(x)(x - a), \tag{V.9}$$

with q continuous at a and $q(a) = f'(a) \neq 0$. It follows that $1/q$ is defined on a small interval I centered at a and is continuous at a as well. In order to complete the proof, we shall assume that f is one-to-one and continuous on I, so that its inverse g is continuous as well on the interval $J = f(I)$. (See Theorem 54 in Section 3.4 — to be precise, in the proof we had assumed that f was strictly monotonic.) Consequently, assuming $x \in I$, by substituting $f(x) = y \in J$, $f(a) = b$, and $x = g(y)$, $a = g(b)$ into the above factorization for f one obtains

$$y - b = q(g(y))(g(y) - g(b)) \text{ for all } y \in J.$$

Now solve this equation for $g(y) - g(b)$ to obtain

$$g(y) - g(b) = \frac{1}{q(g(y))}(y - b). \tag{V.10}$$

Since the composition $(1/q) \circ g$ of continuous functions is continuous at b, equation (V.10) shows that g is differentiable at b, with derivative $D[g](b) = 1/q(g(b)) = 1/q(a) = 1/f'(a)$, as claimed.

This result is another concrete instance of the general principle that whatever is true for linear functions will usually remain correct locally for differentiable functions as well.

Note that the formula for the derivative of the inverse is meaningless if $f'(a) = 0$. In this case the tangent line to the graph of f is horizontal, and hence its reflection on the line $y = x$, which is the tangent to the graph of the inverse, is vertical, so that its slope is not defined. Therefore the inverse g of f is NOT differentiable at points $b = f(a)$ where $f'(a) = 0$.

Let us apply what we just learned to the exponential function $E(x) = e^x$. Here $E'(a) \neq 0$ for all a (why?), and hence the inverse function $g(x) = \ln x$ of E is differentiable at all points $b = e^a$, i.e., at all points of its domain $\{b \in \mathbb{R} : b > 0\}$. Furthermore, by the result we just proved one has

$$D[\ln](b) = \frac{1}{E'(\ln b)} = \frac{1}{E(\ln b)} = \frac{1}{b}.$$

We have thus verified the differentiation formula for the natural logarithm function

$$D[\ln](x) = \frac{1}{x} \text{ for all } x > 0,$$

which we had already obtained in Section 4.6 by a geometric argument.

Completely analogous arguments show that

$$D[\log_b](x) = \frac{1}{x \cdot \ln b}$$

for any base $b \neq 1$ and all $x > 0$.

Example. Let $n \in \mathbb{N}$ and consider the function $f(x) = x^n$ on the interval $\mathbb{R}^+ = (0, \infty)$. f is strictly increasing and continuous, so it has an inverse function $g : \mathbb{R}^+ \to \mathbb{R}^+$ that is also continuous. Since f is differentiable on $\mathbb{R}^+$, and $f'(x) = nx^{n-1} \neq 0$ for $x > 0$, the result we just proved implies that g is also differentiable at each point $y > 0$, and furthermore,

$$D[g](y) = \frac{1}{D[f](g(y))} = \frac{1}{n(g(y))^{n-1}}.$$

Note that $g(y) = \sqrt[n]{y}$ is the nth root of y, and we know that $g(y) = y^{1/n}$ in terms of exponentials. Thus, if $y > 0$, we obtain

$$D[y^{1/n}] = \frac{1}{n[y^{1/n}]^{n-1}} = \frac{1}{n}\frac{1}{y^{1-1/n}} = \frac{1}{n}y^{-(1-1/n)} = \frac{1}{n}y^{1/n-1}$$

by the rules for exponents, that is, by the functional equation of exponential functions and its variations. We note that this is the familiar power rule, initially established for positive integer exponents, now extended to exponents $1/n$. Finally, by applying the chain rule to $G(y) = y^{m/n} = (y^{1/n})^m$, where $m \in \mathbb{Z}$, one obtains

$$D[y^{m/n}] = D[(y^{1/n})^m] = m(y^{1/n})^{m-1}\left[\frac{1}{n}y^{1/n-1}\right]$$
$$= \frac{m}{n}y^{m/n-1/n} \cdot y^{1/n-1} = \frac{m}{n}y^{m/n-1},$$

which shows that the power rule remains valid for any rational exponent m/n.

Power Functions with Real Exponents   We note that the last example is essentially of an algebraic nature. In Section III.2.4 we had already considered the case $n = 2$, i.e., $g(y) = \sqrt{y}$. Appropriate modifications of the arguments in that section prove the analogous result for arbitrary n. In fact, one can handle all this within the rational numbers, provided one restricts the function $g(y) = \sqrt[n]{y}$ to the set $\{r^n : r \in \mathbb{Q} \text{ and } r > 0\}$.

However, once we have introduced exponential functions for arbitrary real numbers as inputs (remember: this required the full theory of real numbers and basic properties of limits), we can consider power functions $f(x) = x^p$ on $\mathbb{R}^+$ with an arbitrary real exponent p, such as $x^{\sqrt{2}}$ or x^π. The definition of such functions requires the exponential function, unless p happens to be rational, but that is the case we just considered in the previous section. In fact, x^p is defined by $x^p = E_x(p)$, so the input variable is in the base, making this a very different type of function than the familiar exponential functions $x \to E_b(x)$, where the base is fixed. To handle this new type of function we use the natural logarithm to write $x = e^{\ln x}$, and thus $f(x) = x^p = (e^{\ln x})^p = e^{p\ln x}$. By writing f in this way, we recognize that f is the composition $E \circ h$ of $h(x) = p\ln x$ with the natural exponential function E. By the chain rule it follows that $f(x)$ is differentiable for all

$x > 0$, and furthermore one obtains (recall $E' = E$!)

$$D[x^p] = D[E \circ h](x) = E'(h(x)) \cdot h'(x)$$

$$= E(p \ln x) \cdot \frac{p}{x} = p \cdot e^{p \ln x} \cdot \frac{1}{x} = p\left(x^p \frac{1}{x}\right)$$

$$= p x^{p-1}.$$

Fortunately, this result matches exactly the power rule that we had obtained earlier in case of *rational* exponent. In other words, the familiar power rule for derivatives applies *to any real exponent*, although the verification of this fact involves significantly different and deeper methods when the exponent is not a rational number. Returning to the examples at the beginning of this section, we thus have the formulas

$$(x^{\sqrt{2}})' = \sqrt{2} \cdot x^{\sqrt{2}-1} \text{ and } (x^{\pi})' = \pi \cdot x^{\pi-1} \text{ for all } x > 0.$$

It is important to remember that this power rule applies only when the variable appears in the base. When the variable x appears in the exponent rather than in the base, one deals with an *exponential* function, whose differentiation formula looks quite different, namely $D[b^x] = b^x \ln b$.

Finally, what if the variable appears both in base and exponent, that is, if we consider $f(x) = x^x$ for $x > 0$? Which rule applies now? Surely both rules should be involved somehow. Recall the old technique to use logarithms to simplify operations. In this case, since $x^x > 0$ for $x > 0$, we can consider

$$\log(f(x)) = x \log x,$$

which implies that

$$f(x) = E(x \log x).$$

This shows that f is the composition of differentiable functions, and hence is differentiable, and furthermore one now obtains

$$D[f](x) = D[E](x \log x) \cdot D[x \log x]$$

$$= E(x \log x) \cdot \left[1 \cdot \log x + x \cdot \frac{1}{x}\right]$$

$$= f(x) \cdot [\log x + 1] = f(x) \cdot \log x + f(x).$$

Note that the first summand is the result obtained by treating the base fixed and applying the differentiation rule for exponential functions, while the second summand is obtained by keeping the exponent fixed and applying the power rule: $D[x^a] = ax^{a-1}$. Since $a = x$, this results in $x \cdot x^{x-1} = x^x = f(x)$.

V.5.4   *Differentiation Rules for Algebraic Combinations*

Finally, we need to generalize differentiation rules that we had established for various algebraic combinations of rational functions to the general class of differentiable functions. Since these rules apply, in particular, to linear functions, by the general principle mentioned in Section 2, these rules should then also apply to differentiable functions. In fact, the precise verification is completely straightforward as a consequence of properties of continuous functions. Let us state these rules in the following theorem.

Theorem 61 *Suppose f and g are differentiable at the point $a \in \mathbb{R}$. Then*

i) if $c, d \in \mathbb{R}$, then $cf \pm dg$ is differentiable at a, and $(cf \pm dg)'(a) = cf'(a) \pm dg'(a)$;

ii) $f \cdot g$ is differentiable at a, and $(f \cdot g)'(a) = f'(a) \cdot g(a) + f(a) \cdot g'(a)$;

iii) if $g(a) \neq 0$, then f/g is differentiable at a and

$$D\left[\frac{f}{g}\right](a) = \frac{D[f](a) \cdot g(a) - f(a) \cdot D[g](a)}{[g(a)]^2}.$$

Proof. We had verified $ii)$ in detail for rational functions. Exactly the same proof applies, provided one notices that the resulting factor q for the product $f \cdot g$ is now a combination of functions that are continuous at a, and hence is continuous at a as well by general results about continuous functions. $i)$ is even easier, and we leave it to the reader. We shall now prove $iii)$ in detail, since we didn't do it earlier in case of rational functions. First, note that since g is continuous at a and $g(a) \neq 0$, there exists an interval $I = I_\delta(a)$ so that $g(x) \neq 0$ for all $x \in I$, so that f/g is defined on I. We shall only consider $x \in I$ in what follows. By hypothesis, we have the factorizations $f(x) - f(a) = q_f(x)(x - a)$ and $g(x) - g(a) = q_g(a)(x - a)$, with q_f and q_g continuous at a. Now consider

$$
\begin{aligned}
\frac{f(x)}{g(x)} - \frac{f(a)}{g(a)} &= \frac{f(x)g(a) - f(a)g(x)}{g(x)g(a)} \\
&= \frac{f(x)g(a) - f(a)g(a) - [f(a)g(x) - f(a)g(a)]}{g(x)g(a)} \\
&= \frac{[f(x) - f(a)]g(a) - f(a)[g(x) - g(a)]}{g(x)g(a)} \\
&= \frac{[q_f(x)(x-a)]g(a) - f(a)[q_g(x)(x-a)]}{g(x)g(a)} \\
&= \frac{q_f(x)g(a) - f(a)q_g(x)}{g(x)g(a)}(x - a).
\end{aligned}
$$

Clearly the last equation gives the relevant factorization $f/g(x) - f/g(a) = q_{f/g}(x)(x-a)$ for f/g, and the factor $q_{f/g}$ is continuous at a by general rules for continuous functions. Consequently, f/g is differentiable at a, and

$$D\left[\frac{f}{g}\right](a) = q_{f/g}(a) = \frac{[q_f(a)g(a) - f(a)q_g(a)]}{g(a)g(a)}$$
$$= \frac{D[f](a)g(a) - f(a)D[g](a)}{[g(a)]^2}. \blacksquare$$

Remark. In case f and g are rational, the proof of iii) uses exactly the same algebraic process to identify the relevant factor $q_{f/g}$. It then continues just using algebra: since q_f and q_g are rational functions in this case, one just notices that $q_{f/g}$ is rational as well, and we are done.

V.5.5   *Periodic Functions*

Now that we have a firm understanding of the concept of a differentiable function, you may wonder what other functions of interest may be differentiable. As we mentioned earlier, aside from the exponential functions and their inverses, that is, the logarithm functions, the most widely used functions in applications are the so-called trigonometric functions that are used to model periodic phenomena. Typical examples include the motion of a pendulum, the bouncing motion of a spring, the rotation of the earth around its axis, the (regular) heart beat of a person, waves in the ocean, sound waves in the air, and electromagnetic waves as they appear in the propagation of light or radio signals. The functions used to model such phenomena must be "periodic", that is, they must exhibit the repetition of basic patterns. More precisely, one says that a function f is *periodic with period* ω if

$$f(x + \omega) = f(x) \text{ for all } x \text{ in the domain of } f.$$

In particular, this implies that if $x \in dom(f)$, then $x + \omega$ must be in $dom(f)$ as well.

The Basic Trigonometric Functions   Except for constant functions, none of the familiar algebraic functions such as polynomials, root functions, and so on are periodic, and neither are the exponential functions. Other mathematical concepts are required to produce concrete examples of periodic functions. We shall now examine two of the most useful periodic functions that have found wide applications in the natural sciences and in mathematics, and that are fundamental for studying general periodic functions. These are the (trigonometric) functions *sine* and *cosine*. They have been used for thousands of years in many practical applications, mainly involving relationships between angles of a triangle and ratios of appropriate sides, in order to solve geometric problems or carry out large scale measurements. The standard high school/pre-calculus curriculum emphasizes numerous formulas and geometric applications of these functions, which may be quite overwhelming, especially if students are expected to learn all this before getting to differentiation. In contrast, in this text we have developed the foundations and general rules of differentiation without ever mentioning trigonometric functions. On the other hand, given their long history and importance for applications, it is now time to introduce these functions and discuss their basic properties. Mainly, we shall focus on the essential features that are most useful for the applications in calculus.

In order to exhibit the periodic behavior of the *sine* and *cosine* functions it is best to describe them in the context of the unit circle in the plane rather than through triangles and ratios of their sides. After all, going around in a circle several times, say on a carousel, is probably one of the most common periodic processes most of us are familiar with. We consider the unit circle $x^2 + y^2 = 1$ in a Cartesian coordinate system. Beginning at the point $(1,0)$, given a real number s we measure the distance $|s|$ along the circle, moving counterclockwise—this is known as *mathematically positive*—around the circle if $s > 0$, and moving clockwise if $s < 0$, thereby reaching a well defined point $P(s)$ on the circle. The precise concept of distance along a curve is actually quite complicated, but we can intuitively visualize the process by using a measuring tape, i.e., a flexible number line, placing its 0 at the point $(1,0)$ on the circle, wrapping it around the circle, and then reading off the distance on the tape. This is visualized in Fig. V.20.

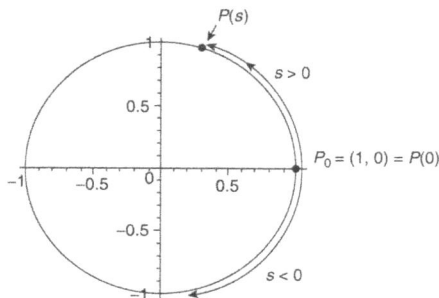

Fig. V.20   Arc of length s on the circle of radius 1.

Naturally, an infinitely long measuring tape can be wrapped around the circle numerous times. One full tour around the circle takes us back to $P(0) = (1, 0)$. This occurs after having moved a distance s corresponding to the circumference of the circle, that is, when $s = 2\pi$ (≈ 6.28). So $P(2\pi) = P(0)$. Similarly, if one moves a distance 2π around the circle clockwise, one obtains $P(-2\pi) = P(0)$. Starting from an arbitrary point $P(s)$ and moving farther along the circle a distance 2π also takes us once around the circle back to $P(s)$, so that

$$P(s + 2\pi) = P(s) \text{ for every } s.$$

We have thus constructed a *periodic* function with period 2π whose values, however, are not real numbers but points in the plane. Writing $P(s) = (x(s), y(s))$, one sees that the coordinate functions $x(s)$ and $y(s)$ satisfy the same periodicity relation

$$x(s + 2\pi) = x(s), \qquad y(s + 2\pi) = y(s).$$

Because of their importance, these periodic functions are given the special names *cosine* and *sine*, abbreviated as

$$\cos s = x(s) \text{ and } \sin s = y(s).$$

The point $P(s)$ is thus described by

$$P(s) = (\cos s, \sin s).$$

Since the reflection of the point $P(s)$ across the x-axis gives the point $P(-s)$, it follows immediately that

$$\cos(-s) = \cos s \text{ and } \sin(-s) = -\sin s.$$

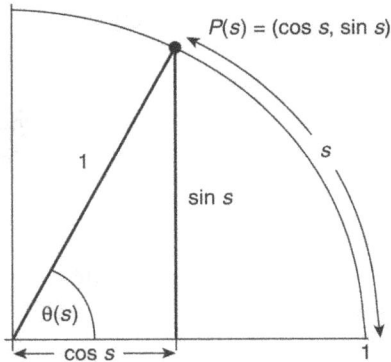

Fig. V.21   Right triangle with legs $\cos s$ and $\sin s$.

Figure V.21 illustrates the connection of these functions for $0 \leq s \leq \pi/2$ with the right triangle with vertices

$$(0,0), (\cos s, 0), \text{ and } P(s) = (\cos s, \sin s),$$

whose hypothenuse has length 1, and the angle $\theta(s)$ at the vertex $(0,0)$. As long as the angle $\theta(s)$ is between 0^0 and 90^0 one sees that

$$\sin s = \frac{\sin s}{1} = \frac{opposite\ leg}{hypotenuse}, \quad \cos s = \frac{\cos s}{1} = \frac{adjacent\ leg}{hypotenuse}.$$

These are the classical formulas that have long been used to define the basic trigonometric functions of an angle in a right triangle. Note that the values of the ratios are independent of the radius of the circle as long as the angle $\theta(s)$ is kept fixed, since the resulting triangles are similar. In order to recognize the periodic nature of the trigonometric functions it is important to consider arbitrary real arguments s as inputs, as introduced here, rather than being restricted to angles in a triangle.

Radian Measure   When the input of the trigonometric functions *sine* and *cosine* is interpreted as an angle, it becomes important to specify how angles are to be measured. The number s (i.e., the distance along the unit circle from the point $(1,0)$ to the point $P(s)$) is known as the *radian measure* of the angle $\theta(s)$ formed by the ray from $(0,0)$ to $P(s)$ and the positive x-axis. (See the last figure.) Note that $(0,1) = P(\pi/2)$, so that the right angle between the (positive) coordinate axis has radian measure $\pi/2$. The familiar *degree measure* for angles is based on dividing a right angle into 90 equal pieces, each piece identifying an angle of 1^0 ($= 1$ degree), so that radian measure $\pi/2$ corresponds to 90^0. In general

$$s \ radians \ correspond \ to \ \left(\frac{180}{\pi}s\right) \ degrees, \ and$$

$$\alpha \ degrees \ correspond \ to \ s = \frac{\pi}{180}\alpha \ radians.$$

While we will occasionally use degree measure for angles in some applications, the input in the trigonometric functions (an arbitrary real number) always refers to the *radian measure* of the relevant (geometric) angle.

Except for very special choices of inputs there is no direct computational procedure to determine exact numerical values of the sine and cosine functions. Of course, approximate values can be obtained from careful graphs of the points $P(s)$ on the unit circle, or from measuring the length of the sides in appropriate right triangles. As we will see later, other analytical approximations can be obtained with the tools of calculus. At a practical level, years ago people had to use tables and slide rules to look up appropriate values of trigonometric functions. Technology has now improved, and the common method to determine (approximate) values for these functions uses scientific calculators. (**Warning:** It is important to set the calculator to the appropriate mode (degree or radian), so that the given input is understood correctly. In most scientific calculators the default mode is degree mode.)

We emphasize that for the purposes of calculus it is more convenient to use the radian measure of an angle as the input in trigonometric functions. Radian measure is based on intrinsic geometric concepts, while degree measure is based on the historical (yet rather arbitrary) partition of a full circle into 360 equal parts that was introduced by the Babylonians almost four thousand years ago.

Simple Trigonometric Identities Since $\cos s$ and $\sin s$ are the coordinates of the point $P(s)$ on the unit circle, one has the obvious fundamental trigonometric identity

$$(\sin s)^2 + (\cos s)^2 = 1 \text{ for all } s \in \mathbb{R}.$$

Other basic relations follow from the geometric observation that $P(s+\frac{\pi}{2}) = (-\sin s, \cos s)$ for all $s \in \mathbb{R}$, as seen in Fig. V.22.

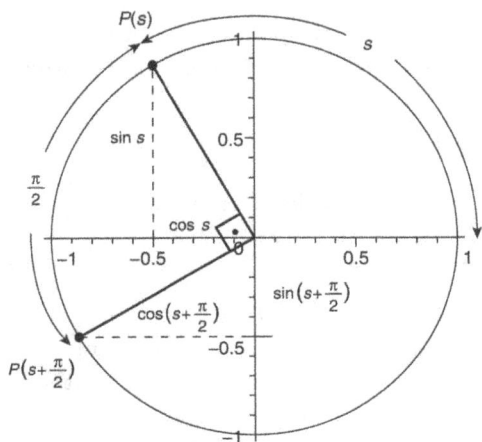

Fig. V.22   Location of $P(s)$ and $P(s + \pi/2)$ on the circle.

In terms of coordinates, this means that

$$\cos\left(s + \frac{\pi}{2}\right) = -\sin s, \text{ and}$$

$$\sin\left(s + \frac{\pi}{2}\right) = \cos s.$$

Replacing s by $-s$ in these formulas one obtains

$$\cos\left(\frac{\pi}{2} - s\right) = -\sin(-s) = \sin s,$$

$$\sin\left(\frac{\pi}{2} - s\right) = \cos(-s) = \cos s.$$

These latter formulas express the trigonometric functions of the complementary angle $\pi/2 - s$ in a right triangle in terms of the opposite trigonometric functions of the original angle.

These formulas are useful in order to translate known statements about one of the trigonometric functions into statements about the other function.

Rather than memorizing all these formulas—it is easy to get mixed up with the minus sign—one should clearly understand the geometric construction that defines the point $P(s) = (\cos s, \sin s)$ on the unit circle.

There are many other identities for trigonometric functions. To keep matters simple, we just mention the *addition formula*

$$\sin(s + t) = \sin s \cos t + \cos s \sin t$$

for the sine function. It can be proved by somewhat involved geometric arguments involving right triangles, but we shall skip the verification at this point. Later, in Section VI.5.3, we shall verify this formula by using the tools of calculus.

From this identity the corresponding addition formula for the cosine function is readily obtained by using the simple formulas we mentioned earlier, as follows.

$$\cos(s + t) = \sin\left(\frac{\pi}{2} - (s + t)\right) = \sin\left(\frac{\pi}{2} - s + (-t)\right)$$

$$\left(\text{use the addition formula with } \left(\frac{\pi}{2} - s\right) \text{ and } (-t)\right)$$

$$= \sin\left(\frac{\pi}{2} - s\right)\cos(-t) + \cos\left(\frac{\pi}{2} - s\right)\sin(-t)$$

$$= \cos s \cos t - \sin s \sin t.$$

Other formulas will be reviewed as needed later on.

In trigonometry courses one usually introduces other functions that are simple algebraic combinations of the two basic functions sine and cosine. For example, the *tangent* function is defined by

$$\tan s = \frac{\sin s}{\cos s} \text{ for all } s \text{ with } \cos s \neq 0.$$

Note that $\cos s = 0$ precisely when $P(s)$ lies on the y-axis, i.e., when $s = \frac{\pi}{2} + k\pi$, k any integer. In terms of the sides of a right triangle with hypothenuse 1 (see Fig. V.23), the tangent of an angle α of s radians is

$$\tan s = \frac{opposite\ side}{adjacent\ side} = \frac{b}{a}.$$

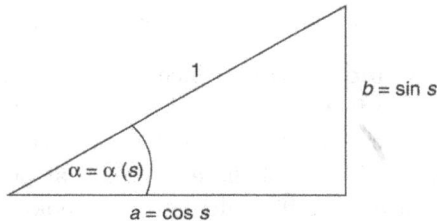

Fig. V.23   Right triangle with angle α of s radians.

We note that for a line in the plane that forms an angle α with the x-axis, its slope m can be described by $m = \tan \alpha$. In this book we shall mainly use the sine and cosine functions.

Graphs   The graphs of the sine and cosine functions are shown in Figs. V.24 and V.25.   Rough approximations of these graphs can be

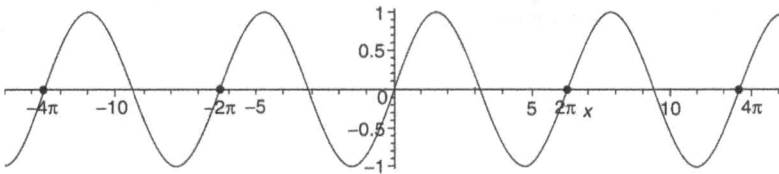

Fig. V.24   Graph of $y = \sin x$.

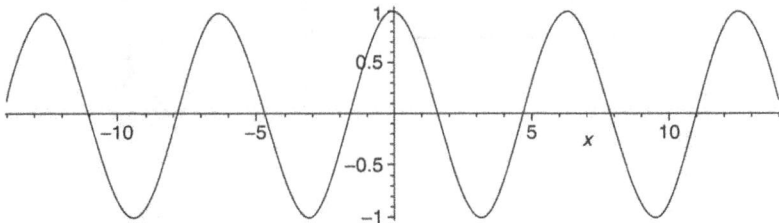

Fig. V.25   Graph of $y = \cos x$.

obtained by inspection of the coordinates of $P(s)$ as the point moves around the circle. Note that because of the periodicity, only points for $0 \leq s \leq 2\pi$ need to be sketched. This part of the graph is then repeated on adjacent period intervals, and so on. Of course, graphing calculators or computers are the most efficient tool for graphing these functions.

V.5.6 *Differentiation of* sine *and* cosine **Functions**

Continuity of *sine* **and** *cosine*   Since *sine* and *cosine* are defined in terms of the unit circle in the x, y-coordinate plane, we shall denote the input variable by a different letter, say t, rather than switching to the standard $y = \sin x$, etc. Values of these functions at points $t \in \mathbb{R}$ are best obtained by means of a scientific calculator. A look at their graphs (see previous section) suggests that these functions are continuous at all real numbers t, so that $\lim_{t \to a} \sin t = \sin a$ for each a, and so on. In particular,

$$\lim_{t \to 0} \sin t = \sin 0 = 0, \qquad \lim_{t \to \pi/2} \sin t = \sin \pi/2 = 1, \text{ and}$$
$$\lim_{t \to 0} \cos t = \cos 0 = 1, \qquad \lim_{t \to \pi/2} \cos t = \cos \pi/2 = 0.$$

The preceding statements follow readily from the geometric definition of *sine* and *cosine* on the unit circle. For example, as shown in Fig. V.26, the length $2 \sin t$ of the secant spanned by the arc of length $2t$ centered at the point $(1, 0)$ surely is shorter than the arc.

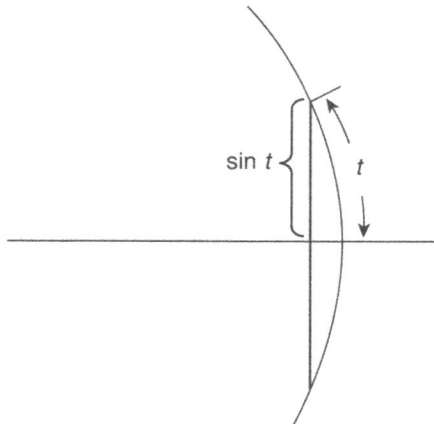

Fig. V.26   The arc t and $\sin t$.

It follows that $\sin t \leq t$ for $t > 0$, which implies that

$$|\sin t| \leq |t| \text{ for all } t \text{ near } 0.$$

This is the same kind of estimate we are familiar with for algebraic functions. It clearly shows that $\sin t \to 0 = \sin 0$ as $t \to 0$. This type of estimate generalizes to prove continuity at an arbitrary point a. Let $t \neq a$ be close to a. Then the distance $dist(P(a), P(t))$ between the points $P(a) = (\cos a, \sin a)$ and $P(t) = (\cos t, \sin t)$ on the unit circle is less than or equal to the length $|t - a|$ of the arc on the circle connecting these two points, i.e., $dist(P(a), P(t)) \leq |t - a|$. By the formula for the distance between these two points one then has

$$|\cos t - \cos a| \leq dist(P(a), P(t)) \leq |t - a| \text{ and}$$
$$|\sin t - \sin a| \leq dist(P(a), P(t)) \leq |t - a|.$$

Just as in the case of algebraic functions, these estimates clearly imply the continuity of *sine* and *cosine* at the point a.

By general rules about continuous functions it then follows, for example, that $\tan t = \sin t / \cos t$ is continuous at all points $t \neq \pi/2 + k\pi$, $k \in \mathbb{Z}$. Also, since $\cos t = \sin(\pi/2 - t)$ and $\sin t = \cos(\pi/2 - t)$ for all t, it is usually enough to verify basic results just for one of the trigonometric functions, and then apply appropriate general principles to extend the results to other functions. For example, once one knows that $\sin t$ is continuous, since $h(t) = \pi/2 - t$ is clearly continuous, the composition $\cos t = \sin(h(t))$ is continuous as well at each point.

The Derivative of $\sin t$ at $t = 0$ The graph of $y = \sin t$ suggests that this function has a *tangent* at every point, that is, $\sin t$ is differentiable everywhere. In order to study its derivative, we shall rely on the definition of the *sine* function on the unit circle in the x, y-coordinate plane.

Let us begin by considering the point $t = 0$. As usual, we need to study the factorization $\sin t = q(t) \cdot t$. The factor $q(t)$ is uniquely determined for $t \neq 0$ by $q(t) = \sin t / t$, but just as in case of the exponential function, the value of q at $t = 0$ is missing, and there is no obvious formula that would produce a suitable value $q(0)$ that would make q continuous at 0. So we must examine directly the behavior of $q(t)$ as $t \to 0$.

Let us first consider numerical approximations for $t_k = 10^{-k}$, $k = 1, 2, 3, \ldots$, as shown in the following table.

k	t_k	$q_a(t_k)$
1	10^{-1}	0.998334166468282
2	10^{-2}	0.999983333416666
3	10^{-3}	0.999999833333342
4	10^{-4}	0.999999998333333
5	10^{-5}	0.999999999983333
6	10^{-6}	0.999999999999833
7	10^{-7}	0.999999999999998
8	10^{-8}	1.00000000000000
9	10^{-9}	1.00000000000000
10	10^{-10}	1.00000000000000

The data provides very strong numerical evidence that

$$\lim_{t \to 0} q(t) = \lim_{t \to 0} \frac{\sin t}{t} = 1.$$

How can we recognize that this statement is indeed correct, without relying on incomplete and possibly misleading numerical "evidence"? There is no obvious method to simplify the expression $\sin t/t$, so we try to argue by using the geometric definition of the *sine* function, as shown in Fig. V.27.

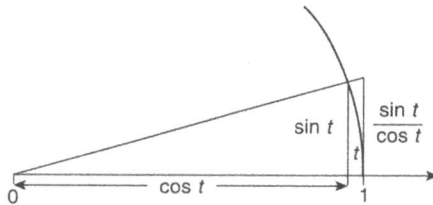

Fig. V.27   The arc t, and the values of $\sin t$ and $\cos t$.

According to Fig. V.27, if the number $t > 0$, which we choose $< \pi/2$, represents the length of the arc on the unit circle, then $\sin t$ measures a line segment that is approximately of length t when t is "small". As $t \to 0$, the approximation appears to improve, so that—geometrically—it is reasonable to expect that

$$\frac{\sin t}{t} \to 1 \text{ as } t \to 0.$$

We can make this precise by comparing the areas of the two similar right triangles shown in Fig. V.27 with the area of the relevant circular sector spanned by the arc of length t. Since the full circle corresponds to arc length 2π, the sector we are looking at represents $t/(2\pi)$ times the area of the full disc, that is, its area is $t/(2\pi) \cdot (\pi 1^2) = t/2.$[4] For $t > 0$ one therefore obtains

$$\frac{1}{2}\sin t \cos t \le \frac{t}{2} \le \frac{1}{2} \cdot 1 \cdot \frac{\sin t}{\cos t},$$

and hence, after dividing the inequalities by $(\sin t)/2 > 0$, it follows that

$$\cos t \le \frac{t}{\sin t} \le \frac{1}{\cos t}.$$

Now take reciprocals (careful with the inequalities!) to obtain

$$\frac{1}{\cos t} \ge \frac{\sin t}{t} \ge \cos t.$$

The geometric argument assumed $t > 0$, but all three expressions above do not change their values if t is replaced by $-t$, so that the latter inequalities hold for all $t \ne 0$ with $|t| < \pi/2$. Since the expression in the middle remains squeezed between the numbers $1/\cos t$ and $\cos t$, both of which approach 1 as $t \to 0$, by the Squeeze Theorem (limit rule iv) in Section 2.5, one obtains

$$\lim_{t \to 0} \frac{\sin t}{t} = 1.$$

This is exactly the result we expected based on the numerical data.

We now define

$$q(0) = \lim_{t \to 0} \frac{\sin t}{t} = 1,$$

thereby extending $q(t)$, defined for $t \ne 0$ by $q(t) = \sin t/t$, to a function that is continuous at 0. Altogether, we have verified that $y = \sin t$ is differentiable at 0 with derivative equal to $q(0) = 1$.

Let us translate the limit result we just obtained to the setting of average slopes. Figure V.28 shows the graph of $y = \sin t$, and the line through the points $(0,0)$ and $(t, \sin t)$ (i.e., secants) for several values $t > 0$. (Replace x by t in the figure.)

Clearly the slope of each secant line is the average rate of change

$$\frac{\sin t - \sin 0}{t - 0} = \frac{\sin t}{t} = q(t)$$

of $y = \sin t$ between 0 and t. As $t \to 0$, these secants turn around the point $(0,0)$ to approximate a line that is "tangential" to the graph of $y = \sin t$ at the point $(0,0)$. The slope of this limiting line is $\lim_{t \to 0} \frac{\sin t}{t}$, which has value 1 as we just determined. We conclude that the line of slope 1 through $(0,0)$, i.e., the graph of

$$y = t,$$

is the tangent to the graph of $y = \sin t$ at the point $(0,0)$.

---

[4]We use the familiar formula that the area of a disc of radius $r > 0$ is $\pi \cdot r^2$.

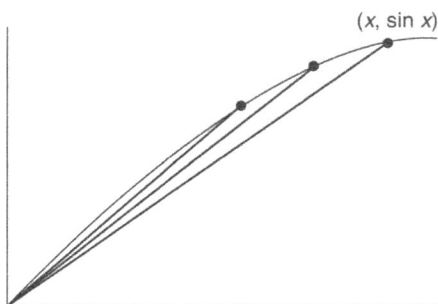

Fig. V.28   Graph of $y = \sin x$ with some secants through $(0, 0)$.

The Derivative of $\sin t$ In order to find the derivative of $\sin t$ at other points $a \neq 0$, we shall use a geometric variation of the argument that we used for the derivative at $a = 0$. It, too, relies on the definition of the *sine* function in terms of the unit circle. Let us first point out a useful consequence of the important formula $\lim_{t \to 0} \sin t/t = 1$ that we established in the preceding section. As seen in Fig. V.29, by reflecting the arc of length t and the corresponding line segment of length $\sin t$ on the x-axis,

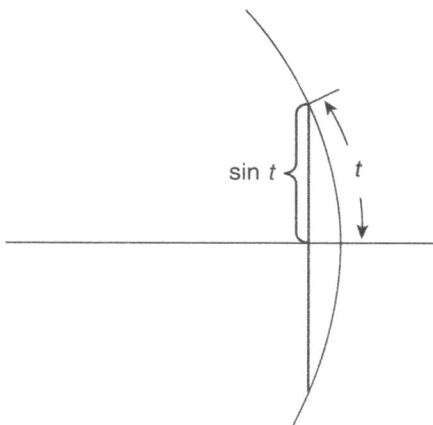

Fig. V.29   Chord spanned by arc of length $2t$.

one obtains that the ratio of the chord $2\sin t$ over the length $2t$ of the corresponding arc also has limit 1 as the length of the arc goes to zero. By rotating the circle, thus moving the point $(1, 0)$ to any other place on the circle, the same relationship between chord and arc persists. Thus we have the general result for arcs of length $h > 0$ centered at any point on the unit circle and the corresponding chord $c(h)$, as shown in Fig. V.30,

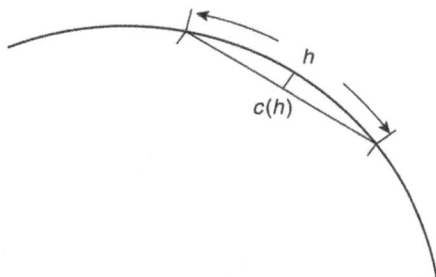

Fig. V.30   Arc of length h spans a chord of length $c(h)$.

that

$$\lim_{h \to 0} \frac{c(h)}{h} = 1.$$

Let us now consider the factorization $\sin(a + h) - \sin a = q_a(a + h)h$. For $h \neq 0$ the factor $q_a(a+h) = (\sin(a+h) - \sin a)/h$. The situation is visualized in Fig. V.31.

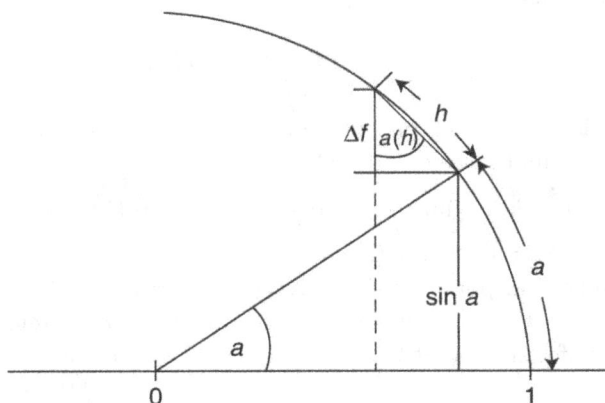

Fig. V.31   Geometric visualization of $q_a(a + h) = \Delta f/h$.

Furthermore, there is a small right triangle whose hypothenuse is the chord $c(h)$ spanned by the arc of length $h \neq 0$ that has the angle $a(h)$ opposite to $P(a)$. Here is an enlargement of that small triangle.

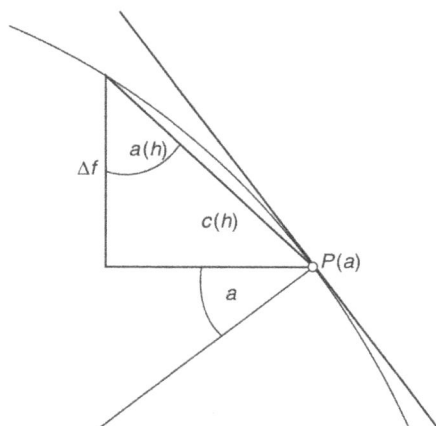

Fig. V.32   Triangle with hypothenuse $c(h)$ and leg Δf.

Since the leg Δf in this right triangle is adjacent to the angle $a(h)$, one has $\Delta f / c(h) = \cos a(h)$. Note that as $h \to 0$ the chord $c(h)$ (i.e., the secant through $P(a)$ and $P(a + h)$) will rotate into the direction of the tangent at $P(a)$, which is perpendicular to the radius at $P(a)$. Therefore the angle $\pi/2 - a(h)$ at the vertex $P(a)$ opposite to the angle $a(h)$ will have limit $\pi/2 - a$; hence the angle $a(h)$ has limit a as $h \to 0$.

It follows that

$$q_a(a + h) = \frac{\Delta f}{h} = \frac{\Delta f \cdot c(h)}{c(h) \cdot h} = \frac{c(h)}{h} \cos a(h).$$

Since $\cos(t)$ is continuous, one has $\lim_{h \to 0} \cos a(h) = \cos[\lim_{h \to 0} a(h)] = \cos a$. By basic limit rules one therefore obtains

$$\lim_{h \to 0} q_a(a + h) = \lim_{h \to 0} \frac{\Delta f}{h} = \lim_{h \to 0} \frac{c(h)}{h} \lim_{h \to 0} \cos a(h) = 1 \cdot \cos a.$$

By defining $q_a(a) = \lim_{h \to 0} q_a(a + h) = \cos a$, we extend q_a to a function that is continuous at point a as well. If one sets $t = a + h$, the factorization $\sin t - \sin a = q_a(t)(t - a)$ with q_a continuous at $t = a$ therefore confirms that $\sin t$ is differentiable at the arbitrary point a, and that its derivative satisfies

$$D[\sin](a) = (\sin)'(a) = q_a(a) = \cos a \text{ for all } a \in \mathbb{R}.$$

Remark. The figures shown above assume that $0 < a < \pi/2$ and $h > 0$. Corresponding figures have to be used in more general cases. However, the crux of the argument remains the same.

Another proof of this differentiation formula, based on the functional equation of the *sine* function, is outlined in Exercise 26 in Section 8.

The *cosine* Function   A geometric argument quite similar to the one we just used for the *sine* function can be used to analyze $\cos t$. (See Exercise 25 in Section 8.) A simpler argument is based on the relation $\cos t = \sin(\pi/2 - t)$ and the Chain Rule. Set $g(t) = \pi/2 - t$; for a fixed point a, set $b = g(a) = \pi/2 - a$. Since sin is differentiable at b, as we just established, the Chain Rule implies that the composition $\sin \circ g$ is differentiable at a and

$$D[\cos](a) = D[\sin \circ g](a) = D[\sin](b) \cdot g'(a) = \cos b \cdot (-1)$$
$$= -\cos\left(\frac{\pi}{2} - a\right) = -\sin a.$$

We have therefore established that $\cos t = (\sin \circ g)(t)$ is differentiable at a and that

$$D[\cos](a) = (\cos)'(a) = -\sin a \quad \text{for all } a \in \mathbb{R}.$$

The derivatives of the *sine* and *cosine* functions are very simple indeed. It is important *not to overlook the minus sign* that appears in the preceding formula in contrast to the earlier one for the derivative of $\sin t$.

Remark. We had mentioned the tangent function $\tan t = \frac{\sin t}{\cos t}$. By the Quotient Rule it follows that this function is differentiable at all points $a \neq \frac{\pi}{2} + k\pi$, $k \in \mathbb{Z}$, with

$$D[\tan](a) = D\left[\frac{\sin}{\cos}\right](a) = \frac{\cos a \cdot \cos a - \sin a \cdot (-\sin a)}{(\cos a)^2}$$
$$= \frac{(\cos a)^2 + (\sin a)^2}{(\cos a)^2} = \frac{1}{(\cos a)^2}, \text{ or } = 1 + (\tan a)^2.$$

A Differential Equation for *sine* and *cosine*   As we just verified, the *sine* and *cosine* functions satisfy the simple differentiation formulas

$$(\sin x)' = \cos x,$$
$$(\cos x)' = -\sin x$$

for all $x \in \mathbb{R}$. In particular, it follows that *sine* and *cosine* are infinitely often differentiable.

By standard differentiation rules it follows that a function f defined by $f(x) = A \sin x + B \cos x$, where A and B are constants, satisfies

$$D[f](x) = f'(x) = (A \sin x + B \cos x)' = A \cos x - B \sin x.$$

Upon differentiating one more time one obtains

$$D[f'](x) = f''(x) = -A \sin x - B \cos x = -f(x).$$

We thus see that f satisfies the differential equation

$$y'' + y = 0.$$

More generally, it is straightforward to verify that if w is any fixed real number, then the function f_w defined by $f_w(x) = f(wx)$, where f is the function we just considered, satisfies $f_w''(x) = -w^2 f_w(x)$, so that it is a solution of the differential equation

$$y'' + w^2 y = 0.$$

In analogy to the differential equation $y' = k \cdot y$ that models processes involving growth and decay, the differential equation $y'' + w^2 y = 0$ appears in many applications that involve periodic processes such as waves or a pendulum, to name just a few. As we shall prove later on in Section VI.5.2, it is indeed the case that *any* solution of this equation, that is, any function $y = f(x)$ that satisfies the equation $f''(x) + w^2 f(x) = 0$ on some interval I is of the form $f(x) = A \sin wx + B \cos wx$ for some constants A, B. This result and related applications will be discussed in the next chapter.

A Differentiable Function with Discontinuous Derivative   We can now give an explicit example of a function as described in the heading, something we had wondered about already in Section 1. Since functions that occur naturally are usually well behaved, such an example must be somewhat "unnatural". Let us define f by

$$f(x) = \begin{cases} x^2 \sin(1/x) & \text{for } x \neq 0 \\ 0 & \text{for } x = 0 \end{cases}.$$

Note that $f(x) = q(x) \cdot x$, where $q(x) = x \sin(1/x)$ clearly satisfies $\lim_{x \to 0} q(x) = 0$ (note $|x \sin(1/x)| \leq |x|$). Hence f is differentiable at 0, and $D[f](0) = 0$. For $x \neq 0$, the rules of differentiation imply that f is differentiable, with

$$D[f](x) = 2x \sin(1/x) + x^2 \cos(1/x) \left(-\frac{1}{x^2} \right)$$
$$= 2x \sin(1/x) - \cos(1/x) \text{ for } x \neq 0,$$

where we applied product rule and chain rule. Here the first term has limit 0 as $x \to 0$, but $\cos(1/x)$ does NOT have a limit as $x \to 0$. So $D[f]$ is NOT continuous at 0.

V.5.7 *Inverse Trigonometric Functions and Their Derivatives*

The Derivative of the Inverse of Sine Finally we also want to consider briefly inverse functions of sine and cosine. While the trigonometric functions are surely not one-to-one, just as we had done for the function $f(x) = x^2$, we may restrict them to suitable intervals to obtain this property, and then consider their inverse. It turns out that the inverse function rule will lead to an intriguing phenomenon.

To be specific, let us consider the function S defined for $x \in [-\pi/2, \pi/2]$ by $S(x) = \sin x$, whose graph is the thick curve shown in Fig. V.33.

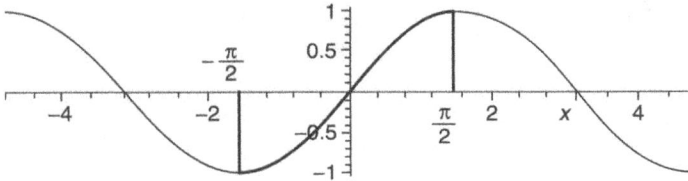

Fig. V.33   Graph of $\sin x$ for $-\pi/2 \le x \le \pi/2$.

Looking at this graph surely suggests that this function is strictly increasing on $[-\pi/2, \pi/2]$ (something that can be verified abstractly by the tools discussed in Section 6.2), and so it has an inverse

$$g : [-1, 1] \to [-\pi/2, \pi/2],$$

that is continuous by Theorem 54. The graph of g is as usual obtained by reflection on the line $y = x$ and is shown in Fig. V.34.

Since S is differentiable with $S'(x) = \cos x$, and $S'(a) \ne 0$ for all $a \in I = (\pi/2, -\pi/2)$ (no endpoints!), the Inverse Function Rule in the preceding section implies that the inverse g of $S(x)$ is differentiable at all points $b = \sin a \in S(I) = (-1, 1)$, and that

$$g'(b) = \frac{1}{S'(a)} = \frac{1}{\cos a}.$$

In order to write this formula in terms of $b = \sin a$, recall that $\sin^2 a + \cos^2 a = 1$ for all a. Therefore

$$\cos^2 a = 1 - \sin^2 a.$$

Since for $a \in (-\pi/2, \pi/2)$ one has $\cos a > 0$, it follows that

$$\cos a = +\sqrt{1 - \sin^2 a} = \sqrt{1 - b^2} \text{ for } a \in (-\pi/2, \pi/2),$$

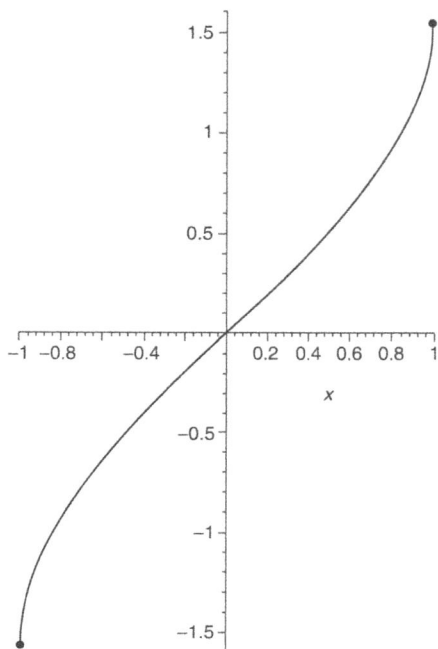

Fig. V.34   Graph of the inverse of $S(x) = \sin x$.

where $b = \sin a$. Hence one obtains

$$g'(b) = \frac{1}{\sqrt{1 - b^2}} \quad \text{for } -1 < b < 1.$$

After introducing the name *arcsin* (or inverse sine)[5] for the function g and replacing b with x, this formula translates into the differentiation formula

$$D[\arcsin x] = \frac{1}{\sqrt{1 - x^2}} \quad \text{for } -1 < x < 1$$

for the inverse of the (partial) *sine* function. Note that the inverse sine is not differentiable at the end points -1 and $+1$, even though $\arcsin x$ is defined and continuous at these points.

By analogous techniques one verifies that the inverse $\arccos x$ of the cosine function restricted to the interval $[0, \pi]$ satisfies

$$D[\arccos x] = -\frac{1}{\sqrt{1 - x^2}} \quad \text{for } -1 < x < 1.$$

See Exercise 34 in Section 8 for details.

---

[5] The terminology $\arcsin x$ for the inverse is a shortened version of *the arc whose sine is x*.

Finally, we also want to consider the inverse of $T(s) = \tan s$ on the interval $I = (-\pi/2, \pi/2)$. Since $\tan s = \frac{\sin s}{\cos s}$ equals the slope of the line through $(0, 0)$ that forms an angle of s radians with the x−axis, it is readily seen that $\tan s$ is strictly increasing on the interval I, and that $T(I) = \mathbb{R}$. Therefore T is invertible, and its inverse $T^{-1} = \arctan : \mathbb{R} \to I$. We had seen that the derivative of T equals, in particular,

$$D[\tan](s) = 1 + (\tan s)^2,$$

and therefore is nowhere equal to zero. If $b = \tan a$ for $a \in I$, the Inverse Function Rule then implies that

$$D[\arctan](b) = \frac{1}{D[\tan](a)} = \frac{1}{1 + (\tan a)^2} = \frac{1}{1 + b^2}.$$

So we have verified that

$$D[\arctan](x) = \frac{1}{1 + x^2} \text{ for all } x \in \mathbb{R},$$

which is a surprisingly simple answer.

A Remarkable Connection Let us conclude this discussion by pointing out a most amazing connection between algebraic and transcendental functions. As we have been learning about these two classes of functions, they seem to live in two completely separate worlds.

To summarize, algebraic functions are built up from the basic algebraic operations we are familiar with from the rational numbers. The identification of their tangents/derivatives involves an elementary application of simple fundamental algebraic techniques. Yes, in general successive differentiations of rational functions and their inverses, where defined, lead to more and more complicated expression, but we still remain within the realm of algebra, and no new concepts and ideas are needed.

In contrast, as we have seen in this chapter, a full understanding of the elementary transcendental functions, which are fundamental for investigations related to natural phenomena, requires extending the rational numbers to the much more mysterious real numbers $\mathbb{R}$, as well as the introduction of the deep limit process. Yet in spite of these complicated foundations, it turns out that the formulas for the derivatives are as simple as could be, almost boring: $D[e^x] = e^x$, or $D[b^x] = k \cdot b^x$, for some constant k, while the derivatives of sine and cosine just jump back and forth between each other, including an occasional minus sign.

On the other hand, differentiation of the corresponding *inverse* functions of these transcendental functions (where defined) often results in *algebraic* functions, i.e., functions of a completely different nature that are, in fact, much more elementary. Just recall the formula $D[\ln x] = \frac{1}{x}$ for $x > 0$, which really is surprisingly simple, and of course we just discovered the (algebraic) formula for the derivative of the inverse $\arcsin x$ of the sine function. The derivative of $\arctan x$ is an even simpler rational function. Differentiation thus establishes a surprising and deep link between transcendental and algebraic objects!

In other words, differentiation provides an amazing connection between what at first appear to be two separate worlds.

We can also look at this process in a different way, namely by considering the reversal of the differentiation process. Given a function f, does there exist a (differentiable) function F, called an *antiderivative* of f, whose derivative $D[F]$ equals the given f? And if so, how do we find F? For example, note that if $f(x) = x^n$, where $n \in \mathbb{Z}$, an antiderivative is easily found by reversing the power rule for derivatives $D[\frac{1}{n+1}x^{n+1}] = x^n$. Note however that we must exclude the case $n = -1$ in this formula. On the other hand, we have discovered that $F(x) = \ln x$ is an antiderivative of $f(x) = 1/x = x^{-1}$ on $\{x > 0\}$. So we see how the search for antiderivatives forces us to introduce new functions, and in particular takes us from the simple rational function $f(x) = x^{-1}$ to the non-algebraic natural logarithm function. We will examine some simple, yet fundamental examples later in Chapter VI.3. However, the systematic investigation of this process will be discussed in Chapter IV of the sequel to this book [WiC?].

V.5.8   *Exercises*

1. a) Show that the function $g(x) = |x|^{3/2}$ is differentiable at 0.

b) What is $g'(0)$?

c) More generally, show that $g_\alpha(x) = |x|^{1+\alpha}$ is differentiable at 0 for any $\alpha > 0$.

2. Let $f(x) = x^2$ and denote its linear approximation at $x = a$ by $L_{f,a}$.

a) Determine $L_{f,a}(x)$ explicitly. (Recall $f'(x) = 2x$.)

b) Evaluate the error $\mathcal{E}_a(x) = f(x) - L_{f,a}(x)$ by algebra and simplify as much as possible.

c) Use the result in b) to verify $\lim_{x \to a} \frac{\mathcal{E}_a(x)}{x-a} = 0$.

3. a) Use techniques of linear approximation to justify the approximation $e^h \approx 1 + h$ for small values of h.

b) Estimate the error $\mathcal{E}_0(h) = k(h) \cdot h$ made by the above approximation in case $h = 0.1$, 0.01 and 0.001 with the help of a calculator.

c) Evaluate $k(h) = \mathcal{E}_0(h)/h$ for the values of h in b). Do the results appear to support the statement that $\lim_{h \to 0} k(h) = 0$?

d) Use additional smaller values of h, if necessary, to estimate $\lim_{h \to 0} k(h)/h$.

4. In order to estimate the area of a circular platform one measures its radius with a measuring tape. The result is $5\ m \pm 0.005$. Use the linear approximation of the area function $A(r) = \pi r^2$ at $r = 5$ to estimate that the area of the platform equals $25\pi \pm \pi(0.05)\ m^2$.

5. Determine the derivatives of the functions

a) $3 \cdot 4^x - 5x^4$, b) $\pi x^\pi - \dfrac{4}{x}$ for $x > 0$, c) $[x^7 - 4 \cdot 5^x]^3$.

6. Suppose that g is differentiable with $g(4) = 2$ and $g'(4) = 1/2$, and that f is differentiable at 2, with $f(2) = 0$ and $f'(2) = 3$. Determine the derivative of the composition $f \circ g$ at $x = 4$.

7. Let $f : \mathbb{R} \to \mathbb{R}$ be differentiable and satisfy the property $f(0) = 0$.

a) Show that $(f \circ f)'(0) = [f'(0)]^2$ and $(f \circ f \circ f)'(0) = [f'(0)]^3$.

b) Let n be a positive integer. Determine the derivative of $(f \circ f \circ \ldots \circ f)$ (n compositions) at 0 in dependence of n.

8. Let $b > 0$ and define $f_1(x) = b^{-x}$ and $f_2(x) = (b^x)^{-1}$.

a) Evaluate the derivatives of f_1 and f_2 by using chain rule and power rule as needed.

b) Do your answers in a) agree? Explain!

9. The function $S^+(x) = x^2$ is one-to-one on the interval $[0, \infty)$, with inverse $g(x) = \sqrt{x}$ also defined on $[0, \infty)$. Use the result in this section to obtain the derivative of g. Where does g fail to be differentiable?

10. If $r = m/n$, with m and n both positive integers and n odd, the function $p(x) = x^r$ is defined in a full neighborhood of 0. Determine the values $r = \frac{m}{n} > 0$ for which p is *not* differentiable at 0, and those values for which p *is* differentiable at 0.

11. Suppose f is a one-to-one differentiable function with differentiable inverse g. Apply the chain rule to the function $f \circ g$ to find the relationship between f' and g'. (Hint: Recall that $(f \circ g)(x) = x$; take derivatives on both sides!)

12. Use the formula for the derivative of the inverse function to verify in detail that

$$D[\log_b x] = \frac{1}{x \ln b}, \quad x > 0$$

for any base $b \neq 1$.

13. Show that if ω is a period for the periodic function f, then 2ω and $-\omega$ are also periods, i.e., $f(x + 2\omega) = f(x)$ and $f(x - \omega) = f(x)$ for all x. More generally, verify that any integer multiple $k\omega$ is also a period.

14. Find
a) $\sin(k\pi)$ and $\cos(k\pi)$ for any integer k, and
b) $\sin(\frac{\pi}{2} + k\pi)$ and $\cos(\frac{\pi}{2} + k\pi)$ for any integer k.

15. Identify the point $P(\pi/4) = (x, y)$ on the unit circle and calculate its coordinates x, y, making use of the fact that $x = y$. Use this to evaluate $\sin\frac{\pi}{4}$ and $\cos\frac{\pi}{4}$.

16. Use the addition formulas for sine and cosine to find formulas for $\sin(2s)$ and $\cos(2s)$ in terms of $\sin s$ and $\cos s$.

17. a) Find the radian measure of 30^0, 45^0, and 60^0.
b) What is the degree measure of the angle of $\pi/12$ radians?

18. a) Determine the mode of your scientific calculator as follows. Enter $\sin(1.57)$. Explain why the result is either close to 1 or close to 0, depending on the mode. Use this fact to determine the mode your calculator is set for.

b) Find sine and cosine of 15^0. (Make sure that the calculator is set to degree mode.)

c) Compare the results in b) to the values of $\sin 15$ and $\cos 15$. (Use radian mode.)

19. Use a graphing calculator to display the graphs of the sine function and of $f(x) = \sin 3x$ and $g(x) = \sin(\frac{x}{3})$ in one window.

20. Use numerical methods to estimate the slope of the tangent line to the graph of $y = \sin(2x)$ at the point $(0, 0)$.

21. Given that $\lim_{x \to 0} \frac{\sin x}{x} = 1$, verify that $\lim_{x \to 0} \frac{\sin(ax)}{x} = a$ for any number a. (Hint: Replace $xa = h$, and consider $h \to 0$.)

22. The tangent function is defined by $\tan x = \frac{\sin x}{\cos x}$ for $x \neq \frac{\pi}{2} + n\pi$, n integer. Determine

$$a) \quad \lim_{s \to 0} \frac{\tan s}{s}, \qquad b) \quad \lim_{s \to 0} \frac{s}{\tan(3s)}.$$

23. a) The graph of the cosine function suggests that its tangent at $(0, 1)$ is the horizontal line $y = 1$. Use numerical data to confirm this geometric conclusion.

b) Use an accurate graph of $y = \cos x$ to estimate the slope of the tangent at the point $(\frac{\pi}{3}, \frac{1}{2})$.

c) Use a calculator to confirm your answer in b) by approximating the slope of the tangent by average rates of change on intervals $[\frac{\pi}{3}, \frac{\pi}{3} + 10^{-k}]$ for $k = 2, 3, ..., 6$.

24. Determine $\lim_{x \to \pi/2} \frac{\cos x}{x - \pi/2}$. (Hint: Use $\sin(\frac{\pi}{2} - t) = \cos t$.)

25. Work out the details of the *geometric* argument analogous to the one used for the sine function to find the derivative of $f(t) = \cos t$ directly. (Hint: Start with a sketch similar to Fig. V.31; note that Δf now corresponds to a different segment than in the case of the *sine* function.)

26. It is known that the *sine* function satisfies the functional equation (i.e., addition formula) $\sin(\alpha + \beta) = \sin\alpha \cos\beta + \sin\beta \cos\alpha$. (This result is usually proved in a course on trigonometry.) Use this equation to give an *analytic* proof of the formula for the derivative of $\sin x$. Use the following outline.

a) Prove that $\lim_{x \to 0} \frac{\cos x - 1}{x} = 0$.

b) Apply the functional equation to $\sin(a + h)$ and use the result to determine a formula for $q_a(a + h) = [\sin(a + h) - \sin a]/h$ for $h \neq 0$.

c) Take the limit as $h \to 0$ in the formula in b).

27. Determine the limits

$$a) \lim_{x \to 0} \frac{\cos x - 1}{x^2}, \quad b) \lim_{t \to 0} \frac{t^{3/2}}{\sin t}.$$

28. Determine whether $\lim_{x \to 0} \sin\frac{1}{x}$ and $\lim_{x \to 0} |x|^{1/2} \sin\frac{1}{x}$ exist, and find the limits, if possible. Make sure to justify your conclusions.

29. Find the equation of the tangent line to the graph of $y = \cos x$ at $x = \frac{\pi}{4}$.

30. Find the derivatives of the functions

a) $5\cos x + 3^x$,   b) $6x^4 - 2\sin x$,   c) $3\sin x + 2x^7 - 3e^x$.

31. Let $g(x) = x + \sin x$.

a) Find the equation of the tangent line to the graph of g at the point where $x = \frac{\pi}{2}$.

b) Are there any points on the graph of g where the tangent is horizontal? If yes, find all such points.

c) Graph g with a calculator to verify your conclusions in b).

32. Find the derivatives of the following functions.

a) $f(x) = \ln(x^2 + 1)$,   b) $g(t) = \sin(2^t)$,   c) $h(u) = 3^{\cos u}$,

d) $p(x) = x^{\sqrt{3}}$,   e) $G(x) = x^e + e^x$,   f) $q(s) = \dfrac{1}{s^4 + 2s^2 + 3}$.

(Hint for part f): Write $q(s)$ as power with negative exponent.)

33. Determine the derivative of $F(x) = \cos(\sin(x^2 + 1))$.

34. a) Verify that $C(x) = \cos x$ restricted to the domain $[0, \pi]$ is one to one with image $[-1, 1]$. (Look at the definition of $\cos x$ on the unit circle.)

b) Let g be the inverse of the function in a). Find the derivative of g at the point $b = \cos a$, where $a \neq 0, \pi$. Modify the argument in the text used for the *inverse sine* to determine the derivative of g. Where does g fail to be differentiable?

V.6   Some Basic Properties of Differentiable Functions

V.6.1   *The Mean Value Inequality*

Obviously any function $f(x) = c = constant$ satisfies $f' = 0$. Are the constants the only functions with this property? We have not encountered any other function whose derivative is always zero, and—intuitively—it is hard to imagine a *nonconstant* function that has zero derivative everywhere. Geometrically, the graph of a function that has always a horizontal tangent appears to necessarily be a horizontal line. Yet attempts to turn these intuitive ideas into precise form run into difficulties. Further analysis reveals that any correct verification of this apparently so "obvious" conclusion ultimately requires the *completeness* of the real numbers. In fact, if all we could "see" were just the rational numbers (realistically, isn't this the case?), one could indeed build such strange non constant functions with zero derivative "everywhere" (that is, at all visible rational points).

We shall now discuss a process that will allow us to extract a precise verification of what seems so obvious to the eye. More significantly, we will obtain an estimate for derivatives that has far reaching applications. In order to motivate our arguments, let us translate the problem into the setting of motion and velocity. So we assume that the function $s = s(t)$ measures the position of a vehicle at time t. Suppose we know that the *instantaneous* velocity $v(t) = \frac{ds}{dt}(t)$ is zero for all t in a time interval I. We want to conclude that $s(t)$ is constant on I, i.e., that there is no motion at all. Alternatively, assume that there is some motion between two points in time t_1 and t_2, i.e., there are $t_1 < t_2 \in I$ with $s(t_1) \neq s(t_2)$; we must then be able to find a time t^* in the time interval $[t_1, t_2]$ at which the *instantaneous* velocity $v(t^*) = ds/dt(t^*) \neq 0$. Our experience tells us that this must indeed be correct, but how do we back up our experience with a solid argument? The only thing we can be absolutely certain about is that the *average* velocity $\frac{s(t_1) - s(t_2)}{t_1 - t_2}$ in the time interval $[t_1, t_2]$ is nonzero.

We want to use this fact to produce a point t^* with $v(t^*) \neq 0$. Since the instantaneous velocity is approximated by average velocities over very short time intervals, we need to find a sequence of *shrinking* intervals, so that the corresponding average velocities will remain nonzero, and will in fact *converge to a nonzero limit*, that is, to a nonzero instantaneous velocity. The following simple observation provides the crux of the construction.

If during each of two successive time periods $[t_0, t_1]$ and $[t_1, t_2]$ the average velocity is less than or equal to v^*, then the average velocity over the combined period $[t_0, t_2]$ is also less than or equal to v^*.

Again, this statement is consistent with our experience and completely "obvious". But in contrast to the earlier situation, no subtle "limits" are involved here, and this makes it very easy to back up our intuition by a convincing argument that only requires simple algebra. In fact, if

$$\frac{v(t_1) - v(t_0)}{t_1 - t_0} \leq v^* \text{ and } \frac{v(t_2) - v(t_1)}{t_2 - t_1} \leq v^*,$$

then

$$v(t_1) - v(t_0) \leq v^* \cdot (t_1 - t_0) \text{ and } v(t_2) - v(t_1) \leq v^* \cdot (t_2 - t_1),$$

so that by adding the two inequalities one obtains

$$v(t_1) - v(t_0) + v(t_2) - v(t_1) \leq v^* \cdot (t_1 - t_0) + v^* \cdot (t_2 - t_1).$$

After rearranging and cancellations, one is left with

$$v(t_2) - v(t_0) \leq v^* \cdot (t_2 - t_0),$$

which implies the desired conclusion after division by $(t_2 - t_0) > 0$.

It is now clear how to proceed. We assume that the average velocity v_0 over a time interval $[c_0, d_0]$ is nonzero, say $v_0 > 0$. Divide the interval in half. By the observation just made, v_0 cannot exceed the maximum of the average velocities over each half interval. In other words, the average velocity v_1 over at least one of these half intervals must be at least as large as v_0, i.e., $v_1 \geq v_0$. Label that half by $[c_1, d_1]$. Now repeat this process starting with v_1 and the interval $[c_1, d_1]$, and then repeat over and over. At the nth step one obtains an interval $[c_n, d_n] \subset [c_{n-1}, d_{n-1}] \subset [c_0, d_0]$ of length $(d_0 - c_0)/2^n$, so that the average velocity v_n over $[c_n, d_n]$ satisfies $v_n \geq v_{n-1} \geq v_0$. The Nested Interval Theorem (Theorem 28 in Section 1.5), which is a consequence of the completeness of $\mathbb{R}$, guarantees that there is at least one point t^* contained in all these intervals. Then the instantaneous velocity $v(t^*)$, being the limit of average velocities $v_n \geq v_0$ over shorter and shorter time intervals shrinking to t^*, must also be greater than or equal to $v_0 > 0$, and hence $v(t^*) \neq 0$ as needed. (If desired, the last (intuitive) argument can be made rigorous by invoking the precise limit definition of derivatives combined with the observation above to pass from average velocities over $[c_n, d_n] = [c_n, t^*] \cup [t^*, d_n]$ to intervals with one endpoint at t^*.)

What if $v_0 < 0$? Then the preceding argument still gives $v(t^*) \geq v_0$, although now this does not imply $v(t^*) \neq 0$. Yet surely the whole argument can be modified to find another $t^{\#}$ with $v(t^{\#}) \leq v_0$. We therefore have verified the desired conclusion: If the average velocity over a time interval is not zero, then at some time during that interval the instantaneous velocity has to be non-zero as well.

Let us recast the conclusion we just obtained in the setting of an arbitrary differentiable function. Given such a function f defined on an interval I, we define the average rate of change of f over $[a, b] \subset I$, where $a < b$, by

$$\Delta(f, [a, b]) = \frac{f(b) - f(a)}{b - a}.$$

The argument we just went through (in the language of velocities) proves the second inequality in the following theorem. The first inequality follows by applying that result to $-f$ in place of f. We refer to this result as the *Mean Value Inequality*.

Theorem 62 *(Mean Value Inequality) Assume that f is differentiable on I. If $[a, b] \subset I$, then there exist x_{low} and $x_{high} \in [a, b]$ such that*

$$D[f](x_{low}) \leq \Delta(f, [a, b]) \leq D[f](x_{high}).$$

The result that prompted the whole discussion follows immediately.

Corollary 63 *If f is differentiable on the interval I with $f' \equiv 0$ on I, then f is constant.*

Proof. By the Theorem, the hypothesis implies $0 \leq \Delta(f, [a, b]) \leq 0$, i.e., $\Delta(f, [a, b]) = 0$, and hence $f(b) = f(a)$ for all $a, b \in I$.

We have thus verified that the only solutions of the differential equation $y' = 0$ are indeed just the constant functions.

Corollary 64 *If two differentiable functions f_1 and f_2 satisfy $D[f_1]) = D[f_2]$ on an interval I, then f_1 and f_2 differ by a constant, that is, there is $C \in \mathbb{R}$ such that $f_2(x) = f_1(x) + C$ for all $x \in I$.*

Proof. The function $f_2 - f_1$ has derivative $(f_2 - f_1)' = f_2' - f_1' = 0$, and hence is a constant C.

We therefore know *all* solutions to the differential equation $y' = g$ as soon as we know *one* solution.

Examples.

i) Find all functions that satisfy $y' = 2x^4$. By the standard differentiation rules, $(ax^5)' = a5x^4$ for any constant a. Choose a so that $5a = 2$, i.e., $a = 2/5$. Then the function $f(x) = \frac{2}{5}x^5$ satisfies $f'(x) = 2x^4$, so is one solution. All other solutions are therefore of the form $\frac{2}{5}x^5 + C$, where C is a constant.

ii) All solutions of $y' = 3^x$ are of the form $g(x) = \frac{1}{\ln 3}3^x + C$.

Another important consequence of Theorem 62 is the following estimate.

Corollary 65 *Suppose f is differentiable on I and that its derivative $D[f]$ is bounded on the interval $[a, b] \subset I$, that is, there exists K such that $|D[f](x)| \leq K$ for $x \in [a, b]$. Then*

$$|f(x_1) - f(x_2)| \leq K\,|x_1 - x_2|$$

for all $x_1, x_2 \in [a, b]$.

Recall that we had proved this kind of estimate for any polynomial already in Section IV.1.1, as well as a local version for rational functions in Section IV.1.2 by elementary arguments. This sort of estimate led us to introduce the concept of continuity, which has become central in our discussion of differentiable functions.

Proof. We apply the Mean Value Inequality to the interval $[a, b]$. Since $|D[f](x_{low})| \leq K$ and $|D[f](x_{high})| \leq K$, that result implies $|\Delta(f, [x_1, x_2])| \leq K$ for any two different $x_1, x_2 \in [a, b]$. The desired estimate then follows by multiplying with $|x_1 - x_2|$.

Finally, the Mean Value Inequality easily implies also the following theorem, which is known as the *Mean Value Theorem*.

Corollary 66 *(**Mean Value Theorem**) Suppose f is differentiable on the interval I and that its derivative $D[f]$ is continuous on I. Given $[a, b] \subset I$, there exists a point $c \in [a, b]$ such that*

$$D[f](c) = \Delta(f, [a, b]) = \frac{f(b) - f(a)}{b - a}.$$

Proof. By the Mean Value Inequality one has

$$D[f](x_{low}) \leq \Delta(f, [a, b]) \leq D[f](x_{high}).$$

Since $D[f]$ is assumed to be continuous, the Intermediate Value Theorem (Theorem 47 in Section 2.6) gives the existence of the desired solution c of the equation $D[f](x) = \Delta(f, [a, b])$. ■

Remark. In the standard calculus curriculum this MVT is usually stated and proved without assuming continuity of the derivative of f. The proof is somewhat more elaborate, and it involves a rather unnatural detour. Differentiable functions whose derivatives are not continuous are mainly of theoretical interest for professional mathematicians. The version stated here is quite sufficient for most of us.

V.6.2 *Differential Characterization of Increasing and Decreasing Functions*

Review of Monotonic Functions   We have seen the importance of the concept of an increasing (or decreasing) function in the context of the exponential function. In particular, if appropriately bounded, such functions have a limit at each point which is precisely given by the LUB or GLB of certain sets. Let us recall that a function f is *increasing on the interval I* if

$$f(x_1) \leq f(x_2) \text{ for all } x_1, x_2 \in I \text{ with } x_1 < x_2.$$

f is said to be *strictly increasing* on I if

$$f(x_1) < f(x_2) \text{ for all } x_1, x_2 \in I \text{ with } x_1 < x_2.$$

Geometrically, this means that the graph of f moves higher as we move towards the right. Correspondingly, one has the concepts of *decreasing* and *strictly decreasing* function, as follows. f is decreasing (strictly decreasing) on the interval I if

$$f(x_1) \geq f(x_2) \ (f(x_1) > f(x_2)) \quad \text{for all } x_1, x_2 \in I \text{ with } x_1 < x_2.$$

In this case the graph moves lower as one moves to the right. Figure V.35 visualizes these concepts.

A function f is said to be *monotonic* on the interval I if f is either decreasing on I or increasing on I.

Note: Common usage does not distinguish between increasing and *strictly* increasing, so one must be careful with the more precise language used here. For example, a constant function, whose graph is a horizontal

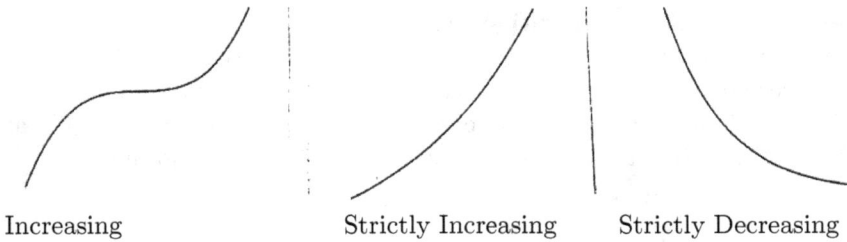

Increasing   Strictly Increasing   Strictly Decreasing

Fig. V.35

line, is both increasing and decreasing according to the definition given here, although it clearly is not *strictly* increasing or *strictly* increasing.[6]

Examples.

i) The functions $f(x) = 2^x$ and $g(x) = x^3$ are strictly increasing on $\mathbb{R}$.

ii) $f(x) = e^{-x}$ is strictly decreasing on $\mathbb{R}$.

iii) $p(x) = x^2$ is strictly decreasing on $(-\infty, 0]$ and strictly increasing on $[0, \infty)$.

iv) $y = \cos x$ is strictly decreasing on $[0, \pi]$.

All these properties are immediately confirmed by looking at the familiar graphs of these functions. (See Fig. V.36.) However, there are abstract proofs for these properties that use information about the derivatives of the functions, as will be discussed below.

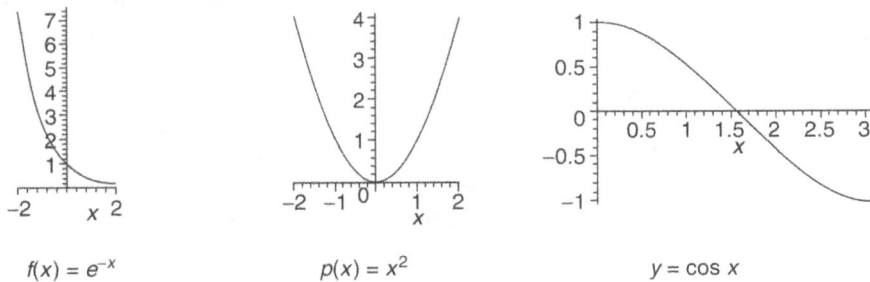

$f(x) = e^{-x}$ $p(x) = x^2$ $y = \cos x$

Fig. V.36

---

[6] An alternative terminology refers to "increasing", as defined here, as *nondecreasing*, and it uses the term *increasing* for what is called "strictly increasing" here. Correspondingly, one then uses the terms *nonincreasing* and *decreasing* in place of decreasing and strictly decreasing. In this terminology a constant function is both nondecreasing and nonincreasing, which may sound more reasonable than to say that such a function is both increasing and decreasing.

Relationship with Derivatives   The derivative of a function allows us to readily characterize the geometric properties illustrated above. From the preceding figures it is clear that an increasing function on I that is differentiable must have derivative ≥ 0. The analytic argument is just as simple. Fix a point $a \in I$. Consider the basic factorization formula

$$f(x) - f(a) = q(x)(x - a) \tag{V.11}$$

at $x = a$. If f is increasing on I, then $x - a > 0$ implies that the left side in (V.11) is ≥ 0, which implies that $q(x) \geq 0$ as well. By continuity of q at $x = a$ it follows that $f'(a) = q(a) = \lim_{x \to a^+} q(x) \geq 0$ as expected.

But one should NOT jump to the conclusion that if f is *strictly* increasing, then $f'(a) > 0$. While it is true that in this case equation V.11 implies that $q(x) > 0$ for all $x > a$, the limit as $x \to a^+$ may very well turn out to be zero. For example, $f(x) = x^3$ is strictly increasing, yet $f'(x) = 3x^2$ has a zero at $x = 0$.

Similarly, if f is decreasing on I, it follows that $f'(a) \leq 0$ for all $a \in I$. The converse of the above conclusion holds as well.

Lemma 67 *If the function f is differentiable on the interval I and satisfies $f'(x) \geq 0$ (or $f'(x) \leq 0$) for all $x \in I$, then f is increasing (resp. decreasing) on I.*

Proof. Pick any two points $x_1, x_2 \in I$ with $x_1 < x_2$. By the Mean Value Inequality (Theorem 62 in Section 6.1) there exists $x_{low} \in [x_1, x_2]$ with

$$f'(x_{low}) \leq \Delta(f, [x_1, x_2]) = \frac{f(x_2) - f(x_1)}{x_2 - x_1}.$$

Since by assumption $f'(x_{low}) \geq 0$, it follows that $f(x_2) - f(x_1) \geq 0$, i.e., $f(x_1) \leq f(x_2)$ as required. The proof for the corresponding result when $f'(x) \leq 0$ uses the existence of x_{high} with $\Delta(f, [x_1, x_2]) \leq f(x_{high}) \leq 0$.

To summarize:

A differentiable function f on an interval I is increasing on I if and only if $f'(x) \geq 0$ for all $x \in I$.

By replacing $\geq$ with $>$ in the Lemma and its proof one also obtains the following result.

If f satisfies $f'(x) > 0$ on I, then f is strictly increasing on I.

As noted earlier, the converse of this last statement is not correct in general.

By completely analogous arguments one sees that decreasing functions on I are characterized by $f'(x) \leq 0$ on I, and that in case $f'(x) < 0$ for all $x \in I$, one gets the stronger conclusion that f is *strictly* decreasing on I.

Example. Determine intervals where the function $p(x) = x^3 + \frac{3}{2}x^2 - 18x + 5$ is strictly increasing or decreasing.

Solution. This is easily done by visual inspection of the graph of p obtained with the aid of a graphing calculator. If no graphing calculator is available, we apply the principles we just discussed and consider the derivative $p'(x) = 3x^2 + 3x - 18$. The set of points where $p'(x) \neq 0$ is the complement of the zeroes of p', i.e., of the solutions of

$$p'(x) = 3x^2 + 3x - 18 = 0.$$

These solutions are -3 and 2. (Use the formula for solving quadratic equations.) The real line is thus separated into the intervals $(-\infty, -3)$, $(-3, 2)$, and $(2, \infty)$, on each of which p' has no zero. Furthermore, on each of these intervals p' is either always positive or always negative, since a change of sign would result in an additional zero by the Intermediate Value Theorem in Section 2.6. Note that for large $|x|$ the polynomial $p'(x)$ is positive; also, $p'(0) = -18 < 0$. It follows that p is strictly increasing on the intervals $(-\infty, -3)$ and $(2, \infty)$, and strictly decreasing on $(-3, 2)$, as seen in Fig. V.37.

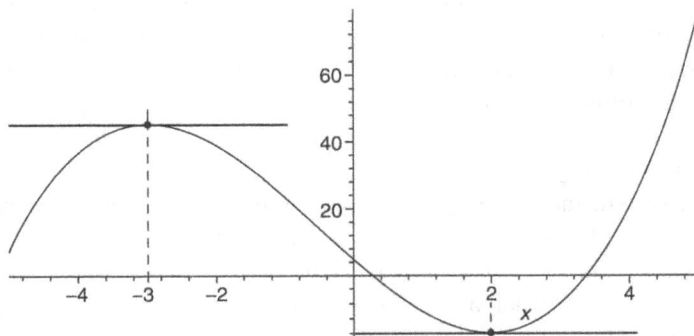

Fig. V.37   Regions where p is increasing or decreasing.

V.6.3   *Local Extrema*

Continuing with the last example, note that the points where $p'(x) = 0$ have a special geometric significance. The tangent line at these places is horizontal. As seen from the last figure, the function has a high point where $x = -3$ and a low point at $x = 2$. These are examples of *local (or relative) extrema*, which are defined as follows.

Definition 68 *The function $y = f(x)$ has a local maximum (or a local minimum) at the point a if there exists a neighborhood U of a, such that*

$$f(x) \leq f(a) \qquad (or\ f(x) \geq f(a)) \qquad for\ all\ x \in U.$$

Geometrically, at a local maximum the graph of the function changes from increasing to decreasing, while at a local minimum the opposite change occurs.

The following result is geometrically evident.

Theorem 69 *Suppose f has a local extremum at a. If f is differentiable at a, then $f'(a) = 0$.*

For completeness' sake we also present the simple analytic proof. It is enough to consider the case that f has a local maximum at $x = a$. The other case follows by an analogous argument. We again consider the factorization

$$f(x) - f(a) = q(x)(x - a).$$

Given that f has a local maximum at a implies that the left side is ≤ 0 for all x near a. If $x < a$ the factorization implies that $q(x) \geq 0$, while for $x > a$ one must have $q(x) \leq 0$. The differentiability of f at a, i.e., the continuity of q at a, implies that $q(a) = \lim_{x \to a} q(x)$. Hence both one-sided limits exist as well and must be equal. Since $\lim_{x \to a^-} q(x) \geq 0$ and $\lim_{x \to a^+} q(x) \leq 0$ by the preceding observations, it follows that $q(a) = \lim_{x \to a} q(x) = 0$. Therefore $f'(a) = q(a) = 0$.

Remark. The problem of finding local extrema of a function has major historical significance. Surely such points are of interest in many applications, for example, economists try to determine the price of a product so as to maximize profits. Already in the 17th century, philosophers and scientists tried to find mathematical techniques to identify such local extrema. For example, Pierre de Fermat (1607–1665) devised a technique that eventually was recognized as being the equivalent of identifying points where the graph of the function has a horizontal tangent. In the language of calculus, developed soon after Fermat, this is a point where the derivative of the function is zero.

We note that a function may have a relative extremum at a point a and not be differentiable at a. For example, $g(x) = |x|$ clearly has a local minimum at 0, and g is not differentiable at that point.

One says that a is a *critical point* of the function f if either f fails to be differentiable at a or else $f'(a) = 0$. So the results we just discussed can be summarized by saying that local extrema are found among the critical points. Note, however, that not every critical point is necessarily a point at which there is a local extremum.

Example. $f(x) = x^3$ has derivative $f'(x) = 3x^2$ for all x. So 0 is the (only) critical point of f. But clearly f takes on values that are less than $f(0) = 0$, as well as values that are greater than $f(0)$ in any neighborhood of 0. So f does not have a local extremum at 0. For the inverse function $g(x) = x^{1/3}$ of f one sees that g fails to be differentiable at 0, while $g'(x) = \frac{1}{3}x^{-2/3} \neq 0$ for all $x \neq 0$. So 0 is the only critical point of g, yet g has no extremum at 0.

Usually a function of one variable has only finitely many critical points on any given finite interval, and these can be determined readily in many cases. However, in general, to find the solutions of the equation $f'(x) = 0$ may require numerical approximations. It turns out that the tools of calculus can be used to set up an efficient approximation process to find zeroes of differentiable functions. We refer the reader to Chapter III.7 in [WiC?].

The discussion in the last two sections gives a brief glimpse into how the tools of derivatives can be used to describe geometric properties of graphs of functions. Here we just focused on monotonicity, given the importance of this concept in our earlier investigations, as well as on local extrema, which are most relevant in applications. More general techniques, that also involve the second derivative of functions (see next section for this concept) are used to describe other important features of graphs of functions. Such techniques were very important before the age of computers and graphing calculators, and the traditional curriculum devoted much time to practice graphing of differentiable functions by using the tools of calculus. Today, while it is useful to consider such additional geometric concepts, such as convexity, for example, their application to graphing functions is no longer so central. Some of the basic techniques are discussed in [WiC?].

V.6.4 *Higher Order Derivatives*

Suppose f is differentiable on an interval I. Then its derivative $D[f]$ is a new function on I that may also be differentiable at each point of I. In that case one can define its derivative $D[D[f]] = D^{(2)}[f]$, or f'', which is called the second derivative of f. Clearly the process may be continued. For example, since the derivative $D[R]$ of a rational function R is again rational, with the same domain as R, one may continue differentiating any number of times. For each $n \in \mathbb{N}$ one may differentiate the rational function R n times, resulting in the nth order derivative of R, denoted by

$$D^{(n)}[R] \text{ or } R^{(n)}.$$

Notice that if one starts with a rational function that is not a polynomial, successive differentiations lead to functions that are more and more complicated, as a consequence of the quotient and product rules.

Example. Let us find the third derivative of $R(x) = \frac{x}{x^2-1}$ for $x \neq 1, -1$. Applying the quotient rule one obtains

$$D[R] = \frac{1(x^2-1) - 2x \cdot x}{(x^2-1)^2} = -\frac{x^2+1}{(x^2-1)^2}.$$

Since we need to take derivatives again, we simplified the numerator. When calculating $D^{(2)}[R]$ we will need to use the chain rule when differentiating the denominator, resulting in

$$D^{(2)}[R] = -\frac{2x(x^2-1)^2 - [2(x^2-1) \cdot 2x](x^2+1)}{(x^2-1)^4}.$$

Here we can cancel the common factor (x^2-1) in numerator and denominator, resulting in

$$D^{(2)}[R] = -\frac{2x(x^2-1) - [2 \cdot 2x](x^2+1)}{(x^2-1)^3} = -\frac{-2x^3 - 6x}{(x^2-1)^3} = 2\frac{x^3+3x}{(x^2-1)^3}.$$

At this point we leave the calculation of $D^{(3)}[R]$ as an exercise.

Polynomials P are a special case: if P has degree n, then $D[P]$ is again a polynomial, now of degree $n-1$. Continuing with taking derivatives leads to $P^{(n)} = $ constant, and hence $P^{(n+1)} = 0$. On the other hand, successive differentiations of exponential functions become pretty boring. In fact, for the natural exponential function E one has $D[E] = E$, so that $D^{(n)}[E] = E$ for each $n = 1, 2, 3, ...$, while for arbitrary base $b > 0$ one has

$$D^{(n)}[E_b] = (\log b)^n E_b.$$

Since we can continue differentiating as often as we wish, one says that rational and exponential functions are infinitely often differentiable. Most of the functions that occur "naturally" in higher mathematics are infinitely differentiable. We noticed that *sine* and *cosine* are also infinitely often differentiable. To find differentiable functions that do not have this property one must make some special constructions. For example, let us consider $f(x) = x \, |x|$. We know that $|x|$ is not differentiable at 0, so we can not apply the product rule to f at 0. Instead, we have to look more carefully. Note that $f(x) - f(0) = q(x)(x - 0)$, where the factor q equals $q(x) = |x|$, which is continuous at 0. Hence f is differentiable at 0, with $f'(0) = q(0) = 0$. To check whether f' is differentiable, we must determine its values also at $x \neq 0$. Note that

$$f(x) = \begin{cases} x^2 & \text{for } x > 0 \\ -x^2 & \text{for } x < 0 \end{cases},$$

so that

$$f'(x) = \begin{cases} 2x & \text{for } x > 0 \\ -2x & \text{for } x < 0 \end{cases}.$$

Hence $f'(x) = 2 \, |x|$ for $x \neq 0$, and we saw earlier that this holds also for $x = 0$, i.e., $f'(x) = 2 \, |x|$ for all x. We thus see that f' is NOT differentiable at 0, so that f is NOT 2 times differentiable at 0. Of course f is 2 times differentiable at all other points, with $f''(x) = 2$ for $x > 0$, and $f''(x) = -2$ for $x < 0$, but we see that $f''(x)$ does not have a limit at 0, and hence cannot be extended to a continuous function there.

This example shows how it is possible to build up strange functions whose properties do not match "natural" expectations that we associate with "natural" functions such as rational or exponential ones. In fact, matters can be really weird! Karl Weierstrass (1815–1897), one of the great mathematicians of the 19th century, who is credited with much of the theoretical foundations of calculus, was able to construct a continuous function on an interval I that is NOT differentiable at any point $a \in I$. It is strange phenomena such as these that forced mathematicians to create and make precise abstract concepts, such as the system of real numbers, the notion of completeness, and limits, as they realized that one could not just rely on one's intuition.

Finally, let us mention that polynomials are the only functions for which successive differentiations eventually result in a function that is identically 0, as made precise in the following Lemma.

Lemma 70 *Suppose f is a function defined on an interval I that can be differentiated infinitely often, and that there is a positive integer n, so that $f^{(n)}(x) = 0$ for all $x \in I$. Then either $f(x) = 0$ for $x \in I$, or f is a polynomial of degree $\leq n - 1$.*

Proof. We shall prove this by induction on n. First, we check this for $n = 1$, that is, we are given that $f^{(1)}(x) = f'(x) = 0$ for $x \in I$. By Corollary 63 in Section 6.1, this implies that $f(x) = c_0$ for some constant $c_0 \in \mathbb{R}$, which proves the Lemma in this case. Next, we prove the inductive step. We assume that we have proved the result for some $n \in \mathbb{N}$, and we want to verify it for $n + 1$. Note that $f^{(n+1)}(x) = [D[f]]^{(n)}(x)$, so that assuming $f^{(n+1)}(x) = 0$ for all $x \in I$ means that we can apply the inductive hypothesis with the integer n to the derivative $D[f]$ of f. Hence $D[f]$ is either 0, or a polynomial of degree $\leq n - 1$, i.e.,

$$D[f] = c_{n-1}x^{n-1} + c_{n-2}x^{n-2} + \ldots + c_1 x + c_0.$$

By reversing the power rule for differentiation, it is easy to find a polynomial g whose derivative equals $D[f]$, namely

$$g(x) = \frac{c_{n-1}}{n}x^n + \frac{c_{n-2}}{n-1}x^{n-1} + \ldots + \frac{c_1}{2}x^2 + c_0 x.$$

Note that if $D[f] = 0$ then $g = 0$ as well. Since $D[f] = D[g]$ on I, Corollary ( ) implies that there is a constant C such that $f = g + C$, which clearly is a polynomial of degree $\leq n$, or $f = 0$. ∎

V.6.5   *Exercises*

1. Let $f(t) = \cos(t)$.

a) Determine the average rate of change $R = \Delta(f, [0, \pi])$ of f over the interval between $t = 0$ and $t = \pi$.

b) Find specific points l and m in $[0, \pi]$ so that $f'(l) \leq R \leq f'(m)$.

c) Is there any subinterval $[a, b] \subset [0, \pi]$, with $\Delta(f, [a, b]) > 0$? Explain your answer!

2. Suppose g satisfies $g'(x) < 0$ for $1 < x < 3$, $g'(3) = 0$, and $g'(x) > 0$ for $3 < x < 4$.

a) Does g have a local extremum at 3? If so, is it a local maximum or minimum?

b) Assume $g(3) = -1$. Sketch a possible graph of a function g that satisfies all the given properties.

3. Verify that $p(x) = x^5$ satisfies $p'(0) = 0$. Does p have a local extremum at 0? If so, is it a local maximum or minimum? Explain!

4. Make a sketch of the graph of a function F that satisfies the following conditions:

$F(0) = -2$, $F(2) = +2$, $F'(x) > 0$ for $0 < x < 2$, and $F(x) > 0$ and $F'(x) < 0$ for all $x > 2$.

5. Figure V.38 shows the graph of the *derivative* $D[g] = g'$ of a function g.

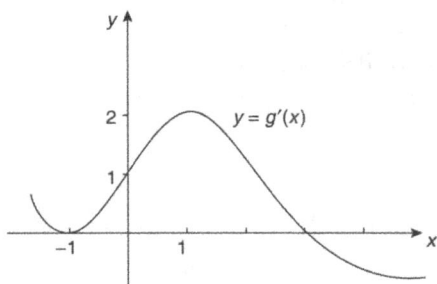

Fig. V.38   Graph of derivative of g.

Suppose that $g(0) = 0$. Make a sketch of a possible graph of g that matches the properties of the derivative shown in Fig. V.38.

6. Let $f(x) = x \cdot 4^{-x}$.

a) Where is f decreasing, where is it increasing?

b) Does f have any local extrema? If yes, find them and describe their type.

7. Answer the same questions as in problem 6 for the function $f(x) = x^2 \cdot e^x$. (Hint: Factor f' as much as possible, and analyze the sign of the various factors.)

8. a) Where is $G(x) = x - 4\sqrt{x}$ increasing, where is it decreasing?

b) Explain why $G'(x)$ is close to 1 when x is getting very large.

c) Use the information in a) and b) to sketch the graph of G.

9. Does $y = x + \sin x$ have any local extrema? Explain.

10. Use the first derivative to determine the intervals where the function $f(x) = 3x^4 - 4x^3 - 12x^2 + 5$ is strictly increasing or strictly decreasing.

11. Find the first three derivatives of the function $g(x) = x^2 |x|$ at all points where they exist.

12. Suppose F is a 3 times differentiable function that satisfies $F^{(3)} = 0$ on $\mathbb{R}$.

a) What is the most general form for F'?

b) Suppose $F'(0) = F''(0) = 0$. What is the most general form for F in this case?

13. Suppose g is 2 times differentiable and $g''(x) > 0$ on the interval $(-1, 1)$.

a) What can you say about the derivative g' on $(-1, 1)$? Prove your answer.

b) Is g increasing or decreasing on $(-1, 1)$? Explain!

c) Give an explicit example of such a function g.

Chapter VI

Applications of Derivatives: A Brief Introduction

In the preceding chapter we systematically investigated and formalized the central new mathematical concepts that are the foundation of what is known as the Differential Calculus, culminating in the definition of differentiable functions and the basic rules that they satisfy. All of more advanced Calculus and, more generally, the vast mathematical area of Analysis, builds on top of these foundations. We shall now take a break from expanding the more theoretical investigations, and instead discuss a few simple, yet central applications of these new tools, so that we may get a better understanding of how they can be used.

VI.1 Acceleration and Motion with Constant Acceleration

We shall begin by returning to the study of motion that was initiated by Galileo early in the 17th century, and that we already touched upon in Section II.3.

VI.1.1 *Acceleration*

The concept of *acceleration* originated with Galileo's discovery that a freely falling object near the surface of the Earth moves at an increasing velocity, and this change in velocity is what is known as the acceleration of the motion. We shall now use differentiation to give a precise formulation of this idea.

We consider the motion of an object along a straight line. The motion is described by a function $s = s(t)$ which measures the distance of the object from a fixed point at the time t. Velocity, in essence, captures the *rate* at which the distance changes with time. The simplest version is the *average velocity* between two moments in time t and $t + h$, where $h > 0$, which is defined by

$$\frac{s(t+h) - s(t)}{(t+h) - t} = \frac{s(t+h) - s(t)}{h}.$$

By shrinking the length of the time interval, i.e., by letting $h \to 0$, the average velocity will approximate what we call the instantaneous velocity at time t. In terms of derivatives, this is just the (first) derivative $D[s](t) = s'(t)$, which, in the present context, is also denoted by $\frac{ds}{dt}(t)$ to remind us that it is defined via quotients that represent average velocities. We shall denote the (instantaneous) velocity at time t by $v(t)$; thus

$$v(t) = D[s](t) = \frac{ds}{dt}(t).$$

We note that common units to measure velocity are *ft/sec, m/sec,* or *km/hour,* and others.

In complete analogy to velocity, which measures the rate of change in distance, one can consider the rate of change of velocity $v(t)$, which is called "acceleration". Again, there is the elementary concept of average acceleration

$$\frac{v(t+h) - v(t)}{h}$$

over the time interval from t to $t+h$, and by letting $h \to 0$ one captures the instantaneous acceleration $a(t)$ at time t. Of course, the precise definition again involves the derivative, that is,

$$a(t) = D[v](t) = \frac{dv}{dt}(t).$$

Based on this definition, one sees that units to measure acceleration are based on *velocity/time,* and therefore equal, for example, *ft/sec²* or *m/sec².*

Of course, for all this to make sense, one must assume that the functions that are used to describe the process are differentiable, and, in particular, the distance function $s(t)$ should be at least two times differentiable, since the acceleration, as we just defined it, is the second derivative of s, namely

$$a(t) = D[v](t) = D[D[s]](t) = D^2[s](t) = s''(t).$$

Example. A boat floating on the sea bounces up and down with the waves. Suppose the motion is described by the function $h(t) = 12\sin(0.25t)$, where $h(t)$ measures the amount in feet that the boat rises above or drops below a fixed level in dependence of time t, measured in seconds. At what times does the acceleration have maximal absolute value? What is that maximal acceleration?

Solution. By differentiation one obtains $h'(t) = 12(.25)\cos(0.25t)$, and by differentiating a second time one obtains the acceleration $h''(t) = -12(.25)^2\sin(0.25t)$. Hence $|h''(t)| = \frac{3}{4}|\sin(0.25t)|$. The maximal value $3/4\,ft/sec^2$ occurs when $|\sin(.25t)| = 1$. The motion is "roughest" at those moments when the boat is on top of a wave or in the valley between two waves.

Acceleration is really the central concept in the study of motion. This is a consequence of Newton's Law of Motion

$$Force = mass \times acceleration,$$

which describes how the underlying forces determine the acceleration, which in turn, determines the motion. The mathematical problem thus can be summarized as follows: Given information about the acceleration of a motion, determine the position function of the motion.

VI.1.2 *Free Fall*

To illustrate how to apply the general relationship between position, velocity, and acceleration established by the process of differentiation, we shall consider the simplest case when the acceleration is *constant*. In particular, as a consequence of basic physical principles, we shall recover the results of Galileo's investigations about falling objects that we had already discussed in Section II.3. We consider an object (say a rock falling off a cliff) falling towards the ground. We denote by $s(t)$ its height above ground at time t. Neglecting all perturbations due to wind or air resistance, the only force that acts on such an object is the gravitational force F_g exerted by the earth. At heights that are very small compared to the radius of the earth, this force can be assumed to be constant, that is, independent of the height of the rock. According to Newton's law of motion which we stated above, this force causes an acceleration a (i.e., a change in velocity) on the object, which therefore must also be constant near the surface of the earth. This acceleration a due to earth's gravity is usually denoted by $-g$ with $g > 0$, where the minus sign makes explicit the fact that the force pulls downwards, so that the height is decreasing. Depending on the units chosen, the numerical value of the constant g is approximately 9.81 m/sec^2 or 32 ft/sec^2.

Given information about the acceleration a, we can now determine the position function s by using the tools of calculus. In fact, we must reverse the process of differentiation, since the acceleration is the second derivative of the position function $s(t)$. Since $s''(t) = -g$ is constant, taking the derivative one more time gives 0. By the result in Section V.6.4, we know that the position function must then be given by a polynomial of degree ≤ 2, and the condition $s''(t) = -g = $ constant implies that we must have

$$s(t) = -\frac{g}{2}t^2 + v_0 t + s_0,$$

where the notation for the two other coefficients is chosen so as to indicate their relationship to the motion. In fact, $s_0 = s(0)$ is the initial position (i.e., at time $t = 0$), and since $v(t) = s'(t) = -gt + v_0$, the constant $v_0 = v(0)$ is the initial velocity of the motion. In particular, it then follows from the above formula that $v(t_2) - v(t_1) = (-g) \cdot (t_2 - t_1)$, that is, the motion of a freely falling body is "uniformly accelerated". This, of course, is the fundamental fact discovered by Galileo early in the 17th century. The formula we determined for the height $s(t)$ of the object at time t is completely determined if one knows the values of the *initial* position and velocity.

Example. *A stone is dropped from the top of a tower, and it hits the ground 4 seconds later. Find the height of the tower.*

The model for the motion under constant acceleration applies. Let h denote the (unknown) height of the tower in meters. This is the initial position at time $t = 0$. Since the stone is simply "dropped" at that time, the initial velocity $v_0 = 0$, so that using meters and seconds for the units, one has

$$s(t) = -\frac{9.81}{2}t^2 + h \text{ meters.}$$

When the stone hits the ground ($t = 4$), the height $s(t)$ is zero, so that $s(4) = -\frac{9.81}{2}4^2 + h = 0$. Solving for h gives

$$h = \frac{9.81}{2} \cdot 16 = 75.48 \text{ meters.}$$

The tower is approximately 75 meters high.

VI.1.3   *Constant Deceleration*

We discuss another situation where constant acceleration occurs.

Example. *A car travels along a highway at 50 miles/hour. The driver sees a washed out bridge approximately 100 ft down the road, and immediately applies the brakes with constant pressure. After one second his speed is down to 35 miles/hour. Will the car stop in time?*

Solution. Based on the given information, we assume that the car decelerates at a constant rate of a ft/sec^2. Let $t = 0$ correspond to the time when the brakes are first applied, measured in seconds. Conversion of the initial speed of 50 miles/hour to ft/sec gives $50 \cdot 5280/3600 \approx 73.3$ ft/sec. Thus $v(t) = -at + 73.3$. (The minus sign reflects the fact that the braking action slows down the car, which we interpret as a negative acceleration.) Since we are told that

$$v(1) = 35\, m/h = 35 \cdot 5280/3600 \approx 51\ 3 ft/sec,$$

we can determine the deceleration rate a from the equation $51.3 = -a \cdot 1 + 73.3$, resulting in $a = 22$ ft/sec^2. Next we can determine the time required for the car to stop (assuming that there is no interruption) from the equation $0 = v(t) = -22t + 73.3$. One obtains $t = 73.3/22 \approx 3.3$ seconds. We can now calculate the distance the car would travel until coming to a stop as follows. The position function $s(t)$ satisfies

$$s(t) = -22\frac{t^2}{2} + 73.3t + 0.$$

Hence $s(3) = -22 \cdot 9/2 + 73.3 \cdot 3 = 120.9$. Unfortunately, the car will not stop in time before falling over the edge...if only the driver had replaced his worn tires the day before!

VI.1.4 *Exercises*

1. The specifications for a shipping box state that the box should withstand an impact against a fixed object up to a speed of 10 ft/sec. What would happen to the box if it is dropped from a balcony 20 ft above the ground?

2. While traveling at 95 miles per hour in your Ferrari on a road where the speed limit is 60 m.p.h. you spot a state trooper in the distance and immediately apply the brakes with constant pressure. After 100 ft your speed is down to 70 m.p.h. Continuing with the same deceleration, what will be your speed when you pass the trooper who is an additional 100 ft away? Will you get a ticket? (60 m.p.h. equals 88 ft/sec.)

3. A car accelerates under constant acceleration from 0 to 100 km/h in 6 seconds. Find the acceleration in m/sec^2.

4. A cannon fires its ammunition straight up with an initial speed of 60 m/sec.
   a) How high will the cannon ball reach?

b) How long will it take for the cannon ball to hit the ground again? (Better move away...).

c) At what speed does the cannon ball hit the ground?

5. A stone dropped from a tower takes 6 seconds to hit the ground. How high is the tower?

6. On a snowy afternoon Joe Q. travels on the very slippery interstate highway at a speed of 60 ft/sec (roughly 40 m.p.h.), when he suddenly sees a big truck ahead losing control, rolling over and blocking the road. He applies the brakes, but because of the snow his deceleration is just 10 ft/sec^2.

a) Find the speed of Joe's car $t > 0$ seconds after he applies the brakes.

b) How many seconds would it take for the car to come to a stop?

c) When Joe applied the brakes, the truck was about 200 ft away. Will Joe be able to stop before hitting the truck? Explain!

VI.2   The Inverse Problem and Antiderivatives

VI.2.1   *Solutions of $y' = ky$*

In order to apply a differential equation such as $y' = ky$, where k is a constant, in studying exponential models in applications, as we shall do in the next section, one needs to know that ALL its solutions are of the form
$$f(t) = Ce^{kt}, \text{ where } C \text{ is a constant.}$$
Recall that in Section V.6.1 we investigated the analogous, but apparently simpler problem of determining all solutions of the differential equation $y' = 0$. The result was that any function that satisfies $f' = 0$ on some interval must be constant. It turns out that the above differential equation, as well as other cases, can be handled by appropriate simple modifications.

In fact, suppose that $y = f(x)$ is a solution of the equation $y' = ky$ on an interval I, that is, f satisfies $D[f](x) = kf(x)$ for all $x \in I$. We want to show that $f(x) = Ce^{kx}$ for some constant C. We therefore consider the function h defined by
$$h(x) = f(x)/e^{kx} = f(x)e^{-kx}.$$
Differentiation (using the product rule, the differential equation for f, and the chain rule) gives
$$h'(x) = f'(x)e^{-kx} + f(x)[e^{-kx}]'$$
$$= kf(x)e^{-kx} + f(x)[e^{-kx}](-k)$$
$$= e^{-kx}[kf(x) - kf(x)] = 0.$$

The result we just recalled implies that h is a constant C and we are done. In particular, it follows that every solution of $y' = ky$ is defined on the whole real line.

Remark. Notice that if a solution $f(x) = Ce^{kx}$ of $y' = ky$ takes on the value 0 at some point x_0, then necessarily $C = 0$, since $e^{kx} \neq 0$ for all x. Therefore $f(x) = 0$ for *all* x. So the only solution that takes on the value zero at some point is the constant function $f \equiv 0$. All other solutions are never zero.

Let us summarize the main conclusions.

Theorem 71 *Let f be a solution of the equation $y' = ky$ on the interval I. Then there exists a constant C, so that $f(x) = Ce^{kx}$ for all $x \in I$. If $f(x_0) = 0$ for some point $x_0 \in I$, then $f(x) = 0$ for all $x \in I$.*

Note that for $k = 0$ this theorem includes the earlier result that a function whose derivative is always zero is necessarily constant.

VI.2.2 *Antiderivatives*

We introduce a new name and notation to describe a generalization of the equation $D[f](x) = 0$ for all x in some interval, as follows. A function F on an interval I is called an *antiderivative of f* if F is differentiable and $F'(x) = f(x)$ for all $x \in I$. An antiderivative of f is denoted by the symbol $\int f(x)dx$. Therefore

$$F = \int f(x)dx \iff F' = f.$$

Sometimes the symbol $\int f(x)dx$, also called *an indefinite integral of f*, is used to denote the collection of *all* antiderivatives of f.

Note that any two antiderivatives of f must differ by a constant. Therefore, if F is any antiderivative of f, then any other antiderivative is of the form $F + C$ for some $C \in \mathbb{R}$.

In Section V.6.4 we had already used the fairly obvious fact that $\frac{1}{n}x^n$ is an antiderivative of $f(x) = x^{n-1}$ for any $n \in \mathbb{N}$. So we can say that

$$\int x^{n-1}dx = \frac{1}{n}x^n + C.$$

By simply reversing known rules of differentiation one obtains other formulas for antiderivatives, as follows.

i) $\int e^x dx = e^x + C$;

ii) $\int b^x dx = \frac{1}{\ln b} b^x + C$, for any $b > 0$ and $\neq 1$;

iii) $\int \sin x \, dx = -\cos x + C$; $\int \cos x \, dx = \sin x + C$;

iv) $\int x^r dx = \frac{1}{r+1} x^{r+1} + C$ for $r \neq -1$ and $x > 0$.

v) $\int \frac{1}{x} dx = \ln x + C$ for $x > 0$; $\int \frac{1}{x} dx = \ln(-x) + C$ for $x < 0$.

These two results can be combined into $\int \frac{1}{x} dx = \ln |x|$ for $x \neq 0$.

vi) $\int [af(x) + bg(x)] dx = a \int f(x)dx + b \int g(x)dx$ for any constants a, b.

It is not immediately obvious if $\ln x$ has an antiderivative on $(0, \infty)$. On the other hand, after some trial and error one may discover that $D[x \ln x - x] = 1 \cdot \ln x + x \cdot \frac{1}{x} - 1 = \ln x$. So we can say that

vii) $\int \ln x \, dx = x \ln x - x + C$ on $(0, \infty)$.

We refer the reader to the book [WiC?] for further results about antiderivatives. In particular, there are general techniques that are helpful in finding antiderivatives such as this last one. However, the general problem is quite complicated, and there are no rules that handle all possible situations, as is the case for taking derivatives. Note, for example, that the antiderivative of the simple algebraic function $f(x) = 1/x$ on $x > 0$ is a new type of function, namely $\ln x$. Also, recall from Section V.5.7 that the antiderivative of $\frac{1}{1+x^2}$ is the inverse tangent function, again a new type of function. So it is a remarkable and important result that *every* continuous function f on an interval I does have an antiderivative on I. The proof of this fact requires a detailed study of the concept of definite integral that is closely related to the abstract process of constructing antiderivates. Again, we refer the reader to [WiC?].

VI.2.3   *Initial Value Problems*

We saw in the preceding section that the solutions of differential equations such as $y' = g(x)$ and $y' = ky$ are determined up to a constant. By prescribing the value y_0 of a solution at one fixed point x_0 one obtains an additional condition that typically will be satisfied by one and only one choice of that constant. In this way one singles out a particular solution. Combining the differential equation with such a choice (x_0, y_0) determines what is called an *initial value problem*.

Example. *Solve the initial value problem* $y' = 2 \cos x$ *with* $y(\pi/2) = 1$.

Solution. The differential equation has solutions $\int 2 \cos x \, dx = 2 \sin x + C$. The initial value condition requires

$$2 \sin \frac{\pi}{2} + C = 1,$$

or $2 + C = 1$, so that $C = 1 - 2 = -1$. So the desired solution f is given by $f(x) = 2 \sin x - 1$.

Geometrically, we see that the graphs of the family of all solutions of $y' = 2 \cos x$ are obtained by parallel translation in the vertical direction (i.e., by adding a constant to the y-coordinate) of the graph of a particular solution. Specifying an initial value (x_0, y_0) therefore selects the one graph that goes through that particular point.

Example. *Describe, for varying times, the number of bacteria in a culture that is of size 2000 at 1 p.m. and which doubles every 5 hours. Use an exponential growth model.*

Solution. Let $P(t)$ denote the size of the culture, where t is measured in hours, so that $t = 0$ corresponds to 1 p.m. Assuming an exponential growth model, one has $P' = kP$ for a constant k. The solutions are of the form $P(t) = Ce^{kt}$. The initial value condition implies $2000 = P(0) = C$. To determine k we use the information $P(5) = 2P(0)$; this implies $e^{k5} = 2$. Hence $k5 = \ln 2$, so that $k = (\ln 2)/5 \approx 0.139$. The desired function that describes the number of bacteria at time t therefore is

$$P(t) = 2000 \, e^{0.139 \, t}.$$

If we want to describe the population in terms of the time T given by the clock, note that $T = t + 1$, or $t = T - 1$. Hence $P(T) = 2000e^{0.139 \, (T-1)}$.

This example generalizes easily to the following result, which is the prototype of the general existence and uniqueness theorem for initial value problems.

Theorem 72 *Given an arbitrary point (x_0, y_0) in the plane, there exists exactly one solution f of the differential equation $y' = ky$ defined on $\mathbb{R}$ that satisfies the initial value condition $f(x_0) = y_0$.*

Proof. We know that any solution f is of the form $f(x) = Ce^{kx}$. The condition $y_0 = f(x_0) = Ce^{kx_0}$ implies that $C = y_0/e^{kx_0} = y_0e^{-kx_0}$. So

$$f(x) = y_0e^{-kx_0}e^{kx} = y_0e^{k(x-x_0)}$$

is the (unique) solution to the initial value problem.

VI.2.4 *Exercises*

1. Let $p(x) = 4x^3 - 2x^2 + 1$.
   a) Find *all* antiderivatives of $p(x)$.

b) Find a function F so that $F(1) = 0$ and $F'(x) = p(x)$.

2. Suppose the functions f and g satisfy $f'(x) = 4g'(x)$ for all x. If $f(0) = g(0) = 1$, and $f(10) = 5$, what is $g(10)$?

3. Solve the following initial value problems.
a) $y' = 2\cos x$, $y(0) = 1$;
b) $y' = 3^x$, $y(0) = 0$;
c) $y' = 2y$, $y(0) = 10$;
d) $y' = 3y$, $y(0) = 0$.
e) $y' = -y$, $y(2) = 3$.

4. A particle moves along the graph of a function f, so that at each point $(x, f(x))$ on the trajectory the tangent has slope $4x$. Assume the particle goes through the point $(1, 2)$. Determine the function f.

5. Use a graphing calculator to sketch in one window the graphs of the antiderivatives of $g(x) = 3x^2$ that go through the points $(0, -1)$, $(0, 0)$, $(0, 1)$, and $(0, 2)$. Describe the geometric relationship of the graphs.

6. Find a function F that satisfies the equation $F' = kF$ for some constant k, and so that $F(0) = 2$ and $F(1) = 5$.

VI.3   Exponential Models

VI.3.1   *Growth and Decay Models*

We had already seen in Section V.4.7 how a simple model for the growth of a population leads to the differential equation $y' = ky$, where k is a constant. We had shown in Section 2.1 that *all* solutions f of this equation (that is, functions f that satisfy $f'(t) = kf(t)$) are functions of the form $f(t) = Ce^{kt}$. This is the result that is the basis for all further analysis of this and similar models. The basic hypothesis underlying this model states that the rate of change $\frac{dP}{dt}(t)$ of a population $P(t)$ is proportional to the size $P(t)$ of the population, that is, there exists a constant k so that

$$\frac{dP}{dt}(t) = kP(t) \text{ for all times } t \text{ under consideration.}$$

It is of course assumed that there is no change in any of the relevant external conditions during the time period that is studied. This is certainly not true in concrete situations. At best, one may say that conditions remain approximately stable over a limited period of time, so that the "exponential growth model" has to be applied with care.

A much more stable, and hence better, situation arises with natural *decay* processes, such as they occur with radioactive substances. For example, the isotope U_{238} of uranium emits radiation that arises from the splitting of the uranium atoms into other elements. Experiments reveal that this radioactive decay process is not affected by any changes in the environment whatsoever, and hence is extremely stable over very long periods of time. The amount $A(t)$ of U_{238} present at time t decays at a rate $\frac{dA}{dt}$ that is proportional to the amount $A(t)$, so

$$\frac{dA}{dt}(t) = -\lambda A(t)$$

for a constant $\lambda > 0$ that is called the *decay constant* (specific to U_{238}).[1] Experimental data leads to the value $\lambda = 0.155 \times 10^{-9}$ when t is measured in years. The differential equation then implies that

$$A(t) = Ce^{-\lambda t}$$

for a constant C that is identified with the amount $A(0) = Ce^0 = C$ that is present at time $t = 0$.

Rather than describing the decay process by the decay constant λ, physicists often use the *half life* T of a radioactive element. This is that time T in which *half* of the initial amount has decayed. So T satisfies the equation

$$A(0)e^{-\lambda T} = A(T) = \frac{1}{2}A(0),$$

or

$$e^{-\lambda T} = \frac{1}{2}.$$

Notice that this last equation does not depend on the initial amount $A(0)$. It follows that

$$e^{\lambda T} = 2, \text{ or } \lambda T = \ln 2.$$

For example, the half life of uranium U_{238} equals $T = \ln 2/\lambda = 4.47 \times 10^9 = 4.47$ billion years. Uranium does indeed decay very slowly.

VI.3.2 *Radiocarbon Dating*

For other radioactive substances the half life is much shorter. For example, the isotope C_{14} of carbon is radioactive (hence called *radiocarbon*) with a

---

[1]In a decay process, dA/dt is negative; it is convenient to write the relevant equation so as to make this clearly visible.

half life of about 5730 years. The corresponding decay constant is $\lambda = \ln 2/5730 = 0.00012$.

Radiocarbon has been used successfully to date ancient objects that have biological origins. The method is based on the fact that radiocarbon occurs naturally in the atmosphere, and that it is continuously created in the upper atmosphere from nitrogen subject to intensive cosmic radiation, at a rate that compensates the loss due to radioactive decay. Consequently, the ratio of radiocarbon to the normal carbon C_{12} in the atmosphere has remained quite stable over very long periods of time. All living organisms assimilate radiocarbon along with normal carbon, but this process stops when the organism dies, and hence the amount of radiocarbon in the remnants will then decrease over time. The amount of radiocarbon in an ancient object derived from living organisms can be measured, and that information can be used to estimate the age of the object.

Example. Archeologists find remnants of a skeleton in a cave. Analysis of a specimen reveals that the ratio of radiocarbon to carbon is about 82% of the ratio q found in the atmosphere. If $A(t)$ is the amount of radio carbon in the specimen at time t (measured in years, with $t = 0$ corresponding to the time the specimen stopped living), then the decay model

$$\frac{dA}{dt} = -\lambda A(t)$$

implies that $A(t) = A(0)\, e^{-\lambda t}$. The value $A(t)$ measured today is $.82\, A(0)$, so that

$$.82\, A(0) = A(0)\, e^{-\lambda t}$$

implies $e^{-\lambda t} = .82$. Hence $-\lambda t = \ln(0.82)$. With $\lambda = 0.00012$, one obtains

$$t = -\ln(0.82)/\lambda \approx 1653.$$

Therefore one may conclude that the cave was inhabited about 1650 years ago.

VI.3.3   *Compound Interest*

An exponential growth model arises also in finance. A principal amount of $A(t)$ dollars at time t (t measured in years) is said to grow under *continuous compounding* at an annual rate r, if the rate of growth of the capital $A(t)$ satisfies

$$\frac{dA}{dt}(t) = rA(t).$$

It then follows that

$$A(t) = A(0)e^{rt}.$$

Other compounding methods are based on dividing the year into n equal compounding periods of length $1/n$ years. At the end of each such period simple interest is added to the principal at the beginning of that period at the rate r/n obtained by dividing the annual interest rate evenly over the n periods. The resulting formula for compound interest after t years reads

$$A_n(t) = A(0)\left(1 + \frac{r}{n}\right)^{nt}.$$

Commonly used methods are annual compounding ($n = 1$), quarterly compounding ($n = 4$), and daily compounding ($n = 360$).

The question arises about the relationship of these compound interest formulas with *continuous compounding*. Let us investigate what happens with $A_n(t)$ as n gets increasingly larger, thereby making the compounding periods increasingly shorter. Since

$$A_n(t) = A(0)\left(1 + \frac{r}{n}\right)^{nt} = A(0)\left(\left(1 + \frac{r}{n}\right)^{n}\right)^{t},$$

we are led to consider

$$\lim_{n \to \infty} (1 + r/n)^{n}.$$

For simplicity, consider $r = 1$ first, so we need to analyze $(1 + 1/n)^{n}$. Let us set $1/n = h$, so that $n = 1/h$. Then $n \to \infty$ is equivalent to $h \to 0$, and therefore

$$\lim_{n \to \infty}\left(1 + \frac{1}{n}\right)^{n} = \lim_{h \to 0}(1 + h)^{1/h}.$$

Now, for h close to 0, yet $h \neq 0$, use the equation $(1+h) = e^{\ln(1+h)}$ to write

$$(1 + h)^{1/h} = [e^{\ln(1+h)}]^{1/h} = e^{[\ln(1+h)]/h}.$$

Recall that since $\ln 1 = 0$, one obtains

$$\lim_{h \to 0} \frac{\ln(1 + h)}{h} = \lim_{h \to 0} \frac{\ln(1 + h) - \ln 1}{h}$$
$$= \text{ derivative of } \ln x \text{ at the point } x = 1.$$

Since $(\ln x)' = 1/x$, the value of this limit is $1/1 = 1$. The continuity of $E(x) = e^{x}$ then implies that

$$\lim_{n \to \infty}(1 + 1/n)^{n} = \lim_{h \to 0}(1 + h)^{1/h}$$
$$= \lim_{h \to 0} e^{[\ln(1+h)]/h} = \lim_{u \to 1} e^{u} = e^{1} = e.$$

We have thus discovered the representation

$$e = \lim_{n \to \infty} \left(1 + \frac{1}{n}\right)^n$$

for the base e of the natural logarithm.

A minor modification of this argument shows that

$$e^r = \lim_{n \to \infty} \left(1 + \frac{r}{n}\right)^n \quad \text{for any real number } r.$$

Consequently, we obtain that continuous compounding of interest at the annual rate r, given by the function $A_c(t) = A(0)e^{rt}$, arises as the limiting case $n \to \infty$ of compounding over n equal periods per year. Continuous compounding may be viewed as "instantaneous" compounding, i.e., interest is added to the capital at every moment.

It is of interest to compare the growth of capital under various compounding methods. Suppose the annual interest rate is 6%, i.e., $r = 0.06$, and that $A(0) = \$100,000$. After 10 years, annual compounding results in

$$A_1(10) = 100000 * (1 + 0.06)^{10} = 1.79084\,77 \times 10^5 = \$179,848.$$

Quarterly compounding gives

$$A_4(10) = 100000 * (1 + 0.06/4)^{40} = 1.81401\,84 \times 10^5 = \$181,402,$$

while daily compounding results in

$$A_{360}(10) = 100000 * (1 + 0.06/360)^{3600} = 1.82202\,99 \times 10^5 = \$182,023.$$

Finally, continuous compounding gives

$$A_c(10) = 100000 * \exp(0.06 * 10) = 1.82211\,88 \times 10^5 = \$182,212.$$

Notice that continuous compounding yields the best result. While the difference to daily compounding is perhaps insignificant, continuous compounding yields almost $\$2,500$ more than annual compounding. For this reason, banks often state the "yield" of continuous compounding, i.e., that annual rate that produces the same result by annual compounding. For example, for continuous compounding at 6%, the yield is determined by solving

$$e^{0.06} = (1 + r)$$

for r. The result is $r = e^{0.06} - 1 = 1.06183\,65 - 1 = 0.0\,61836\,5$, i.e., the yield is 6.18%.

VI.3.4 *Exercises*

1. A bank offers a 5-year certificate of deposit which pays interest annually at a rate of 7.5% compounded continuously. Determine the effective yield of the certificate.

2. Bank ABC offers a certificate of deposit at 6.125% compounded monthly, while its competitor Bank QRS offers 6.1% compounded continuously. Which bank would you choose? Justify your choice by comparing yields.

3. A population of bacteria in a laboratory grows exponentially at the rate of 5% per day. If the initial size is 1000, after how many days will the population have grown to 2000?

4. Radon 222 is a radioactive gas that is found to be harmful to humans if they are exposed to it in excessive amounts. Its half life is about 3.8 days. Because of a leak, the basement of a factory has reached dangerously high levels of Radon 222, and the health inspector recommends that no one should enter the basement until the radioactive level has decreased to 10% of the original level. How many days should people stay out of the basement?

5. Newton's Law of Cooling states that the temperature $T(t)$ of an object placed in an environment at constant temperature A changes at a rate that is proportional to the difference $A - T(t)$, i.e., there is a constant $k > 0$ so that

$$\frac{dT}{dt} = k(A - T).$$

a) Explain the meaning of the sign of dT/dt. Consider the cases $A > T$ and $A < T$ separately.

b) Show that if $T(0) > A$, then $T(t) = A + (T(0) - A)e^{-kt}$. (Hint: Set $y = T(t) - A$ and show that $y' = -ky$.)

c) Will the temperature $T(t)$ ever be equal to A? Explain.

6. At 5 p.m. police find the body of the victim of a murder in a room whose temperature was maintained at $20^0 C$. At that time the temperature of the body was measured to be $30^0 C$, and two hours later it had decreased to 25^0. Assuming that the normal body temperature of a living human is $37^0 C$, determine how many hours ago the murder was committed. (Hint: Use Newton's Law of Cooling (Exercise 5). Use the initial condition $T(0) = 30^0$ and $T(2) = 25^0$ to determine k. Then set $T(0) = 37^0$ to determine the time t at which $T(t) = 30^0$.)

7. An ice cube tray with water at 12^0C is placed in a freezer kept at -10^0C. An hour later the temperature of the water is measured at 6^0C. Estimate how much longer it will take until the ice cubes are ready? (Assume this will happen when the water has reached the freezing temperature 0^0C. Use Newton's Law of Cooling (Exercise 5).

VI.4   "Explosive Growth" Models

VI.4.1   *From Polynomial to Exponential Growth*

We say that a function $f(t)$ is growing like a polynomial (or has polynomial growth), if there are a constant $C > 0$ and a natural number n, so that $f(t)$ is very close to $f_n(t) = Ct^n$ for all sufficiently large t. Clearly as n gets larger, the corresponding polynomial growth functions f_n will eventually grow faster. Recall that we had seen in Section V.3.3 that an exponential function eventually grows faster than any polynomial function. We can describe this relationship more clearly by considering the differential equation satisfied by f_n, as follows. Note that

$$f'_n(t) = Cnt^{n-1} = Cn(t^n)^{\frac{1}{n}(n-1)} = Cn\left(\frac{f_n}{C}\right)^{1-\frac{1}{n}} = k \cdot (f_n)^{1-\frac{1}{n}}$$

for some constant k. So f_n satisfies the differential equation

$$y' = ky^{1-\frac{1}{n}}.$$

Compare this to the differential equation for the exponential function $y' = ky$, and we see that the growth for the derivative here is an upper bound for polynomial growth. The obvious question, discussed in the next section, is then: What sort of growth goes beyond exponential growth?

VI.4.2   *Beyond Exponential Growth*

The basic model underlying exponential growth is described by the differential equation $y' = ky$, where $k > 0$ is constant. It implies that both $y(t)$ and the rate of growth $dy/dt = y'$ increase in time. The differential equation requires that dy/dt and y are just a fixed constant multiple of each other. The rate of growth of dy/dt is measured by its derivative $D[dy/dt] = D[ky] = kD[y]$, so the rates of growth $D[dy/dt]$ and $D[y]$ of dy/dt and y are still proportional, with the same factor k. Furthermore, $D^{(2)}[y] = D[ky] = k^2y$. Continuing to differentiate, one sees that for each positive integer n one has

$$D^{(n)}[y] = kD^{(n-1)}[y] = k^2D^{(n-2)}[y] = \ldots = k^ny.$$

Thus there is a *linear* relationship between any two derivatives of a solution. Each derivative of y still satisfies an exponential growth model in relationship to y, although the relevant constant k^n changes according to the number of differentiations involved. This situation expresses a deep regularity of the underlying process. As we know, the solutions are given by exponential functions that are defined for all values of t, and unless external conditions change, the growth process continues indefinitely, leading to $\lim_{t\to\infty} y(t) = \infty$ as soon as $y(t_0) \neq 0$ at some moment in time.

The situation changes dramatically if one considers a *non-linear* growth model described, for example, by the differential equation $y' = ky^2$, where k is again constant. Here y' grows much faster than y. The relationship is not linear but quadratic. To simplify, let us choose $k = 1$. We want to differentiate the equation again. Since the solution $y(t)$ is assumed to be differentiable, it follows that $y'(t) = [y(t)]^2$ is differentiable as well, and therefore $D[y'] = D[[y(t)]^2] = 2y(t) \cdot D[y]$ by the chain rule. The differential equation then implies that

$$D[y'] = 2y \cdot y' = 2y \cdot y^2 = 2y^3.$$

Thus $D[y'] = D^2[y]$ grows very much faster in comparison to $D[y] = y'$ as y increases. By differentiating again, and so on, one obtains that

$$D^{(n)}[y] = n \cdot (n-1) \cdot \ldots \cdot [y]^{n+1}$$

for $n = 2, 3, \ldots$. We see that the nth derivative of any solution $y(t)$ must grow like the power y^{n+1}. In contrast to the exponential model $y' = y$, the derivatives in this simple non-linear model grow progressively much faster, far exceeding the rate of natural growth of exponential models. This suggests a most unusual behavior of the solutions.

As we will show, there exists a critical point in time T_c that depends on the initial conditions, such that the corresponding solution is defined only for $t < T_c$. Furthermore, the process literally blows up as t approaches the critical point T_c.

VI.4.3   *An Explicit Solution of $y' = y^2$*

It turns out that we can quite easily determine a formula for the solution of any initial value problem related to the non-linear differential equation $y' = y^2$. To be specific, suppose that $f(t)$ is a solution defined near the point $t = 0$, and that $f(0) = 1$. By continuity of f it follows that $f(t) > 0$ on a sufficiently small interval I centered at 0. Hence the differential equation $D[f] = f^2$ can be written in the form

$$\frac{D[f](t)}{f(t)^2} = 1 \text{ for } t \in I.$$

By the reciprocal rule for derivatives we see that $h(t) = -1/f(t)$ is an antiderivative of the left side on the interval I. Since $\int 1 dt = t$ is also an antiderivative of $D[f]/f^2$, it follows that there exists a constant C such that

$$h(t) = t + C \text{ for } t \in I,$$

that is, we have verified that $-1/f(t) = t + C$. The initial value condition implies that $C = -1/f(0) = -1$, and therefore $1/f(t) = 1 - t$. Since $f(t) > 0$ for $t \in I$, we must have $t < 1$ for $t \in I$. It follows that

$$f(t) = \frac{1}{1 - t} \tag{VI.1}$$

is the unique solution of $y' = y^2$ with $f(0) = 1$ on the interval I. The formula (VI.1) shows that the solution f initially defined on the interval I has a natural extension to the interval $\{t : t < 1\}$ which satisfies the differential equation for all $t < 1$. In fact, the expression on the right side of (VI.1) is the *only* extension of the solution f from the interval I to the interval $(-\infty, 1)$ that continues to satisfy the differential equation $y' = y^2$. (See Exercise 2 for more details.) Furthermore, (VI.1) shows that $f(t)$ cannot be extended in any meaningful way to $t = 1$, since $f(t) \to \infty$ as $t \to 1$ from the left side. Thus the solution $f(t)$ "blows up" as t approaches the critical value $T_c = 1$. Figure VI.1 shows the "explosive" behavior of the graph of $f(t)$ as $t \to 1^-$.

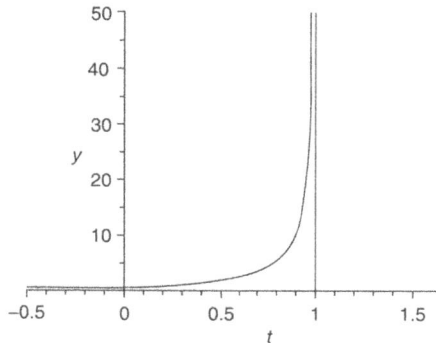

Fig. VI.1   The solution $f(t)$ blows up as t approaches 1.

Simple modifications of these techniques show that all solutions of $y' = ky^r$ exhibit similar explosive behavior at corresponding critical points as soon as the exponent r is greater than 1. The (linear) exponential model $y' = ky$ thus plays a very essential role indeed: we can say that it is the borderline case between polynomial growth that is defined for all t, and growth that literally blows up when approaching a certain moment in time. The implications of this mathematical truth are potentially quite shocking. For example, suppose that the stock market has been growing for a while, and that a careful analysis of the recorded data shows that the S&P 500 index SP has been growing at a rate proportional to $(SP)^{1.001}$. A knowledgeable investor would be able to determine when the index would literally "go through the roof", and adjust his investment strategy accordingly.

VI.4.4   *Exercises*

1. Modify the arguments given in the text for the initial value $f(0) = 1$ to show that if $b \neq 0$, then the initial value problem $y' = y^2$ with $y(t_0) = b$ has a unique solution on some interval I centered at t_0.

   2. a) Verify that $g(t) = 1/(1 - t)$ satisfies $g'(t) = g(t)^2$ for $t < 1$ and $g(0) = 1$.

   b) Let f be any solution of the initial value problem in a) on the interval $(-\delta, 1)$ for some $\delta > 0$. Let $\Lambda = \{\lambda \in (-\delta, 1) : f(t) = g(t) \text{ for all } t \in (-\delta, \lambda)\}$. Show that Λ is not empty. (Hint: Modify the argument in the text.)

   c) Let $\lambda^* = \sup \Lambda$. Suppose that $\lambda^* < 1$. Show that $f(\lambda^*) = g(\lambda^*) \neq 0$ and that any solution $h(t)$ of the initial value problem $y' = y^2$ with $h(\lambda^*) = g(\lambda^*)$ must agree with $g(t)$ for all t in some interval $I = (\lambda^* - \varepsilon, \lambda^* + \varepsilon)$ centered at λ^*. (Hint: Use Exercise 1).

   d) Show that c) implies that $f(t) = g(t)$ for all $-\delta < t < \lambda^* + \varepsilon$.

   e) Show that the result in d) implies that $\lambda^* = 1$, so that $f(t) = g(t)$ for all $t < 1$. (Hint: Explain why the conclusion in d) contradicts the assumption $\lambda^* < 1$ made in c).)

   3. a) Given $r > 1$, find a solution $f_r(t)$ for the initial value problem $y' = y^r$ with $y(0) = 1$ on some interval centered at 0.

   b) Determine the critical point T_r for f_r, so that $f_r(t)$ is defined for all $t < T_r$ and $\lim_{t \to T_r^-} f_r(t) = \infty$.

   4. Follow the steps below to show that if $f(t)$ is a solution of $y' = y^2$ on some interval I centered at 0 which satisfies $f(0) = 0$, then there exists an interval $[-\delta, \delta] \subset I$ such that $f(t) = 0$ for all t with $|t| \leq \delta$. Thus this initial value problem has the *unique* solution $f(t) = 0$ for all t near 0.

a) Show that there exists $0 < \delta < 1/2$, such that $|f(t)| < 1/2$ for all t with $|t| \leq \delta$.

b) Show that if for some $k \in \mathbb{N}$ one has $|f(t)| \leq (1/2)^k$ for all t with $|t| \leq \delta$, then $|f(t)| \leq (1/2)^{2k+1}$ for all t with $|t| \leq \delta$. (Hint: Use the differential equation to first estimate $|f'(t)|$, and then apply the Mean Value Inequality to estimate $|f(t)|$ for $|t| \leq \delta < 1/2$.)

c) Show that a) and b) imply that $|f(t)| = 0$ for $|t| \leq \delta$.

5. Consider the differential equation $y' = y^{1/2}$. Show that there are two *different* solutions of this equation on $\mathbb{R}$ that satisfy the initial value condition $y(0) = 0$. Hence *uniqueness* of solutions fails for this initial value problem. (Hint: Clearly $g(t) = 0$ for all t solves the initial value problem. Show that the function f defined by $f(t) = (1/4)t^2$ for $t \geq 0$ and $f(t) = 0$ for $t < 0$ is differentiable at $t = 0$ and (trivially) at all other points, and that it also solves the initial value problem.)

VI.5 Periodic Motions

VI.5.1 *A Model for a Bouncing Spring*

Suppose a steel ball of mass m is attached to a spring that hangs from the ceiling. The weight will stretch the spring by an amount s_0, at which point the ball will be at rest at the equilibrium position. Suppose the ball is now pulled down an additional amount c_0 and released; it will then bounce up and down around the equilibrium point, as seen in Fig. VI.2. We want to find a mathematical model to describe the motion of the ball under the action of the spring. As the motion appears to involve some periodicity, we expect that the model will involve trigonometric functions or perhaps other more complicated periodic functions.

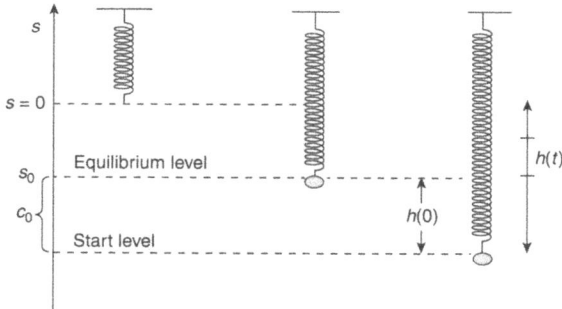

Fig. VI.2   The mass stretches the spring by $|s_0|$.

At the equilibrium level $s = s_0$ the ball is at rest. (Recall that according to our choice of orientation, s_0 is a *negative* number.) So the sum $F(s_0)$ of all forces acting on the ball at that position is zero. One force is the gravitational force F_G, which equals $-mg$. The other force F_S comes from the spring; according to Hooke's Law from physics, that force is a multiple $-ks$ of the amount s that the spring has been stretched, where $k > 0$ is the so-called spring constant. In particular, $F_S(s_0) = -ks_0$. Since $s_0 < 0$, one has $F_S(s_0) > 0$, which is consistent with the fact that the spring pulls the ball upwards. The equation $F(s_0) = F_G(s_0) + F_S(s_0) = -mg + (-ks_0) = 0$ then implies that $-mg = ks_0$. Incidentally, this result gives a practical method to determine the spring constant k: If a mass m stretches the spring by the amount $-s_0$, then $k = -mg/s_0 = mg/(-s_0)$. If one uses the standard metric units kilograms for mass, meters for length, and seconds for time, the spring constant k is measured in kg/sec^2. For example, if a mass of 1 kg stretches the spring by 10 cm ($= 0.1$ m), then $k = 1 \cdot 9.81/0.1 = 98.1$ kg/sec^2.[2]

Rather than focusing on the level $s(t)$ of the mass, we shall consider the (signed) distance $h(t) = s(t) - s_0$ of the mass from the equilibrium level given by s_0. Since $h(t)$ and $s(t)$ differ by a constant, one has $h''(t) = s''(t)$. According to Newton's law of motion, *Force = mass x acceleration.* Together with Hooke's law one then obtains

$$mh''(t) = ms''(t) = F_G + F_S(s(t)) = -mg - ks(t) = ks_0 - ks(t)$$
$$= -k(s(t) - s_0) = -k \cdot h(t)$$

at time t. Therefore the function $h = s - s_0$ satisfies

$$h''(t) + \frac{k}{m}h(t) = 0.$$

If we set $\omega = \sqrt{k/m}$, the position function $h(t)$ relative to the equilibrium point of the bouncing ball must satisfy the differential equation $y'' + \omega^2 y = 0$. We have already seen this differential equation in Section V.5.6.

The main result proved in the next section implies that

$$h(t) = A\sin(\omega t) + B\cos(\omega t)$$

for some constants A and B. The constants are determined by the initial position and velocity of the ball at time $t = 0$. In fact, the above equation

---

[2]The unit kg/sec^2 for the spring constant k may appear quite unnatural. If one introduces a separate unit to measure forces, one obtains a different description for k. More precisely, one defines 1 *Newton* (1 N) to be the size of a force that accelerates a mass of 1 kg by 1 m/sec^2. The unit kg/sec^2 translates to $kg \cdot (m/sec^2)/m = N/m$. This latter unit for k more directly reflects Hooke's law $F = -k\,s$, which implies $k = -F/s$.

implies that $h(0) = B$ and $h'(0) = A\omega$. Concretely, suppose the experiment is started by pulling the ball down from the equilibrium level by an amount $c_0 > 0$ and then releasing it at time $t = 0$, so that its velocity right at that moment is zero. This means that $h(0) = -c_0$ and $h'(0) = 0$. Consequently, given these initial conditions, the motion of the ball is described by the function

$$h(t) = -c_0 \cos(\omega t) = -c_0 \cos(\sqrt{k/m}\, t).$$

Figure VI.3 qualitatively visualizes the vertical displacement $h(t)$ of the ball as a function of time t in seconds.

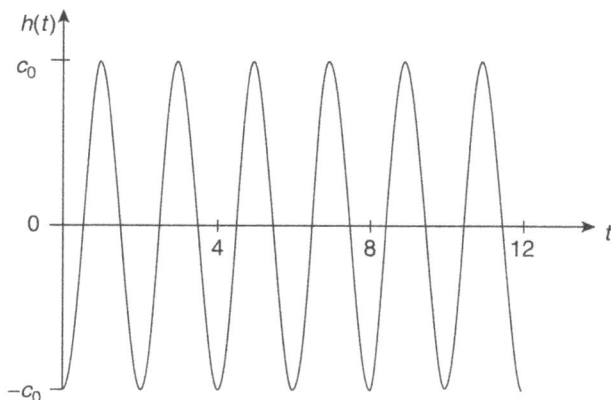

Fig. VI.3   The motion of the spring over time t.

The constant c_0 in this equation is called the *amplitude* of the (periodic) motion. It measures the maximum displacement from the equilibrium position. The other quantity that is often used to describe periodic motions is the so-called *frequency* ν, which gives the number of cycles per unit of time. The frequency thus is an aggregate measure of the speed by which the ball bounces up and down, sort of like an average speed.

The reciprocal $T = 1/\nu$ of the frequency is the *period* of the motion, i.e., T is that time it takes the ball to complete a full cycle to return to the initial position at the bottom. In Figure VI.3, the period T equals 2 seconds and hence the frequency ν is equal to $1/2$ cycles per second. Since the cosine function has a period 2π, the period T of the bouncing mass must satisfy the equation $\sqrt{k/m}\,T = 2\pi$, that is, $T = 2\pi\sqrt{m/k}$. The result shows, in particular, the effect of the spring constant k and the mass m on the motion of the ball. The period increases if the mass m of the ball is increased. Similarly, if one uses balls of equal weights with two springs of different stiffness, the equation shows that the period for the stiffer spring (i.e., larger k) will be shorter than the one for the softer spring. Stated differently, a stiffer spring produces a much quicker bounce, that is, higher frequency, than a softer spring.

Remark. A more accurate model of the bouncing spring needs to take into account other factors, such as the resistance encountered by the motion of the ball in the surrounding medium and internal friction of the spring. The main new effect is that energy is lost over time, thereby slowing down the motion until the ball eventually comes to a stop. The resistance gives an additional force F_R acting on the ball. The simplest model assumes that this force is proportional to the velocity y' and acts in the direction opposite to the direction of the motion, i.e., $F_R = -Ry'$ for some constant $R > 0$. The total force $F = my''$ now satisfies $F = -mg + F_R + F_S$. The resulting differential equation for the displacement $y(t)$ from equilibrium then is

$$my'' + Ry' + ky = 0,$$

where all constants are positive. The analysis of the solutions for this equation when $R > 0$ is more involved than the case when $R = 0$. The reader interested in more details should consult a basic text on Differential Equations.

VI.5.2 The Solutions of $y'' + \omega^2 y = 0$

We saw in the preceding section that the function h that describes the motion of a ball attached to a spring must be a solution of the differential equation $y'' + \omega^2 y = 0$. Recall that $\sin t$ and $\cos t$ satisfy the differential equation $y'' + y = 0$. More generally, we had already remarked in Section V.5.6 that all functions f of the form $f(t) = A\sin(\omega t) + B\cos(\omega t)$, where A and B are constants, satisfy the corresponding equation

$$y'' + \omega^2 y = 0. \tag{VI.2}$$

We also had mentioned that in fact ALL solutions of the above differential equation are of this form. We shall now prove this important fact. The analogous problem of identifying ALL solutions for the differential equation $y' = ky$ was handled in Section 2.1 by reducing the problem to the simpler equation $y' = 0$. Similarly, the equation $y'' + \omega^2 y = 0$ can be reduced to that form by a simple procedure, as follows.

We multiply the equation (VI.2) by the derivative y' to obtain
$$y''y' + \omega^2 yy' = 0$$
Note that by the chain rule one has $D[[y(t)]^2] = 2y(t)D[y(t)] = 2yy'$. Similarly, $D[[y'(t)]^2] = 2y'(t)D[y'(t)] = 2y'y''$. Combining these two equations results in
$$D[[y'(t)]^2 + \omega^2 [y(t)]^2] = 2y'y'' + \omega^2 2yy'$$
$$= 2y'[y'' + \omega^2 y].$$
Therefore, if the function y is a solution of (VI.2), then $D[[y'(t)]^2 + \omega^2 [y(t)]^2] = 0$. This easily implies the following statement.

Lemma 73 *Let y be any solution of the differential equation $y'' + \omega^2 y = 0$ on the interval I. Then*
$$[y'(t)]^2 + \omega^2 [y(t)]^2 \text{ is constant for all } t \in I.$$
In particular, any such solution y that satisfies $y(t_0) = y'(t_0) = 0$ at some point $t_0 \in I$ must satisfy $y(t) = 0$ for all $t \in I$.

Proof. We saw that for any solution y of (VI.2) the function $[y'(t)]^2 + \omega^2 [y(t)]^2$ has zero derivative on I. Hence it must be a constant C on I. The conditions $y(t_0) = y'(t_0) = 0$ imply that $C = 0$. Since $[y'(t)]^2 \geq 0$ and $\omega^2 [y(t)]^2 \geq 0$, their sum can be zero only if the terms are zero individually. In particular, it follows that $\omega^2 [y(t)]^2 = 0$, and hence $y(t) = 0$ for all $t \in I$. ∎

We can now readily prove the main result of this section.

Proposition 74 *If the function f satisfies the differential equation $y'' + \omega^2 y = 0$ on the interval I, then there exist constants A and B such that*
$$f(t) = A\sin(\omega t) + B\cos(\omega t).$$

Proof. Let us assume first that $0 \in I$. The key idea is to find an explicit solution of the form given above on the right that satisfies the same initial conditions at 0 as f, so that subtracting it from f gives a solution of the differential equation with initial value conditions 0. A straightforward computation shows that
$$h(t) = f(t) - [f'(0)/\omega]\sin(\omega t) - f(0)\cos(\omega t)$$
does the job. More precisely, it follows that h also satisfies $h'' + \omega^2 h = 0$, and furthermore, $h(0) = f(0) - 0 - f(0) = 0$, and since $h' = f' - f'(0)\cos(\omega t) + f(0)\omega\sin(\omega t)$, it also follows that $h'(0) = f'(0) - f'(0) + 0 = 0$. By Lemma 73, $h(t) = 0$ for all $t \in I$, and clearly this implies the desired conclusion, with $A = f'(0)/\omega$ and $B = f(0)$. An appropriate modification of this argument works in case $0 \notin I$. See Exercise 3 in Section 5 for details.

VI.5.3 *Addition Formulas for* sine *and* cosine *Functions*

In Section V.5.6, we had briefly mentioned the addition formulas for the trigonometric functions, that is, formulas to express $\sin(t+s)$ and $\cos(t+s)$ in terms of $\sin t$, $\cos t$, $\sin s$, $\cos s$. We can now give simple proofs for these formulas, based on the knowledge of the differential equation satisfied by these functions.

We fix $c \in \mathbb{R}$ and consider the function $h(t) = \sin(t + c)$. By using the Chain Rule it follows that $h'(t) = \cos(t+c)$ and $h''(t) = -\sin(t+c) = -h(t)$. Hence h satisfies the differential equation $y'' + y = 0$. We proved in the previous section that any solution of this equation is of the form

$$h(t) = A \sin t + B \cos t$$

for suitable constants A and B. It follows that $B = h(0) = \sin(0 + c) = \sin c$. By taking derivatives, we see that $h'(t) = A \cos t - B \sin t$, and hence $A = h'(0) = \cos(0 + c) = \cos c$. Therefore

$$\sin(t + c) = h(t) = \sin t \cos c + \cos t \sin c.$$

The same sort of proof applied to $\cos(t + c)$ shows that

$$\cos(t + c) = \cos t \cos c - \sin t \sin c.$$

By choosing $c = t$ one obtains the formulas for the double angle

$$\sin(2t) = 2 \sin t \cos t$$

and

$$\cos(2t) = (\cos t)^2 - (\sin t)^2.$$

VI.5.4 *The Motion of a Pendulum*

A pendulum, such as found, for example, in big wall clocks, provides another familiar example of a periodic motion. We shall now investigate the corresponding mathematical model. We consider a pendulum consisting of a weight of mass m attached to the bottom of a rigid rod of length L, whose mass we assume to be negligible compared to m. (See Fig. VI.4.) The rod swings from a hinge at the top. Neglecting, as usual, factors such as resistance, etc., the only force acting on the pendulum is the gravitational force $F_G = -mg$ that pulls the weight vertically down.

The motion of the pendulum is described by the arc $s(t)$ on the circle of radius L that measures the distance the weight has moved from the central position at the bottom. The orientation is chosen so that $s(t) > 0$

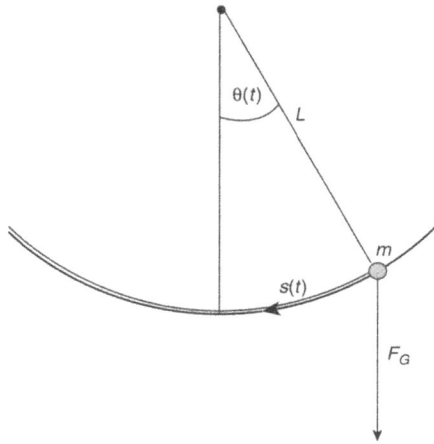

Fig. VI.4   Motion of a pendulum of length L.

corresponds to a position on the right of the center, while $s(t) < 0$ means that the weight is on the left side. As shown in Fig. VI.4, the position is also identified by the angle $\theta(t)$ that the rod forms with the vertical line. If $\theta(t)$ is measured in radians, one has $s(t) = L \cdot \theta(t)$.

In order to apply Newton's law of motion we need to identify the force that acts in the direction of the motion of the weight along the circle. According to basic physical principles, and as shown in Fig. VI.5, at any position $s(t)$ the gravitational force F_G may be decomposed into

$$F_G = F_T + F_N,$$

where F_T is tangential to the circle and F_N is normal, i.e., perpendicular to the circle at the point corresponding to $s(t)$. Clearly F_N has no effect on the motion of the pendulum: it simply tries to stretch the rod, which is assumed to be rigid. So the only force relevant to the motion of the pendulum is the tangential component F_T. Consequently one has $ms'' = F_T$.

Figure VI.5 shows that $F_T = F_G \sin \theta(t) = -mg \sin \theta(t)$. Note that the sign of F_T indicates that this force pulls the weight towards the center, regardless of whether $\theta(t)$ (i.e., $\sin \theta(t)$) is positive or negative. Newton's law of motion thus takes the form

$$ms''(t) = F_T = -mg \sin \theta(t).$$

Since $s''(t) = L\theta''(t)$, after rearranging and dividing by mL, one obtains

$$\theta''(t) + \frac{g}{L} \sin \theta(t) = 0.$$

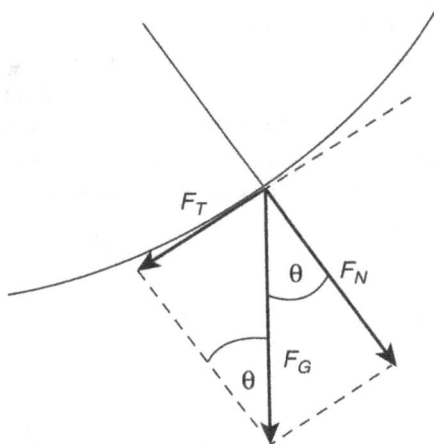

Fig. VI.5   Decomposition of the force F_G.

We notice that this equation is more complicated than the corresponding equation for the motion of the bouncing spring studied in Section 5.1, as it involves $\theta(t)$ composed with the sine function. In order to simplify, we rely on the basic principle that differentiability means that on sufficiently small neighborhoods of a point P the graph of a function is very well approximated by the tangent line at P. Applied to the function $\sin \theta$ at $\theta = 0$, whose tangent at that point is given by $l(\theta) = \theta$, this principle implies that $\sin \theta(t) \approx \theta(t)$, and that the approximation improves the smaller θ gets. Similarly, physical principles suggest that if relevant quantities in a process are changed just by small amounts, the corresponding motion will also change by appropriately small amounts. Altogether, if we approximate the tangential component F_T of the gravitational force by

$$F_T = -mg \cdot \sin \theta(t) \approx -mg \cdot \theta(t)$$

the solutions of the resulting differential equation

$$\theta''(t) + \frac{g}{L}\theta(t) = 0$$

will provide an approximation for the actual motion of the pendulum provided the angle θ is sufficiently small.

Let us set $\omega = \sqrt{g/L}$. It then follows from Section 5.2 that

$$\theta(t) = A \sin \omega t + B \cos \omega t.$$

If we assume that the initial position satisfies $\theta(0) = 0$, it follows that $B = 0$, so that

$$\theta(t) = A \sin \omega t = A \sin(\sqrt{g/L}\,t).$$

The *amplitude* A measures the maximal size of the angle in the motion of the pendulum. As indicated above, we need to assume that A is rather small to be assured that this solution gives a good approximation of the motion of the pendulum. Since $s(t) = L\theta(t)$ and $\theta'(t) = A\omega\cos\omega t$, the amplitude is related to the initial velocity $v_0 = s'(0)$, that is, to the velocity of the pendulum right when the weight is at the bottom, by the equation $A = \theta'(0)/\omega = s'(0)/L\omega$. Since $L\omega = L\sqrt{g/L} = \sqrt{Lg}$, it follows that

$$A = v_0\sqrt{1/(Lg)}.$$

Of course, for a swinging pendulum, the most interesting piece of information is the length of a period T. Just as in case of the bouncing spring, the period T of the pendulum is determined by $\omega T = 2\pi$, so that

$$T = 2\pi/\omega = 2\pi/\sqrt{g/L} = 2\pi\sqrt{L/g}.$$

What is perhaps surprising is that—in contrast to the bouncing spring— the mass m of the weight attached to the pendulum does not appear in this formula for the period. In other words, changing the weight of a pendulum does not affect its period. On the other hand, the above formula for T clearly shows the effect of the *length* L of the pendulum, that is, of the distance of the weight from the hinge at the top. By increasing that length, the period increases, i.e., the motion of the pendulum is slowed down. This is a phenomenon familiar to anyone who ever attempted to adjust the accuracy of a wall clock.

VI.5.5   *Exercises*

1. Consider the spring model discussed in the text. Suppose the displacement c_0 from the equilibrium level is doubled, i.e., c_0 is replaced by $2c_0$. Determine the effect on i) the velocity $v(t)$ of the ball, ii) the *average* velocity of the ball between a low point and the following high point, and iii) the frequency of the motion.

2. A weight of 5 *kg* is attached to a spring, which causes the spring to stretch by 15 *cm*. Determine the period of the motion that results after the weight has been given an initial push.

3. This exercise completes the proof of Proposition 74 in Section 2 in case $0 \notin I$. Let $f(t)$ be a solution of $y'' + \omega^2 y = 0$ on the interval I, and choose any point $t_0 \in I$.

   a) Show that $h(t) = f(t) - \frac{f'(t_0)}{\omega}\sin\omega(t-t_0) - f(t_0)\cos\omega(t-t_0)$ satisfies the equation $y'' + \omega^2 y = 0$.

   b) Use Lemma 73 to show that $h(t) = 0$ for all $t \in I$.

c) Show that $f(t) = A\sin(\omega t) + B\cos(\omega t)$ for suitable constants A and B. (Hint: Use a) and b), expand $\sin\omega(t - t_0)$ and $\cos\omega(t - t_0)$ by means of trigonometric addition formulas obtained in Section 3, and rearrange.)

4. Use the relations $\sin(\pi/2 - t) = \cos t$ and $\cos(\pi/2 - t) = \sin t$ to prove the addition formula for the cosine function assuming the addition formula for the sine function proved in Section 3.

5. Use the result of problem 3 to describe the function f that satisfies the equation $f'' + 9f = 0$ and the initial conditions $f(\pi/6) = 0$ and $f'(\pi/6) = 4$.

6. In order for a large wall clock to give accurate time the frequency of its pendulum needs to be exactly $1/2$ cycles per second. Determine the distance in *cm* from the hinge at which the weight needs to placed in order for the clock to be accurate.

Epilogue

Let us take a look back at what we have achieved. Starting with a simple algebraic process, we have introduced tangents to the graphs of polynomials and rational functions. Their slopes, that is, the derivative of the relevant function, has been interpreted as an instantaneous rate of change, something that has numerous applications in the physical sciences, in economics, biology, and much more. Along the way we recognized the importance of the idea of limit, which can only be fully understood in the framework of the real numbers, a sophisticated extension of the more familiar rational numbers. With these more advanced tools we have then been able to investigate the tangent problem for exponential functions, leading up to the general notion of a differentiable function. We then discussed basic rules and properties of such functions, and we concluded with a very brief look at some applications of these new mathematical tools.

As we mentioned at the end of the Preface, the story does continue. In essence, one may say that we covered one half of the story. The other half involves reversing the process of differentiation, something that leads us to *definite integrals*, the other central concept in calculus. Such integrals have numerous applications to topics such as (geometric) areas and volumes, the work done by variable forces, income streams, probability distributions, length of curves, and much more. But rather than adding these topics and other significant applications to this book, your author has chosen to refer you to a book that he has published a few years ago: *What is Calculus? From Simple Algebra to Deep Analysis,*[1] to which we have occasionally referred to as [WiC?]. Having come so far, you, the reader, are well equipped to continue with Chapter IV in that book, titled *The Definite Integral.*

---

[1] World Scientific Publishing, Singapore, London and New York, 2015.

However, I would highly recommend that you continue your studies by first looking at Chapter III of that book, which covers some of the applications of derivatives that we did not discuss here. In particular, you definitely should read Section III.9, *Higher Order Approximations and Taylor Polynomials*, where you will discover some amazing features of the elementary transcendental functions. Alternatively, you may choose to study this section only later, just before reading Section IV.9, where this topic is brought to a fascinating peak. In any case, you should work through Chapter IV in order to get a fairly complete overview of the basics of calculus of functions of *one* variable.

Lastly, after you have mastered Taylor series of exponential and trigonometric functions, you should definitely work through Section 10 in Chapter IV. Here you will find a brief introduction to *complex* numbers, something that we mentioned earlier on several occasions, for example in the context of finding all roots of the quadratic equation even when the discriminant is negative. But the goal at this point is much deeper and far-reaching, namely, to enlarge the domain of the exponential and trigonometric functions from the familiar real numbers to the more general complex numbers in a most natural way. This leads us to discover a most amazing connection between these apparently so different fundamental functions, that is, a connection between growth and periodic processes, that remains hidden as long as one only considers these functions for *real* numbers. In particular, it will reveal and explain one of the most amazing formulas in mathematics, the famous *Euler Formula*,

$$e^{\sqrt{-1}\pi} + 1 = 0,$$

that connects in a simple and mysterious way the fundamental numbers $0, 1, e, \pi$, and the imaginary unit $i = \sqrt{-1}$.

Index